AF678774

Molecular Targeted Therapy of Lung Cancer

Yuichi Takiguchi
Editor

Molecular Targeted Therapy of Lung Cancer

Springer

Editor
Yuichi Takiguchi
Department of Medical Oncology
Graduate School of Medicine
Chiba University
Chiba, Japan

ISBN 978-981-10-2000-1 ISBN 978-981-10-2002-5 (eBook)
DOI 10.1007/978-981-10-2002-5

Library of Congress Control Number: 2017930607

© Springer Science+Business Media Singapore 2017
This work is subject to copyright. All rights are reserved by the Publisher, whether the whole or part of the material is concerned, specifically the rights of translation, reprinting, reuse of illustrations, recitation, broadcasting, reproduction on microfilms or in any other physical way, and transmission or information storage and retrieval, electronic adaptation, computer software, or by similar or dissimilar methodology now known or hereafter developed.
The use of general descriptive names, registered names, trademarks, service marks, etc. in this publication does not imply, even in the absence of a specific statement, that such names are exempt from the relevant protective laws and regulations and therefore free for general use.
The publisher, the authors and the editors are safe to assume that the advice and information in this book are believed to be true and accurate at the date of publication. Neither the publisher nor the authors or the editors give a warranty, express or implied, with respect to the material contained herein or for any errors or omissions that may have been made.

Printed on acid-free paper

This Springer imprint is published by Springer Nature
The registered company is Springer Nature Singapore Pte Ltd.
The registered company address is: 152 Beach Road, #22-06/08 Gateway East, Singapore 189721, Singapore

Preface

In 2004, a randomized study involving patients with advanced non-small cell lung cancer (NSCLC) demonstrated a 1.7 months' benefit in median overall survival time (OS) with the combination chemotherapy consisting of docetaxel and cisplatin (median OS of 11.3 months) over the combination chemotherapy consisting of vindesine and cisplatin (median OS of 9.6 months) [1]. Although this study changed the 20-year standing standard of care for patients with advanced NSCLC that was established at 1981 [2], the survival time advantage was very small. From that time up to now, however, by adding an antiangiogenesis agent to the standard combination chemotherapy [3], by including newer agents [4, 5], and by incorporating maintenance therapy [6], a small advantage in each landmark study had steadily accumulated up to 16.9 months of median OS [6], which was almost twice long compared to the one with vindesine plus cisplatin. Although limited to patients with cancer harboring particular driver mutations, drastic prolongation of median OS has been observed by incorporating molecular-target agents [7–9]. Further improvement in prolonging OS is expected by incorporating newly developed immune checkpoint therapy [10, 11]. The advancement in the treatment for advanced NSCLC in the latest decade has been accompanied by advancement of technologies for molecular-marker detection and in bioinformatics. It has also involved issues of regulatory science and companion diagnostics. Because of the large patient number available and existence of definitive driver oncogenes, the recent clinical study on advanced NSCLC has led the recent clinical studies of many other cancers.

The current topics on treatment of lung cancer, its related translational research, and regulatory science are discussed in this book. Future directions especially focusing on the epithelial-to-mesenchymal transition, cancer stem-cell nature, and the interaction between cancer and its microenvironment are also discussed.

The editor and most of the authors are the faculty members of the integrated and interuniversity educational organization "International Training Program for Co-operative Experts in Clinical Oncology" of "the Cultivation Plan for Cancer Care Professionals (Gann Pro)" (http://kanto-kokusai-ganpro.md.tsukuba.ac.jp) consisting of Chiba University, Tsukuba University, Gunma University, Nippon Medical School, Saitama Medical School, and Dokkyo Medical School. Besides expecting wide-ranging readership, we also provide the book with a role as a textbook for graduated students participating in this program.

The authors are hoping that further development in lung cancer treatment will further relieve the pain of patients and families.

Chiba, Japan Yuichi Takiguchi

References

1. Kubota K, Watanabe K, Kunitoh H, Noda K, Ichinose Y, Katakami N, Sugiura T, Kawahara M, Yokoyama A, Yokota S, Yoneda S, Matsui K, Kudo S, Shibuya M, Isobe T, Segawa Y, Nishiwaki Y, Ohashi Y, Niitani H (2004) Phase III randomized trial of docetaxel plus cisplatin versus vindesine plus cisplatin in patients with stage IV non-small-cell lung cancer: the Japanese Taxotere Lung Cancer Study Group. J Clin Oncol 22(2):254–261. doi:10.1200/jco.2004.06.114
2. Gralla RJ, Casper ES, Kelsen DP, Braun DW, Jr., Dukeman ME, Martini N, Young CW, Golbey RB (1981) Cisplatin and vindesine combination chemotherapy for advanced carcinoma of the lung: a randomized trial investigating two dosage schedules. Ann Int Med 95(4):414–420
3. Sandler A, Gray R, Perry MC, Brahmer J, Schiller JH, Dowlati A, Lilenbaum R, Johnson DH (2006) Paclitaxel-carboplatin alone or with bevacizumab for non-small-cell lung cancer. N Engl J Med 355(24):2542–2550. doi:10.1056/NEJMoa061884
4. Schiller JH, Harrington D, Belani CP, Langer C, Sandler A, Krook J, Zhu J, Johnson DH, the Eastern Cooperative Oncology Group (2002) Comparison of four chemotherapy regimens for advanced non-small-cell lung cancer. N Engl J Med 346(2):92–98. doi:10.1056/NEJMoa011954
5. Ohe Y, Ohashi Y, Kubota K, Tamura T, Nakagawa K, Negoro S, Nishiwaki Y, Saijo N, Ariyoshi Y, Fukuoka M, for the FACS Cooperative Group (2007) Randomized phase III study of cisplatin plus irinotecan versus carboplatin plus paclitaxel, cisplatin plus gemcitabine, and cisplatin plus vinorelbine for advanced non-small-cell lung cancer: Four-Arm Cooperative Study in Japan. Ann Oncol 18(2):317–323. doi:10.1093/annonc/mdl377
6. Paz-Ares LG, de Marinis F, Dediu M, Thomas M, Pujol JL, Bidoli P, Molinier O, Sahoo TP, Laack E, Reck M, Corral J, Melemed S, John W, Chouaki N, Zimmermann AH, Visseren-Grul C, Gridelli C (2013) PARAMOUNT: final overall survival results of the phase III study of maintenance pemetrexed versus placebo immediately after induction treatment with pemetrexed plus cisplatin for advanced nonsquamous non-small-cell lung cancer. J Clin Oncol 31(23):2895–2902. doi:10.1200/jco.2012.47.1102
7. Maemondo M, Inoue A, Kobayashi K, Sugawara S, Oizumi S, Isobe H, Gemma A, Harada M, Yoshizawa H, Kinoshita I, Fujita Y, Okinaga S, Hirano H, Yoshimori K, Harada T, Ogura T, Ando M, Miyazawa H, Tanaka T, Saijo Y, Hagiwara K, Morita S, Nukiwa T (2010) Gefitinib or chemotherapy for non-small-cell lung cancer with mutated EGFR. N Engl J Med 362(25):2380–2388. doi:10.1056/NEJMoa0909530
8. Mitsudomi T, Morita S, Yatabe Y, Negoro S, Okamoto I, Tsurutani J, Seto T, Satouchi M, Tada H, Hirashima T, Asami K, Katakami N, Takada M, Yoshioka H, Shibata K, Kudoh S, Shimizu E, Saito H, Toyooka S, Nakagawa K, Fukuoka M (2010) Gefitinib versus cisplatin plus docetaxel in patients with non-small-cell lung cancer harbouring mutations of the epidermal growth factor receptor (WJTOG3405): an open label, randomised phase 3 trial. Lancet Oncol 11(2):121–128. doi:10.1016/s1470-2045(09)70364-x
9. Solomon BJ, Mok T, Kim DW, Wu YL, Nakagawa K, Mekhail T, Felip E, Cappuzzo F, Paolini J, Usari T, Iyer S, Reisman A, Wilner KD, Tursi J, Blackhall F (2014) First-line crizotinib versus chemotherapy in ALK-positive lung cancer. N Engl J Med 371(23):2167–2177. doi:10.1056/NEJMoa1408440

10. Brahmer J, Reckamp KL, Baas P, Crino L, Eberhardt WE, Poddubskaya E, Antonia S, Pluzanski A, Vokes EE, Holgado E, Waterhouse D, Ready N, Gainor J, Aren Frontera O, Havel L, Steins M, Garassino MC, Aerts JG, Domine M, Paz-Ares L, Reck M, Baudelet C, Harbison CT, Lestini B, Spigel DR (2015) Nivolumab versus docetaxel in advanced squamous-cell non-small-cell lung cancer. N Engl J Med 373(2):123–135. doi:10.1056/NEJMoa1504627
11. Borghaei H, Paz-Ares L, Horn L, Spigel DR, Steins M, Ready NE, Chow LQ, Vokes EE, Felip E, Holgado E, Barlesi F, Kohlhaufl M, Arrieta O, Burgio MA, Fayette J, Lena H, Poddubskaya E, Gerber DE, Gettinger SN, Rudin CM, Rizvi N, Crino L, Blumenschein GR, Jr., Antonia SJ, Dorange C, Harbison CT, Graf Finckenstein F, Brahmer JR (2015) Nivolumab versus docetaxel in advanced nonsquamous non-small-cell lung cancer. N Engl J Med 373(17):1627–1639. doi:10.1056/NEJMoa1507643

Contents

Part I
Diagnosis

Chapter 1
Classification of Adenocarcinoma of the Lung, with a Special Reference to Prognosis

Yukio Nakatani, Yoko Yonemori, Jun Matsushima, and Takuya Yazawa

Abstract Classification of lung adenocarcinoma was largely revised in the 4th edition of WHO classification of tumors of the lung, pleura, thymus, and heart published in 2015. This chapter deals with the major changes in the adenocarcinoma classification, briefly describing the definition, gross and histopathological findings, genetic profiles and clinical features of each subtype, and variants of lung adenocarcinoma. Special reference was also made to the prognostic aspects. The new concepts of adenocarcinoma in situ and minimally invasive adenocarcinoma are especially important from the prognostic point of view because of their virtual connotation as 100% curable cancers if resected completely. Each subtype of invasive adenocarcinoma may be categorized into either good, intermediate, or poor prognostic group. Much progress has been made regarding the genetic profiles as well, such as the occurrence of *EGFR* and *KRAS* mutations, *ALK* fusion genes and recently discovered alterations, and *NRG1* fusion genes in association with adenocarcinomas with certain characteristics. A brief overview of the major changes in the lung adenocarcinoma classification in this chapter will help physicians, radiologists, and pathologists grasp the significance and meaning of the histopathological diagnosis according to the new WHO classification.

Keywords Lung adenocarcinoma • WHO classification

Y. Nakatani (✉) • Y. Yonemori • J. Matsushima
Department of Diagnostic Pathology, Chiba University Graduate School of Medicine, 1-8-1 Inohana, Chuo-ku, Chiba 260-8670, Japan
e-mail: nakatani@faculty.chiba-u.jp

T. Yazawa
Department of Pathology, Dokkyo Medical University, Tochigi, Japan

© Springer Science+Business Media Singapore 2017
Y. Takiguchi (ed.), *Molecular Targeted Therapy of Lung Cancer*,
DOI 10.1007/978-981-10-2002-5_1

1.1 Classification of Adenocarcinoma of the Lung in New WHO Classification

1.1.1 Introduction: Major Changes in the Classification

The 4th edition of *WHO Classification of Tumours of the Lung, Pleura, Thymus, and Heart* was published in 2015 [1]. In this new edition, the significant changes in the lung adenocarcinoma classification include (1) introduction of the new term "adenocarcinoma in situ (AIS)" as a preinvasive lesion in addition to atypical adenomatous hyperplasia (AAH), discarding the old and ambiguous term "bronchioloalveolar carcinoma (BAC)"; (2) introduction of the new term "minimally invasive adenocarcinoma (MIA)"; (3) classification of invasive adenocarcinomas according to the predominant subtype with additional description of minor subtypes; (4) introduction of the new term "invasive mucinous adenocarcinoma" (roughly corresponding to the former mucinous BAC) as a variant of adenocarcinoma; (5) refining the category of adenocarcinoma variants as including invasive mucinous adenocarcinoma, colloid adenocarcinoma, fetal adenocarcinoma (low- and high-grade), and enteric adenocarcinoma; (6) introduction of immunohistochememically defined "solid adenocarcinoma", i.e., diagnosing the former large cell carcinoma as solid adenocarcinoma if tumor cells are immunopositive for pneumocyte markers (TTF1 and/or napsin A); and (7) avoiding the noncommittal diagnosis of non-small cell carcinoma in small biopsy/cytology samples as much as possible by introduction of the new immunohistochemically defined diagnostic category of "non-small cell carcinoma, favor adenocarcinoma" [1–4] (Table 1.1).

It should be emphasized that these major changes in adenocarcinoma classification are deeply related to the ever-growing recognition that a multidisciplinary approach is mandatory for the classification to be clinically relevant: (1) recent advance in molecular biology/oncology has led to the discovery of epidermal growth factor receptor (*EGFR*) mutations and *ALK* gene translocations almost exclusively in lung adenocarcinomas, and targeted therapy with tyrosine kinase inhibitors (TKIs) has become available for these tumors; (2) progress in treatment requires discrimination of squamous cell carcinoma from non-squamous, non-small cell carcinomas such as in application of certain drugs including pemetrexed and bevacizumab; and (3) advancement in knowledge of the intimate correlation between the developmental stages of adenocarcinoma and corresponding CT images has led to its utility in prediction of prognosis and choice of treatment in lung adenocarcinomas [3].

1.1.2 Preinvasive Lesions

1.1.2.1 Atypical Adenomatous Hyperplasia (AAH) (Fig. 1.1)

AAH, by definition, is a small, localized proliferation of mildly to moderately atypical type II pneumocytes and/or club cells (formerly named as Clara cells) lining alveolar walls and sometimes respiratory bronchioles [1]. This lesion is usually

Table 1.1 Lung adenocarcinoma and its precursor. WHO classification [1]

Adenocarcinoma	8140/3
Lepidic adenocarcinoma	8250/3
Acinar adenocarcinoma	8551/3
Papillary adenocarcinoma	8260/3
Micropapillary adenocarcinoma	8265/3
Solid adenocarcinoma	8230/3
Invasive mucinous adenocarcinoma	8253/3
Mixed invasive mucinous and non-mucinous adenocarcinoma	8254/3
Colloid adenocarcinoma	8480/3
Fetal adenocarcinoma	8333/3
Enteric adenocarcinoma	8144/3
Minimally invasive adenocarcinoma	
Non-mucinous	8250/2
Mucinous	8257/3
Preinvasive lesions	
Atypical adenomatous hyperplasia	8250/0
Adenocarcinoma in situ	8140/2
Non-mucinous	8410/2
Mucinous	8257/3

The morphology codes are from the International Classification of Diseases for Oncology (ICD-O)

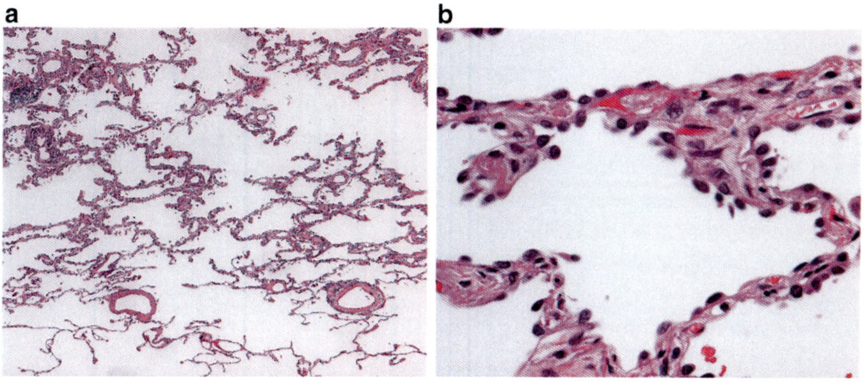

Fig. 1.1 Atypical adenomatous hyperplasia. (**a**) *Low-power view*. Note the slightly thickened alveolar septa with lining cells that show a sharp demarcation from the normal lung parenchyma occupying the lowermost quarter of the field. (**b**) *High-power view*. Cuboidal to somewhat flattened cells with mildly atypical nuclei and scant cytoplasm are growing along alveolar septa

found incidentally in lung specimens resected for cancer or may incidentally be detected as a pure ground-glass nodule (GGN) on high-resolution CT scans during medical examination for some reasons.

In gross examination, AAH typically is a few millimeter-sized, barely discernable gray-white nodule in the peripheral lung. Histopathologically, the distinction between AAH and AIS is sometimes difficult because both show the lepidic pattern or growth along the alveolar wall throughout the lesion, but AAH typically is up to 5 mm in size, and the constituent cells show less nuclear atypism and are less densely populated along alveolar walls than those of AIS [5, 6]. Somewhat paradoxically, the cell shape in AAH is more various with cuboidal, pyramidal, or flat appearances than that of AIS.

AAH is considered to be a precursor lesion of peripheral lung adenocarcinoma. Clinicopathological and clonality/mutational studies have demonstrated that AAH is a clonal lesion with the potential for progression to adenocarcinoma [5–7], harboring *KRAS* and *EGFR* mutations in up to 33 % and 35 %, respectively [1, 8–11]. There is some evidence that *KRAS*-mutated AAH may not progress to AIS or invasive adenocarcinoma as frequently as *EGFR*-mutated AAH and that major driver genes (*EGFR/KRAS/ALK/HER2*) mutation-negative AAH/AIS may not progress to invasive adenocarcinoma so frequently [8, 12]. A recent genetic analysis of AAH/AIS/MIA utilizing next-generation sequencing (NGS) [13] showed an average mutation rate of 2.2 non-synonymous mutations (range 0–6 mutations) per one lesion among 25 AAHs, the most frequently mutated genes being *BRAF* and *ARID1B*. Genes associated with DNA repair and chromatin remodeling network such as *ATM* and *ATRX* were also mutated in multiple lesions, suggesting AAH may be predisposed to the acquisition of secondary genetic aberrations. Mutations in *TP53*, *EGFR*, and *IGFR1* were noted in all developmental stages of AAH/AIS/MIA, but *BRAF* mutation was rarely found in MIA or invasive adenocarcinoma, again suggesting the inequity in the progression potential among various mutations.

The natural history of AAH is not fully elucidated, but a recent radiographic study [14] showed that solitary pure GGNs 5 mm or smaller in CT images, the majority of which presumably represented AAH, grew in 10 % of the cases and developed into MIA or invasive adenocarcinoma in 1 % with the mean period of 3.6 years. This observation appears to corroborate the aforementioned genetic inequity in the progression potential of AAH.

1.1.2.2 Adenocarcinoma In Situ (AIS) (Fig. 1.2)

AIS is a newly introduced entity in the current WHO classification [1]. It is a small (<=3 cm), localized adenocarcinoma with neoplastic cell growth restricted along alveolar walls (pure lepidic growth), lacking stromal, vascular, or pleural invasion. The constituent cells are mostly non-mucinous, but mucinous in rare cases as well. AIS is usually found incidentally as a pure GGN or part-solid nodule on CT scan [14, 15]. Mucinous AIS tends to present as a solid or part-solid nodule with air-containing spaces [16].

Grossly, AIS is an ill-defined, gray-white to tan-colored nodule with somewhat spongy consistency. Histopathologically, type II pneumocyte/club cell-like cuboidal to columnar cells with mild to moderately atypical nuclei are seen along alveolar walls. The alveolar walls are almost normal to moderately thickened with collapse-type fibroelastosis [17]. In the rare mucinous AIS, the lining cells have mucinous

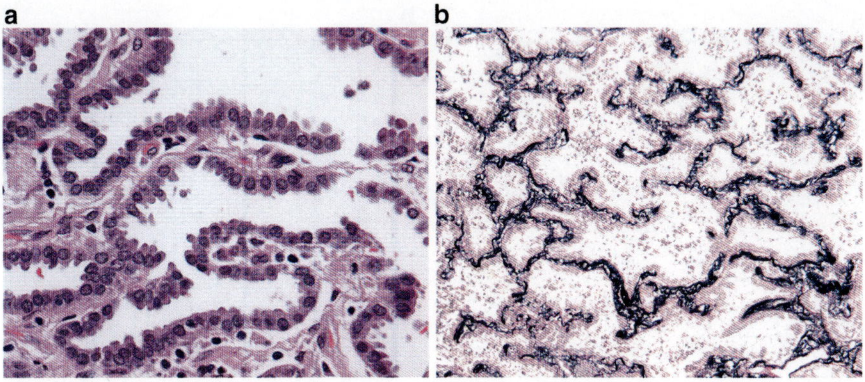

Fig. 1.2 Adenocarcinoma in situ (non-mucinous type). (**a**) The alveolar septa are lined by atypical type II pneumocyte/club cell-like cells. (**b**) The elastic framework of the alveolar septa in AIS is completely preserved. (Elastic van Gieson stain)

cytoplasm, resembling gastric foveolar epithelium or goblet cells. Non-mucinous AIS expresses TTF1 and napsin A, whereas mucinous AIS is often negative for these immunohistochemical markers of alveolar pneumocytes and positive for gastric epithelium-associated mucin such as MUC5AC and MUC6 [18, 19].

Genetically, non-mucinous AIS harbors *EGFR* mutations frequently (40–86 %), but *KRAS* mutations rarely (0–4 %) [12, 19–23]. A recent NGS analysis of AIS in five patients showed an average mutational rate of 6.2 non-synonymous mutations per patient; the mutational landscape varied widely, most mutations including *EGFR* and *TP53* mutations found only in one patient [13]. The lower mutational rate of *EGFR* compared with those of the aforementioned studies [12, 19–23] may be related to different ethnic backgrounds of the cohorts. *EGFR* mutations are rare in mucinous AIS [19, 24].

The clinical significance of diagnosing AIS lies in its connotation as a neoplasm with 100 % disease-free survival if it is resected completely [1, 17, 19–24] (Fig. 1.3) (Table 1.2). It is noteworthy that most of these data are from Japan, where *EGFR* mutation-related adenocarcinomas are common and CT-based examination is part of routine clinical practice. The frequency of AIS among resected lung adenocarcinomas has been 4.5–8.4 % in Japanese cohorts [19–21, 23], whereas it has been less than 1 % in Western countries [24]. The clinical behavior of mucinous AIS is less well elucidated but may also be good [15, 19, 20, 24, 25]. Thus, the most recent article on the eighth TNM classification of lung cancer has proposed the code Tis in place of T1 for AIS [15].

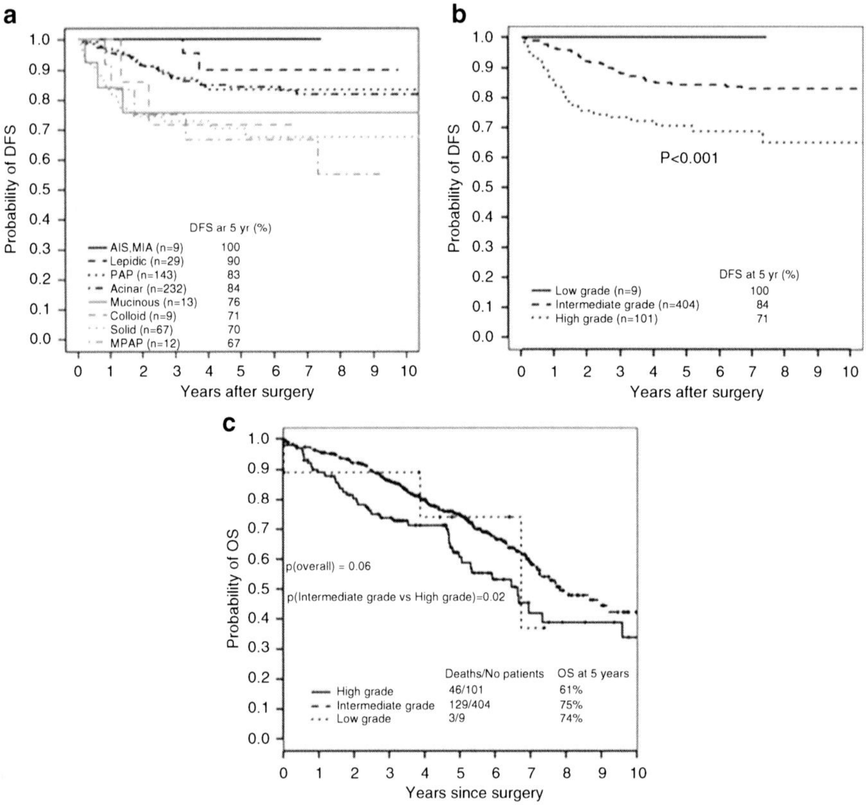

Fig. 1.3 Pulmonary adenocarcinoma subtypes and prognosis. Stage I ($n = 514$). (**a**) Disease-free survival (DFS) for all histological categories ($P < 0.001$). The favorable group includes adenocarcinoma in situ (AIS) and minimally invasive adenocarcinomas (MIA) with 100 % 5-year disease-free survival. Disease-free survival for the intermediate group was 90, 83, and 84 % for lepidic predominant, papillary (PAP) predominant and acinar predominant, and adenocarcinomas, respectively. Disease-free survival for the unfavorable group was 70, 67, 71, and 76 % for solid predominant, micropapillary (MPAP) predominant, colloid predominant, and mucinous and mixed adenocarcinomas, respectively. (**b**) Disease-free survival according to combined histological groupings according to *low*-, intermediate-, and *high-grade* clinical aggressiveness. (**c**) Overall survival (OS) according to combined histological groupings according to *low*-, intermediate-, and *high-grade* clinical aggressiveness (Adopted from Fig. 1.4 of reference [20]). (**b**) Stages I–III ($n = 440$). (**A**) Disease-free survival curves and (**B**) overall survival curves, for the groups, separated by the IASLC/ATS/ERS classification of lung adenocarcinomas (Adopted from Fig. 1.6 of reference [33]). *AIS* adenocarcinoma in situ, *MIA* minimally invasive adenocarcinoma, *Lepidic* lepidic predominant adenocarcinoma, *Aci* acinar predominant adenocarcinoma, *Pap* papillary predominant adenocarcinoma, *Solid* solid predominant adenocarcinoma, *MP* micropapillary predominant adenocarcinoma, *IMA* invasive mucinous adenocarcinoma, *IASLC* International Association for the Study of Lung Cancer, *ATS* American Thoracic Society, *ERS* European Respiratory Society

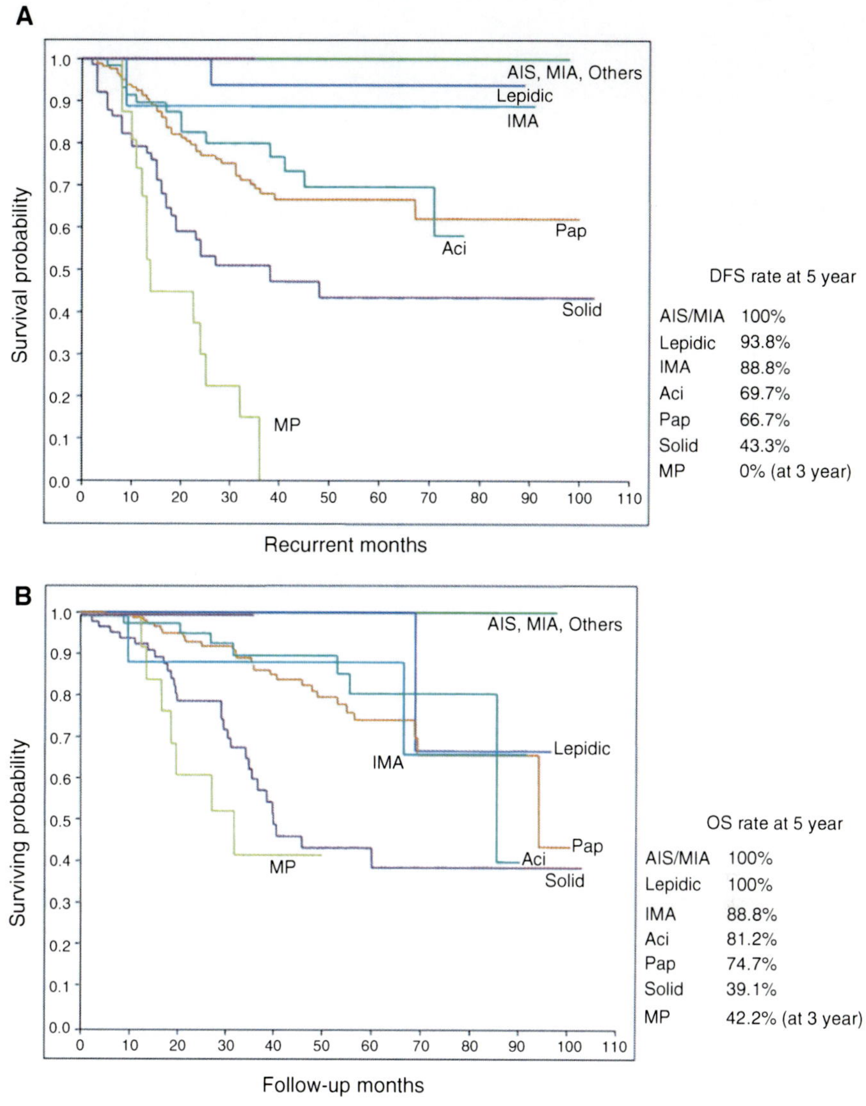

Fig. 1.3 (continued)

1.1.3 Minimally Invasive Adenocarcinoma (MIA) (Fig. 1.4)

MIA is another new entity incorporated into the current WHO classification. It defines the solitary adenocarcinoma (<=3 cm) with a predominantly lepidic pattern and <=5 mm invasion in greatest dimension [1]. MIA should lack lymphatic/vascular/pleural/air space invasion or spread. MIA is non-mucinous in most cases but

Table 1.2 Pulmonary adenocarcinoma subtypes and prognosis

Reporter/pattern		AIS	MIA	Lepidic	Acinar	Papillary	Micropapillary	Solid	IMA	Colloid	Others
Yoshizawa et al. (2011) [33]											
Stage I AC n=514	n (%)	1 (0.2)	8 (1.6)	29 (5.6)	232 (45.1)	143 (27.8)	12 (2.3)	67 (13)	13 (2.5)	9 (1.8)	
	5Y DFS: %	100	100	90	84	83	67	70	76	71	
Russell et al. (2011) [34]											
Stages I–III AC n=210	n (%)	1(0.5)	7 (3)	10 (5)	84 (40)	26 (12)	14 (7)	49 (23)	10 (5)	9 (4)	
	5Y OS: %	100	100	86	68	71	38	39	51	51	
Warth et al. (2012) [40]											
Stages I–IV AC n=500	n (%)	0	0	41 (8.4)	207 (42.5)	23 (4.7)	33 (6.8)	183 (37.6)	12 (2.4)	0	1/enteric
	OS: mean Mo	NA	NA	78.5	67.3	48.9	44.9	58.1	88.7	NA	NA
	DSS: mean Mo	NA	NA	80.3	79.2	56.3	50.4	66.7	All survived	NA	NA
	DFS: mean Mo	NA	NA	72.6	61.7	37.7	33.8	51.2	88.1	NA	NA
Yoshizawa et al. (2013) [20]											
Stages I–III AC n=440	n (%)	20 (4.5)	33 (7.5)	36 (8.1)	61 (13.8)	179 (40.7)	19 (4.3)	78 (17.7)	10 (2.2)	3 (0.7)	1/fetal(0.2)
	5Y OS: %	100	100	100	81.2	74.7	42.2 (at 3Y)	39.1	88.8	NA	NA

	5Y DFS: %	100	100	93.8	69.7	66.7	0 (at 3 Y)	43.3	88.8	NA	NA
Tsuta et al. (2013) [21]											
Stages I–IV AC *n*=904	*n* (%)	69 (8)	33 (4)	136 (15.1)	98 (10.8)	338 (37.4)	61 (6.7)	124 (13.7)	45 (5.0)	0	0
	5Y/10Y OS: %	98/94[a]	98/94[a]	93/85	67/47	74/57	62/47	58/41	76/63	NA	NA
	5Y/10Y DSS: %	100[a]	100[a]								
Gu et al. (2013) [35]	*n* (%)	1(0.3)	14 (4.8)	31 (10.6)	112 (38.4)	36 (12.3)	30 (10.3)	52 (17.8)	10 (3.4)	2(0.7)	4/enteric 81.4)
Stages I–III AC *n*=292	5Y OS: %	100	100	91.4	72.2	71.1	46.6	57.9	73.1	73.1	73.1
	5Y DFS: %	100	100	71.9	54	56.1	25.7	45.7	62.5	62.5	62.5

AC adenocarcinoma, *DFS* disease-free survival, *OS* overall survival, *DSS* disease-specific survival, *5Y* 5 year, *Mo* months, *AIS* adenocarcinoma in situ, *MIS* minimally invasive adenocarcinoma

[a]Subtypes combined

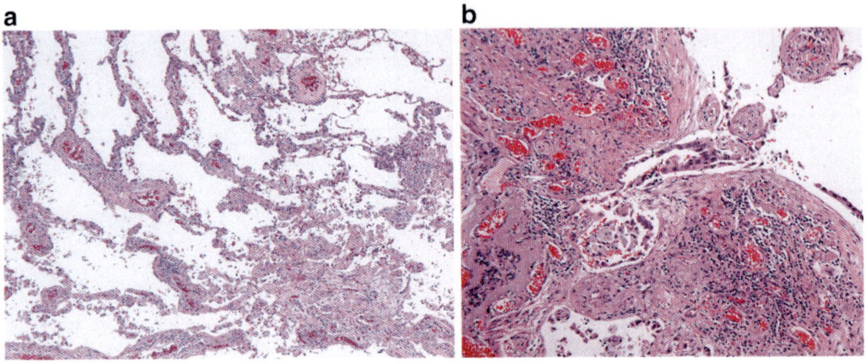

Fig. 1.4 Minimally invasive adenocarcinoma. (**a**) The *left upper field* shows the lepidic pattern of tumor growth with preserved alveolar framework, whereas the *right lower field* shows a fibrotic focus with an invasive growth of neoplastic cells. (**b**) Note an invasive neoplastic acinar structure within the fibrous stroma

may rarely be mucinous as well. This lesion is usually discovered incidentally as a part-solid nodule, pure GGN, or rarely as a solid nodule on CT [15].

Historically, the criteria for this entity were searched after the epoch-making publication of an article on AIS by Noguchi et al. in 1995 [17], and several pioneering studies contributed to its establishment [25–32]. Validation studies [19–23, 33–37] suggested the prognosis of MIA is virtually equal to that of AIS, supporting its recognition as a distinct entity (Fig. 1.3, Table 1.2). The code T1mi is proposed for MIA in the latest TNM system [15].

Histopathologically, the invasive focus may take one of the basic patterns of invasive adenocarcinoma, i.e., papillary, acinar, solid, or micropapillary pattern or tumor cells infiltrating myofibroblastic stroma [1].

Genetically, MIA shows high rates of *EGFR* mutation similar to AIS [20–23]. An NGA analysis of MIA in five patients revealed an average mutation rate of 10.8 non-synonymous mutation per patient with *EGFR* and *TP53* being the most frequently mutated genes [13].

1.1.4 Invasive Adenocarcinoma (Fig. 1.3) (Table 1.2)

Invasive adenocarcinoma is a carcinoma with glandular differentiation, mucin production, or pneumocyte marker expression [1]. The growth pattern includes acinar, papillary, micropapillary, and solid. These patterns often appear admixed with and in transition to one another within the same tumor, and therefore the tumor is classified according to the predominant pattern in proportion with additional description of each component present in 5–10 % increment. Invasive adenocarcinoma is typically localized in the periphery of the lung. Pleural indentation is common due to the retraction caused by central collapse and fibrosis in the tumor. CT images of

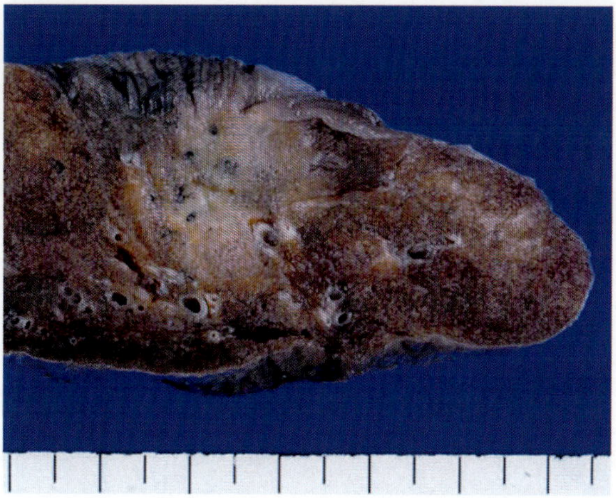

Fig. 1.5 Lepidic adenocarcinoma. Gross appearance: Note the peripheral tumor with a pleural indentation. The central portion of the tumor appears *grayish white* and solid, whereas the peripheral portion is tan in color with a somewhat spongy appearance and unclear margin

invasive non-mucinous adenocarcinoma appear solid or part solid depending on the proportion of lepidic growth vs. invasive growth as well as on the extent of alveolar collapse [1, 3].

1.1.4.1 Lepidic Adenocarcinoma (Fig. 1.5)

In this tumor, the predominant pattern is lepidic with type II pneumocyte/club cell-like atypical cells growing along alveolar walls, but also present is an invasive component of various patterns such as papillary and acinar larger than 5 mm in greatest dimension. Grossly, part of the tumor, often centrally located, is grayish white in color with carbon dust deposition and solid in consistency, whereas the peripheral portion is somewhat ill defined, tan in color, and soft in consistency (Fig. 1.5). The former roughly corresponds to the invasive component with fibrosis and the latter the lepidic component with preserved airspace. This feature is usually reflected as a part solid image at CT scan. The frequency of this subtype among invasive adenocarcinomas varies from 5 % [34] to 18.3 % [23], probably reflecting different ethnic and clinical backgrounds of these cohorts.

Genetically, *EGFR* mutation is frequent [20, 21, 23]. Adenocarcinoma of lepidic pattern with type II pneumocyte/ club cell-like cells (bronchioloalveolar features) has been termed terminal respiratory unit (TRU)-type adenocarcinoma and known to be intimately associated with *EGFR* mutation [38].

Prognostically, this tumor lies intermediate between the good prognostic group of AIS/MIA and the poor prognostic group of micropapillary adenocarcinoma/solid adenocarcinoma [20, 21, 23, 33, 35] (Fig. 1.3, Table 1.2). The prognosis of lepidic adenocarcinoma is related to the proportion of the lepidic growth within the entire tumor, tumors with >50 % to >75 % lepidic pattern showing good prognosis similar to those of AIS/MIA [24, 27]. Adenocarcinomas even with a non-predominant lepidic component show a better outcome than adenocarcinomas without the component [39]. This tendency in prognosis will be more accurately reflected in the 8th

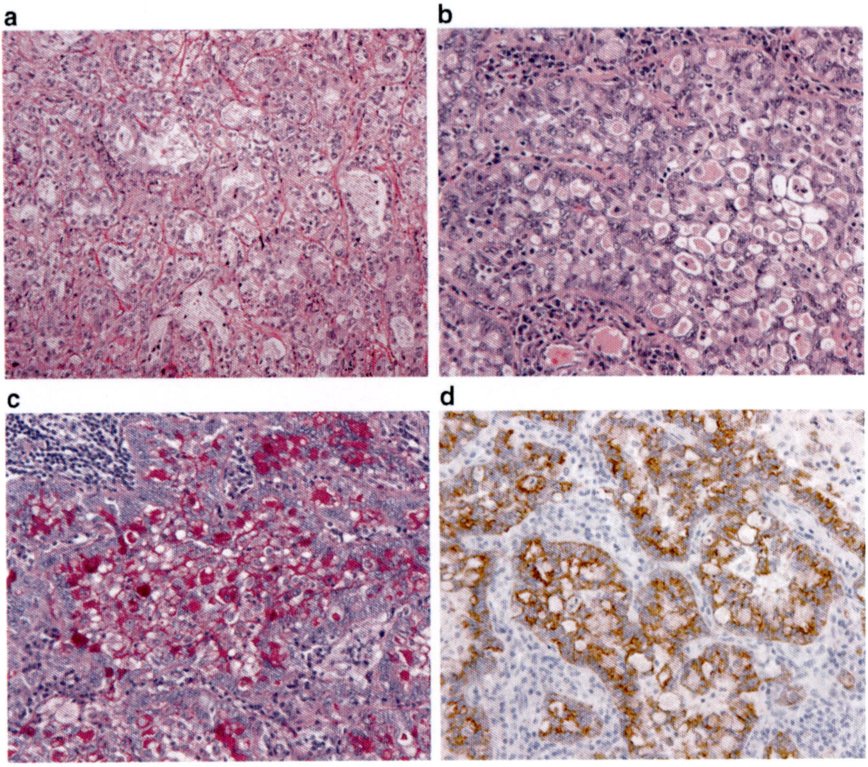

Fig. 1.6 Acinar adenocarcinoma. (**a**) Neoplastic cells are arranged in acinar or tubular structures. (**b**) *ALK*-rearranged adenocarcinoma. The neoplastic cells are arranged in a so-called mucinous cribriform pattern. (**c**) *ALK*-rearranged adenocarcinoma. The luminal space and cytoplasmic vacuoles of tumor cells are abundant in mucin. (PAS reaction). (**d**) *ALK*-rearranged adenocarcinoma. The neoplastic cells are diffusely positive for ALK protein (immunostaining)

edition of the TNM classification of lung cancer in which the invasive tumor size, excluding the lepidic growth, will be used as the T descriptor size [15]. Risk factors for recurrence in lepidic adenocarcinoma may include limited resection with a close margin, lymphovascular invasion, and a substantial component of high-grade pattern such as micropapillary [24].

1.1.4.2 Acinar Adenocarcinoma (Fig. 1.6)

Acinar adenocarcinoma is composed predominantly of acinar or glandular structures with cuboidal to columnar neoplastic cells forming central lumina of various size. Of all subtypes of pulmonary adenocarcinoma, acinar adenocarcinoma is less common (10.8–20.4 %) in Japan [20, 21, 23] than in Western countries (40–45.1 %) [33, 34, 40].

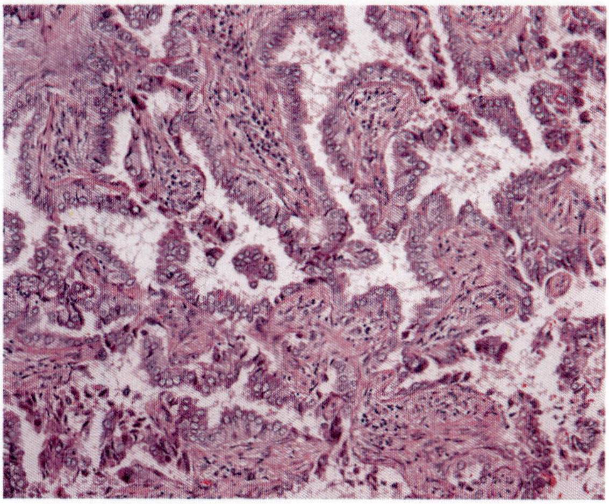

Fig. 1.7 Papillary adenocarcinoma. Neoplastic columnar cells are arranged in a papillary configuration along with the central fibrovascular cores

Genetically, acinar adenocarcinoma shows *EGFR* mutation less frequently and *ALK* rearrangement more frequently than AIS/MIA and lepidic and papillary adenocarcinomas [20, 21, 41]. Mucinous cribriform pattern has been reported as a variant of acinar pattern intimately associated with *ALK*-rearranged pulmonary adenocarcinoma (Fig. 1.6) [41, 42].

Prognostically, acinar adenocarcinoma together with lepidic adenocarcinoma and papillary adenocarcinoma belongs to the intermediate group between the groups of AIS/MIA and micropapillary/solid adenocarcinoma [20, 21, 23, 33, 35] (Fig. 1.3, Table 1.2). In a study of stage I pulmonary adenocarcinomas [43], however, a cribriform pattern-predominant adenocarcinoma has been proposed as a distinct subtype of acinar adenocarcinoma with a poor prognosis compatible to those of the high-grade adenocarcinomas. This needs further validation.

1.1.4.3 Papillary Adenocarcinoma (Fig. 1.7)

Papillary adenocarcinoma shows a predominant papillary pattern with neoplastic cuboidal to columnar cells growing along fibrovascular cores in papillary configuration.

Genetically, papillary adenocarcinoma is among the subtypes with most frequent *EGFR* mutations, revealing the mutation in 50–68.5 % of cases [20, 21, 23]. This corroborates with the observation that this subtype is quite frequent (28–40.7 %) among various subtypes of adenocarcinoma in Japan [20, 21, 23], where *EGFR* mutation-related adenocarcinoma is prevalent, but is less common (12–27.8 %) in Western countries [33, 34, 44].

Prognostically, most studies placed papillary adenocarcinoma in the intermediate prognostic group [20, 21, 23, 33, 35] (Fig. 1.3, Table 1.2), but papillary adenocarcinoma belonged to the poor survival group together with micropapillary and solid adenocarcinomas in a study on a German cohort [40]. The reason for this

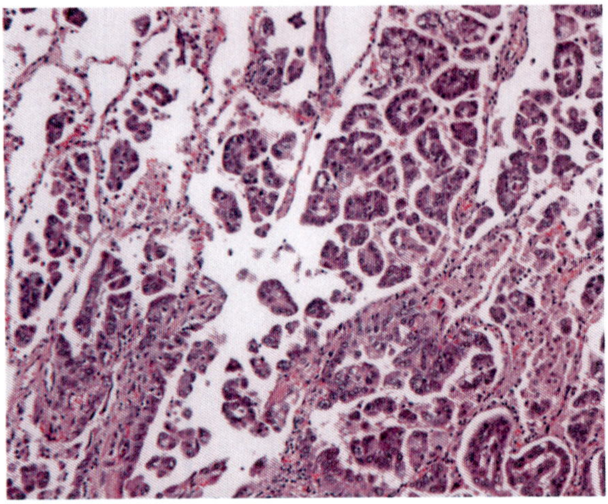

Fig. 1.8 Micropapillary adenocarcinoma. Neoplastic cells are arranged in a micropapillary pattern and show STAS in alveoli surrounding the tumor

discrepancy appears to be the presence of a range of papillary growth from the type architecturally close to lepidic pattern (type I) to the type showing the highest degree of architectural aberrations (type III) [44]: any presence of the type III papillary pattern was associated with poor overall and disease-free survivals, the aforementioned study having applied the most strict criteria (type III) to the recognition of the papillary pattern [41]. Tumors with any type I papillary growth were significantly more likely to harbor *EGFR* mutations than cases with any type II or type III papillary growth [44].

1.1.4.4 Micropapillary Adenocarcinoma (Fig. 1.8)

This is a newly introduced subtype in the current WHO classification [1]. This adenocarcinoma shows the predominant growth of neoplastic cells in micropapillary configuration, i.e., cells forming florets that lack fibrovascular cores, either connected to or detached from alveolar walls. This subtype frequently shows lymphatic permeation and spread through air spaces (STAS) [1, 45]. Micropapillary adenocarcinoma is relatively uncommon, constituting 2.3–19.5 % of all resected pulmonary adenocarcinomas [20, 21, 23, 24, 34–37, 40], most of the cohorts showing the frequency of less than 10 % [20, 21, 23, 24, 34, 37, 40]. However, the presence of micropapillary component itself is not uncommon, any presence (=>1 %) and =>5 % of this component representing 43.6 % and 21.7 % of 525 resected invasive adenocarcinomas, respectively, in one study [46].

Genetically, micropapillary adenocarcinoma shows relatively high rates of *EGFR* mutation (39.7–43 %) next to the adenocarcinomas with predominant lepidic and papillary patterns [20, 21, 23].

Prognostically, there is an agreement that this subtype belongs to the poor prognostic group together with solid adenocarcinoma [20, 21, 23, 24, 33–37, 40] (Fig. 1.3, Table 1.2). The presence of a micropapillary component of 5 % or greater may

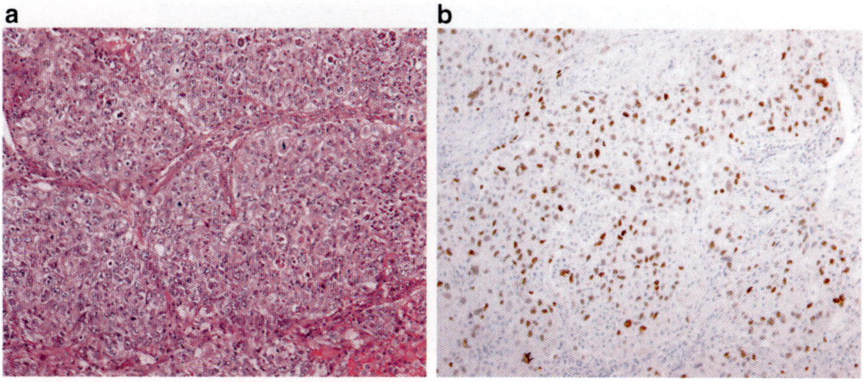

Fig. 1.9 Solid adenocarcinoma. (**a**) The neoplastic cells are arranged in a sheet with no keratinization or acinar formation. (**b**) The neoplastic cells of this tumor express TTF1, qualifying as solid adenocarcinoma in the new WHO classification

be significantly associated with increased risk of local recurrence in patients treated with limited resection [47]. A recent study demonstrated overall survival was significantly better in patients without the micropapillary pattern (<1 %) than in those with the micropapillary pattern (<5 % of the entire tumor), emphasizing the recognition and description of this pattern even in a smallest proportion (=>1 %) [46].

1.1.4.5 Solid Adenocarcinoma (Fig. 1.9)

1. Solid adenocarcinoma shows the predominant growth of neoplastic polygonal cells in a sheetlike arrangement without any recognized pattern of adenocarcinoma described above. In tumors entirely with the solid pattern, intracellular mucin should be present in =>5 tumor cells in each of two high-power fields histochemically, or tumor cells should be positive for pneumocyte markers, i.e., TTF1 and/or napsin A immunohistochemically [1]. The latter immunohistochemically defined solid adenocarcinoma is a newly introduced entity in the current WHO classification. This represents the incorporation of a subset of former large cell carcinomas, the rationale for which is that these immunomarker-defined large cell carcinomas had a distinct adenocarcinoma-related spectrum of therapeutically relevant-driver mutations, including *EGFR*, *KRAS*, and *ALK* [48–50]. The frequency of solid adenocarcinoma among resected lung adenocarcinomas (based on the 2011 IASLC/ATS/ERS international lung adenocarcinoma classification) varies widely from 13 to 37.6 % [21, 22, 34–37, 40].

Genetically, the frequency of *KRAS* mutation is especially high in solid adenocarcinoma, which parallels the observation that *KRAS* mutations are enriched in poorly differentiated adenocarcinomas with a solid component [20, 48–53].

Prognostically, solid adenocarcinoma belongs to the poor prognostic group [20, 21, 23, 33–37, 40] (Fig. 1.3, Table 1.2). Patients who had adenocarcinomas with a

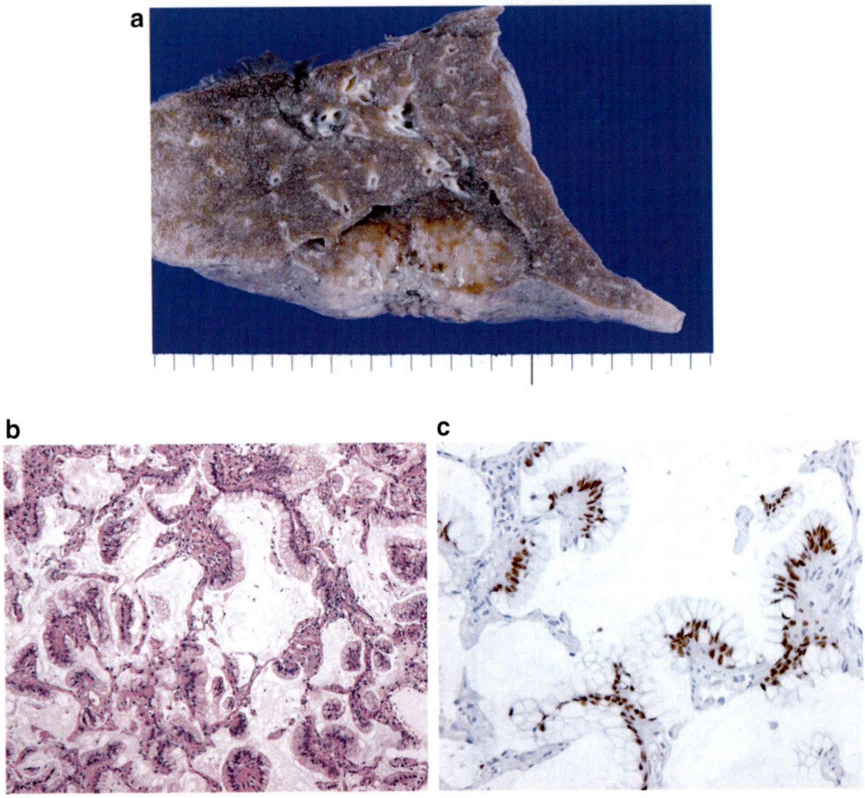

Fig. 1.10 Invasive mucinous adenocarcinoma. (**a**) Gross appearance: A mucinous *grayish-white* nodule with an ill-defined border. (**b**) Columnar cells with mucinous cytoplasm are growing in lepidic and papillary patterns. (**c**) The neoplastic cells express HNF4α in their nuclei (immunostaining)

solid component had significantly lower overall survival and recurrence-free survival rates than patients who had adenocarcinomas with nonsolid components [53]. In patients with stage I pulmonary adenocarcinomas, solid adenocarcinoma recurred significantly earlier than nonsolid adenocarcinomas and was associated with worse post recurrence survival [54].

1.1.5 Variants of Adenocarcinoma

The new WHO classification lists invasive mucinous adenocarcinoma, fetal adenocarcinoma, colloid carcinoma, and enteric adenocarcinoma as variants of pulmonary adenocarcinoma [1]. These variants are all rare but should always be kept in mind as differential diagnoses for appropriate treatment of the patients.

1.1.5.1 Invasive Mucinous Adenocarcinoma (IMA) (Fig. 1.10)

IMA shows growth of neoplastic columnar cells with goblet cell-like or gastric foveolar epithelium-like morphology. The growth pattern can be various but predominantly lepidic in most cases. Tumors solely with the lepidic growth pattern, however, are rare and diagnosed as mucinous AIS. Most of the tumors formerly diagnosed as mucinous bronchioloalveolar carcinoma fall into the category of IMA in the current classification. The CT findings of IMA are variable, including consolidations, air bronchograms, and multifocal and sometimes multilobar solid or subsolid nodules or masses [56]. The frequency of IMA in resected lung adenocarcinomas ranges from 2.2 to 5 % [21, 22, 24, 33–37, 40].

In gross examination, IMA typically displays a somewhat ill-defined, mucinous grayish-white nodule. It may sometimes show a multinodular pattern or a broad lobar consolidation [1]. Histopathologically, the neoplastic columnar cells have basally situated, relatively small and round to oval nuclei with mild atypism. Alveolar spaces within and surrounding the tumor area are often filled with mucin.

Various growth patterns such as papillary and acinar can be seen in addition to the lepidic growth. Frankly invasive areas may show desmoplastic fibrosis.

Immunohistochemically, IMA cells express CK7 and MUC5AC in most cases, and sometimes CK20 as well, but TTF1 only in 11–27.5 % of the cases [18, 56]. Recently, HNF4α was reported as a new immunohistochemical marker for IMA, which was expressed in 92 % of IMA but was negative in normal lung tissues [57]. This transcription factor, however, is expressed in all gastrointestinal adenocarcinomas, pancreatic adenocarcinomas, and mucinous adenocarcinomas of the ovary and uterine cervix, precluding its utility for differentiating lung metastases of these tumors from IMA, which is a major challenge in the histopathological diagnosis of this variant [57].

Genetically, IMA is intimately associated with *KRAS* mutation, disclosing the gene mutation 40–86 % of the examined cases [21, 58–65]. The distribution of *KRAS* amino acid changes more resembled that of colorectal and pancreatobiliary adenocarcinomas than that of pulmonary non-mucinous adenocarcinomas [58, 60, 63, 65]. Smoking status may not be related to *KRAS* mutations in IMA [63]. In addition, *NRG1* fusion genes were recently discovered as novel driver mutations in

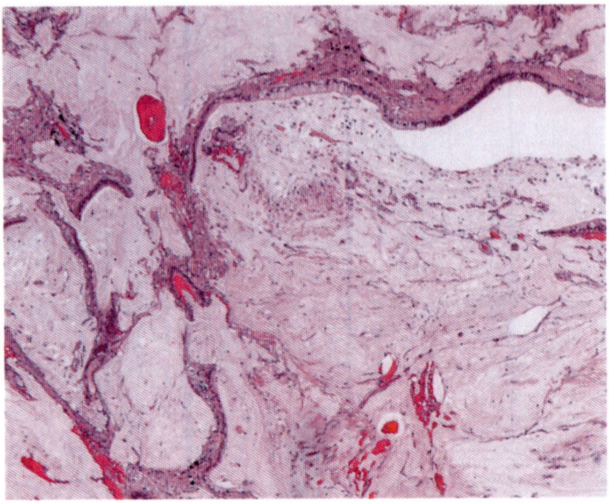

Fig. 1.11 Colloid adenocarcinoma. Note abundant mucinous pool destroying the alveolar framework and incomplete fibrous tissue walls partially lined by neoplastic columnar cells

6.7–27 % of IMA [62–64]. Interestingly, NRG1is known as a regulator of goblet cell formation with MUC5AC/ MUC5B expression in primary cultures of bronchial epithelial cells, suggesting a possible relationship between *NRG1* gene mutation and goblet cell-like morphology/phenotype of IMA [64, 66]. *EGFR* mutations are rare in IMA, ranging 0–22 % in reported studies [21, 22, 24, 60–65]. *KRAS* and *EGFR* mutations are mutually exclusive in IMA but for a few exceptional cases [61, 65]. Rarity of *TP53* mutations in IMA was noted in one study [63].

Prognosis of IMA is somewhat controversial. Some studies found IMA in the poor prognostic group [33, 34], others in the intermediate group [20, 21, 23], while another in the good prognostic group [40] (Fig. 1.3, Table 1.2). Some recent studies show there is no statistically significant difference in prognosis between IMA and non-mucinous invasive adenocarcinomas [58, 63]. Recurrence of IMA after surgical resection was limited to the lungs in one study, suggesting a nonaggressive nature of IMA [63].

1.1.5.2 Colloid Adenocarcinoma (Fig. 1.11)

Colloid adenocarcinoma is an adenocarcinoma in which abundant mucin pools replace air spaces, destroying alveolar framework [1]. This variant may be seen in a pure form or in association with conventional adenocarcinomas.

In gross examination, this variant typically shows a well-demarcated solid or cystic tumor filled with abundant gelatinous material. Histopathologically, the neoplastic cells constitute a relatively small portion of the tumor, columnar cells growing along incompletely developed fibrous tissue septa or small neoplastic cell clusters floating within mucinous pool. Immunohistochemically, the neoplastic cells, especially of goblet cell morphology, often express intestinal markers such as CDX2, MUC2, and CK20, whereas pneumocyte markers such as TTF1and napsin A are variably expressed [56, 67, 68]. Expression of CK7 is usually retained [67, 68].

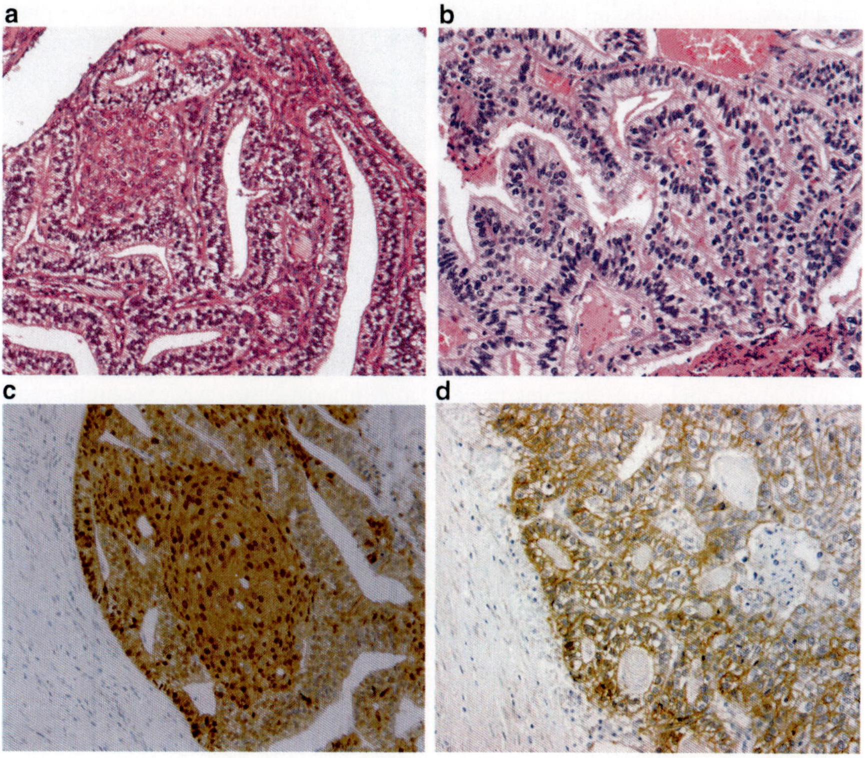

Fig. 1.12 Fetal adenocarcinoma. (**a**) *Low-grade* fetal adenocarcinoma. Note complex glandular structures lined by columnar cells with relatively small and regular nuclei and supra- and subnuclear vacuoles resembling fetal airway epithelium. A characteristic morular formation is also seen. (**b**) *High-grade* fetal adenocarcinoma. The histology resembles the *low-grade* form, but nuclear atypism is more obvious and morular formation is absent. (**c**) *Low-grade* fetal adenocarcinoma. The neoplastic cells show aberrant nuclear/cytoplasmic localization of β-catenin, especially in the morular area (immunostaining). (**d**) High-grade fetal adenocarcinoma. The localization of *β*-catenin is predominantly membranous (immunostaining)

The genetic profile of colloid adenocarcinoma is not well known. *KRAS* mutations were identified in a few cases, while *EGFR* mutation and *ALK* fusion genes were so far not found [68]. Prognostically, a few recent studies suggest this variant may belong to the poor prognostic group [33, 34] in contrast to the previous notion of a relatively favorable prognosis for this tumor [67] (Fig. 1.3a, Table 1.2).

1.1.5.3 Fetal Adenocarcinoma (Fig. 1.12)

Fetal adenocarcinoma is an adenocarcinoma resembling fetal lung [1]. Low-grade and high-grade tumors exist, and they are considered histogenetically different despite their morphologic similarities [69, 70]. Low-grade fetal adenocarcinoma is considered as the epithelial prototype of pulmonary blastoma and occurs in a pure form, whereas high-grade fetal adenocarcinoma frequently coexists with other conventional adenocarcinomas and requires at least 50 % fetal morphology for its diagnosis.

Clinically, low-grade fetal adenocarcinoma occurs in relatively young population with a peak incidence in the fourth decade of life and with a slight female preponderance, whereas high-grade fetal adenocarcinoma occurs predominantly in male heavy smokers [69–71]. High-grade fetal pattern as a minor component of a tumor, however, can be seen more widely in age and gender [72].

Histopathologically, both low-grade and high-grade tumors are characterized by neoplastic columnar cells with glycogen-rich clear cytoplasm in complex papillotubular structures. Low-grade tumors have characteristically small and round nuclei of mild atypia and show morules or cell balls in most cases, whereas high-grade tumors show more obvious nuclear atypia and lack morular formation. Neuroendocrine cells are often admixed with the glandular component. Other types of carcinoma such as large cell neuroendocrine carcinoma, hepatoid adenocarcinoma, and choriocarcinoma may be seen in association with high-grade fetal adenocarcinoma [71, 72]. TTF1 is expressed in low-grade tumors, whereas its expression is often diminished or absent in high-grade tumors [71, 72].

Genetically, low-grade fetal adenocarcinoma is characterized by frequent *β-catenin* gene mutations with aberrant nuclear/cytoplasmic localization of the protein [68, 73], whereas high-grade fetal adenocarcinoma lacks the mutation, rarely showing major driver mutations of conventional pulmonary adenocarcinomas such as *EGFR*, *KRAS*, and *PIK3CA* mutations [71–73]. Somewhat surprisingly, *DICER 1* mutation, which is a characteristic genetic feature of pleuropulmonary blastoma, has recently been reported in a case of low-grade fetal adenocarcinoma occurring in a patient with *DICER1* syndrome [74, 75].

The prognosis of fetal adenocarcinoma is not fully elucidated because of the rarity of the tumors. Low-grade fetal adenocarcinomas are usually found at stage I and show an indolent behavior with approximately 10 % tumor death rate [67], whereas high-grade fetal adenocarcinomas are often found at more advanced stages and show much higher mortality rates [69, 71, 72] .

1.1.5.4 Enteric Adenocarcinoma (Fig. 1.13)

This variant is simply defined as an adenocarcinoma that resembles colorectal adenocarcinomas [1]. Adenocarcinomas may partially take this form, and tumors that show this component at least 50 % of the whole are diagnosed as this variant. This is a very rare tumor; all previous studies on this tumor have been based on a single case or a series of less than ten cases [76–86]. Clinically, this tumor occurs in both sexes almost equally with a median age of 66 [81]. Smoking may be related to the development of this variant [81, 82].

Histopathologically, enteric adenocarcinoma shows acinar, cribriform, or papillotubular structures lined by columnar cells with eosinophilic cytoplasm and brush

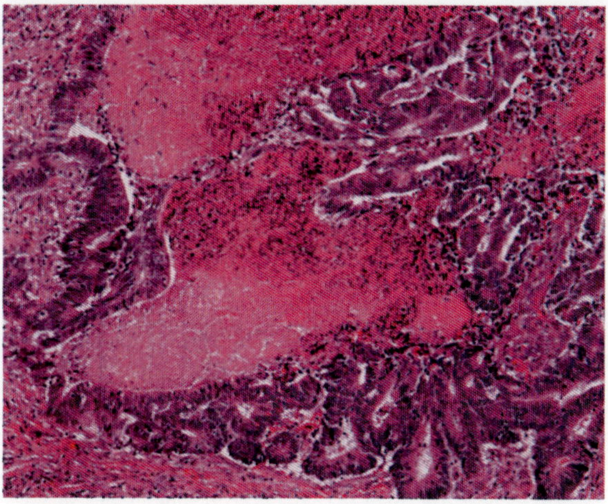

Fig. 1.13 Enteric adenocarcinoma. The neoplastic cells form glandular structures resembling colorectal adenocarcinoma. Central necrosis is prominent

borders just like conventional colorectal adenocarcinomas [1]. Central necrosis is common. Thus, it is mandatory to rule out the possibility of a metastasis of colorectal origin, especially if the tumor is entirely enteric in morphology. Immunohistochemically, the expression of CK7 is retained in the majority of the reported cases, and TTF1 over half of the cases, but the expression of intestinal markers such as CK20 and CDX2 is also noted approximately in one third and a half of the cases, respectively [79, 80]. Rare cases have also been reported in which tumor cells revealed a completely intestinal immunophenotype, i.e., CK7–, TTF1–, CK20+, and CDX2+ [80, 82, 83].

The genetic profile of this variant is not well known. A few cases revealed *KRAS* mutations [83, 85, 86] and *EGFR* mutation [83]. A rare *KRAS* Q22K mutation with concomitant *KRAS* polysomy was noted in one case, which could be related to the aggressive clinical course [85]. A recent MicroRNA profiling of this tumor disclosed similarities to non-small cell lung carcinoma and some overlap with pancreatic ductal adenocarcinoma [86].

Prognostically, it is not certain if this variant behaves differently from conventional invasive adenocarcinomas of the lung [82, 84].

References

1. Travis WD, Brambilla E, Burke AP et al (2015) WHO classification of tumours of the lung, pleura, thymus and heart, 4th edn. International Agency for Research on Cancer Press, Lyon
2. Travis WD, Brambilla E, Nicholson AG (2015) The 2015 World health organization classification of lung tumors: impact of genetic, clinical and radiologic advances since the 2004 classification. J Thorac Oncol 10:1243–1260. doi:10.1097/JTO.0000000000000630
3. Travis WD, Brambilla E, Noguchi M et al (2011) The new IASLC/ATS/ERS international multidisciplinary lung adenocarcinoma classification. J Thorac Oncol 6:244–285

4. Travis WD, Brambilla E, Noguchi M et al (2013) Diagnosis of lung cancer in small biopsies and cytology: implications of the 2011 International Association for the Study of Lung Cancer/ American Thoracic Society/ European Respiratory Society classification. Arch Pathol Lab Med 137:668–684
5. Nakayama H, Noguchi M, Tsuchiya R et al (1990) Clonal growth of atypical adenomatous hyperplasia of the lung: cytofluorometric analysis of nuclear DNA content. Mod Pathol 3:314–320
6. Kitamura H, Kameda Y, Ito T et al (1990) Atypical adenomatous hyperplasia of the lung. Implications for the pathogenesis of peripheral lung adenocarcinoma. Am J Clin Pathol 111:610–622
7. Takamochi K, Ogura T, Suzuki K et al (2001) Loss of heterozygosity on chromosomes 9q and 16p in atypical adenomatous hyperplasia concomitant with adenocarcinoma of the lung. Am J Pathol 159:1941–1948
8. Sakamoto H, Shimizu J, Horio Y et al (2007) Disproportionate representation of KRAS gene mutation in atypical adenomatous hyperplasia, but even distribution of EGFR gene mutation from preinvasive to invasive adenocarcinomas. J Pathol 212:287–294
9. Soh J, Toyooka S, Ichihara S et al (2008) Sequential molecular changes during multistage pathogenesis of small peripheral adenocarcinomas of the lung. J Thorac Oncol 3:340–347. doi:10.1097/JTO.0b013e318168d20a
10. Yoo SB, Chung JH, Lee HJ, Lee CT et al (2010) Epidermal growth factor receptor mutation and p53 overexpression during the multistage progression of small adenocarcinoma of the lung. J Thorac Oncol 5:964–969. doi:10.1097/JTO.0b013e3181dd15c0
11. Yoshida Y, Shibata T, Kokubu A et al (2005) Mutations of the epidermal growth factor receptor gene in atypical adenomatous hyperplasia and bronchioloalveolar carcinoma of the lung. Lung Cancer 50:1–8
12. Kobayashi Y, Mitsudomi T, Sakao Y et al (2015) Genetic features of pulmonary adenocarcinoma presenting with ground-glass nodules: the differences between nodules with and without growth. Ann Oncol 26:156–161. doi:10.1093/annonc/mdu505
13. Izumchenko E, Chang X, Brait M et al (2015) Targeted sequencing reveals clonal genetic changes in the progression of early lung neoplasms and paired circulating DNA. Nat Commun 16:8258. doi:10.1038/ncomms9258
14. Kakinuma R, Muramatsu Y, Kusumoto M et al (2015) Solitary pure ground-glass nodules 5 mm or smaller: frequency of growth. Radiology 276:873–882. doi:10.1148/radiol.2015141071
15. Travis WD, Asamura H, Bankier AA et al (2016) The IASLC Lung Cancer Staging Project: proposals for coding T categories for subsolid nodules and assessment of tumor size in part-solid tumors in the forthcoming eighth edition of the TNM classification of lung cancer. J Thorac Oncol 11:1204–1223. doi:10.1016/j.jtho.2016.03.025
16. Miyata N, Endo M, Nakajima T et al (2015) High-resolution computed tomography findings of early mucinous adenocarcinomas and their pathologic characteristics in 22 surgically resected cases. Eur J Radiol 84:993–997. doi:10.1016/j.ejrad.2015.01.014
17. Noguchi M, Morikawa A, Kawasaki M et al (1995) Small adenocarcinoma of the lung. Histological characteristics and prognosis. Cancer 75:2844–2852
18. Tsuta K, Ishii G, Nitadori J et al (2006) Comparison of the immunophenotypes of signet-ring cell carcinoma, solid adenocarcinoma with mucin production, and mucinous bronchioloalveolar carcinoma of the lung characterized by the presence of cytoplasmic mucin. J Pathol 209:78–87
19. Sato S, Motoi N, Hiramatsu M et al (2015) Pulmonary adenocarcinoma in situ: analyses of a large series with reference to smoking, driver mutations, and receptor tyrosine kinase pathway activation. Am J Surg Pathol 39:912–921. doi:10.1097/PAS.0000000000000458
20. Yoshizawa A, Sumiyoshi S, Sonobe M et al (2013) Validation of the IASLC/ATS/ERS lung adenocarcinoma classification for prognosis and association with EGFR and KRAS gene mutations: analysis of 440 Japanese patients. J Thorac Oncol 8:52–61

21. Tsuta K, Kawago M, Inoue E et al (2013) The utility of the proposed IASLC/ATS/ERS lung adenocarcinoma subtypes for disease prognosis and correlation of driver gene alterations. Lung Cancer 81:371–376
22. Nakagiri T, Sawabata N, Morii E et al (2014) Evaluation of the new IASLC/ATS/ERS proposed classification of adenocarcinoma based on lepidic pattern in patients with pathological stage IA pulmonary adenocarcinoma. Gen Thorac Cardiovasc Surg 62:671–677. doi:10.1007/s11748-014-0429-3
23. Yanagawa N, Shiono S, Abiko M et al (2014) The correlation of the International Association for the Study of Lung Cancer (IASLC)/American Thoracic Society (ATS)/European Respiratory Society (ERS) classification with prognosis and EGFR mutation in lung adenocarcinoma. Ann Thorac Surg 98:453–458. doi:10.1016/j.athoracsur.2014.04.108
24. Kadota K, Villena-Vargas J, Yoshizawa A et al (2014) Prognostic significance of adenocarcinoma in situ, minimally invasive adenocarcinoma, and nonmucinous lepidic predominant invasive adenocarcinoma of the lung in patients with stage I disease. Am J Surg Pathol 38:448–460
25. Eto T, Suzuki H, Honda A et al (1996) The changes of the stromal elastotic framework in the growth of peripheral lung adenocarcinomas. Cancer 77:646–656
26. Suzuki K, Yokose T, Yoshida J et al (2000) Prognostic significance of the size of central fibrosis in peripheral adenocarcinoma of the lung. Ann Thorac Surg 69:893–897
27. Yokose T, Suzuki K, Nagai K et al (2000) Favorable and unfavorable morphological prognostic factors in peripheral adenocarcinoma of the lung 3 cm or less in diameter. Lung Cancer 29:179–188
28. Terasaki H, Niki T, Matsuno Y et al (2003) Lung adenocarcinoma with mixed bronchioloalveolar and invasive components: clinicopathological features, subclassification by extent of invasive foci, and immunohistochemical characterization. Am J Surg Pathol 27:937–951
29. Sakurai H, Maeshima A, Watanabe S et al (2004) Grade of stromal invasion in small adenocarcinoma of the lung: histopathological minimal invasion and prognosis. Am J Surg Pathol 28:198–206
30. Minami Y, Matsuno Y, Iijima T et al (2005) Prognostication of small-sized primary pulmonary adenocarcinomas by histopathological and karyometric analysis. Lung Cancer 48:339–348
31. Kurokawa T, Matsuno Y, Noguchi M et al (1994) Surgically curable "early" adenocarcinoma in the periphery of the lung. Am J Surg Pathol 18:431–438
32. Maeshima AM, Niki T, Maeshima A et al (2002) Modified scar grade: a prognostic indicator in small peripheral lung adenocarcinoma. Cancer 95:2546–2554
33. Yoshizawa A, Motoi N, Riely GJ et al (2011) Impact of proposed IASLC/ATS/ERS classification of lung adenocarcinoma: prognostic subgroups and implications for further revision of staging based on analysis of 514 stage I cases. Mod Pathol 24:653–664
34. Russell PA, Wainer Z, Wright GM et al (2011) Does lung adenocarcinoma subtype predict patient survival?: a clinicopathologic study based on the new International Association for the Study of Lung Cancer/American Thoracic Society/European Respiratory Society international multidisciplinary lung adenocarcinoma classification. J Thorac Oncol 6:1496–1504. doi:10.1097/JTO.0b013e318221f701
35. Gu J, Lu C, Guo J et al (2013) Prognostic significance of the IASLC/ATS/ERS classification in Chinese patients-a single institution retrospective study of 292 lung adenocarcinoma. J Surg Oncol 107:474–480
36. Hung JJ, Yeh YC, Jeng WJ et al (2014) Predictive value of the international association for the study of lung cancer/American Thoracic Society/European Respiratory Society classification of lung adenocarcinoma in tumor recurrence and patient survival. J Clin Oncol 32:2357–2364
37. Cha MJ, Lee HY, Lee KS et al (2014) Micropapillary and solid subtypes of invasive lung adenocarcinoma: clinical predictors of histopathology and outcome. J Thorac Cardiovasc Surg 147:921–928

38. Yatabe Y, Kosaka T, Takahashi T et al (2005) EGFR mutation is specific for terminal respiratory unit type adenocarcinoma. Am J Surg Pathol 29:633–639
39. Mäkinen JM, Laitakari K, Johnson S (2015) Nonpredominant lepidic pattern correlates with better outcome in invasive lung adenocarcinoma. Lung Cancer 90:568–574. doi:10.1016/j.lungcan.2015.10.014
40. Warth A, Muley T, Meister M et al (2012) The novel histologic International Association for the Study of Lung Cancer/American Thoracic Society/European Respiratory Society classification system of lung adenocarcinoma is a stage-independent predictor of survival. J Clin Oncol 30:1438–1446
41. Inamura K, Takeuchi K, Togashi Y (2009) EML4-ALK lung cancers are characterized by rare other mutations, a TTF-1 cell lineage, an acinar histology, and young onset. Mod Pathol 22:508–515. doi:10.1038/modpathol.2009.2
42. Yoshida A, Tsuta K, Nakamura H et al (2011) A comprehensive histologic analysis of ALK-rearranged lung carcinomas. Am J Surg Pathol 35:1226–1234. doi:10.1097/PAS.0b013e3182233e06
43. Kadota K, Yeh YC, Sima CS et al (2014) The cribriform pattern identifies a subset of acinar predominant tumors with poor prognosis in patients with stage I lung adenocarcinoma: a conceptual proposal to classify cribriform predominant tumors as a distinct histologic subtype. Mod Pathol 27:690–700
44. Warth A, Muley T, Harms A et al (2016) Clinical relevance of different papillary growth patterns of pulmonary adenocarcinoma. Am J Surg Pathol 40:818–826. doi:10.1097/PAS.0000000000000622
45. Kadota K, Nitadori J, Sima CS et al (2015) Tumor spread through air spaces is an important pattern of invasion and impacts the frequency and location of recurrences after limited resection for small stage I lung adenocarcinomas. J Thorac Oncol 10:806–814
46. Lee G, Lee HY, Jeong JY et al (2015) Clinical impact on minimal micropapillary pattern in invasive lung adenocarcinoma: prognostic significance and survival outcomes. Am J Surg Pathol 39:660–666
47. Nitadori J, Bograd AJ, Kadota K et al (2013) Impact of micropapillary histologic subtype in selecting limited resection vs lobectomy for lung adenocarcinoma of 2cm or smaller. J Natl Cancer Inst 105:1212–1220
48. Rekhtman N, Tafe LJ, Chaft JE et al (2013) Distinct profile of driver mutations and clinical features in immunomarker-defined subsets of pulmonary large-cell carcinoma. Mod Pathol 26:511–522. doi:10.1038/modpathol.2012.195
49. Rossi G, Mengoli MC, Cavazza A et al (2014) Large cell carcinoma of the lung: clinically oriented classification integrating immunohistochemistry and molecular biology. Virchows Arch 464:61–68. doi:10.1007/s00428-013-1501-6
50. Hwang DH, Szeto DP, Perry AS et al (2014) Pulmonary large cell carcinoma lacking squamous differentiation is clinicopathologically indistinguishable from solid-subtype adenocarcinoma. Arch Pathol Lab Med 138(5):626–635. doi:10.5858/arpa.2013-0179-OA
51. Rekhtman N, Ang DC, Riely GJ et al (2013) KRAS mutations are associated with solid growth pattern and tumor-infiltrating leukocytes in lung adenocarcinoma. Mod Pathol 26(10):1307–1319. doi:10.1038/modpathol.2013.74
52. Driver BR, Portier BP, Mody DR et al (2016) Next-generation sequencing of a cohort of pulmonary large cell carcinomas reclassified by World Health Organization 2015 Criteria. Arch Pathol Lab Med 140(4):312–317. doi:10.5858/arpa.2015-0361-OA
53. Solis LM, Behrens C, Raso MG et al (2012) Histologic patterns and molecular characteristics of lung adenocarcinoma associated with clinical outcome. Cancer 118:2889–2899
54. Ujiie H, Kadota K, Chaft JE et al (2015) Solid predominant histologic subtype in resected stage I lung adenocarcinoma is an independent predictor of early, extrathoracic, multisite recurrence and of poor postrecurrence survival. J Clin Oncol 33:2877–2884
55. Austin JH, Garg K, Aberle D et al (2013) Radiologic implications of the 2011 classification of adenocarcinoma of the lung. Radiology 266:62–71. doi:10.1148/radiol.12120240

56. Wu J, Chu PG, Jiang Z et al (2013) Napsin a expression in primary mucin-producing adenocarcinomas of the lung: an immunohistochemical study. J Clin Pathol 139(2):160–166. doi:10.1309/AJCP62WJUAMSZCOM
57. Sugano M, Nagasaka T, Sasaki E et al (2013) HNF4α as a marker for invasive mucinous adenocarcinoma of the lung. Am J Surg Pathol 37:211–218. doi:10.1097/PAS.0b013e31826be303
58. Geles A, Gruber-Moesenbacher U, Quehenberger F et al (2015) Pulmonary mucinous adenocarcinomas: architectural patterns in correlation with genetic changes, prognosis and survival. Virchows Arch 467(6):675–686
59. Kakegawa S, Shimizu K, Sugano M et al (2011) Clinicopathological features of lung adenocarcinoma with KRAS mutations. Cancer 15(117):4257–4266. doi:10.1002/cncr.26010
60. Hata A, Katakami N, Fujita S et al (2010) Frequency of EGFR and KRAS mutations in Japanese patients with lung adenocarcinoma with features of the mucinous subtype of bronchioloalveolar carcinoma. J Thorac Oncol 5:1197–1200. doi:10.1097/JTO.0b013e3181e2a2bc
61. Finberg KE, Sequist LV, Joshi VA et al (2007) Mucinous differentiation correlates with absence of EGFR mutation and presence of KRAS mutation in lung adenocarcinomas with bronchioloalveolar features. J Mol Diagn 2007(9):320–326
62. Fernandez-Cuesta L, Plenker D, Osada H et al (2014) CD74-NRG1 fusions in lung adenocarcinoma. Cancer Discov 4:415–422. doi:10.1158/2159-8290.CD-13-0633
63. Shim HS, Kenudson M, Zheng Z et al (2015) Unique genetic and survival characteristics of invasive mucinous adenocarcinoma of the lung. J Thorac Oncol 10:1156–1162. doi:10.1097/JTO.0000000000000579
64. Nakaoku T, Tsuta K, Ichikawa H et al (2014) Druggable oncogene fusions in invasive mucinous lung adenocarcinoma. Clin Cancer Res 20:3087–3093. doi:10.1158/1078-0432.CCR-14-0107
65. Ichinokawa H, Ishii G, Nagai K et al (2013) Distinct clinicopathologic characteristics of lung mucinous adenocarcinoma with KRAS mutation. Hum Pathol 44:2636–2642. doi:10.1016/j.humpath.2013.05.026
66. Kettle R, Simmons J, Schindler F et al (2010) Regulation of neuregulin 1beta1-induced MUC5AC and MUC5B expression in human airway epithelium. Am J Respir Cell Mol Biol 42:472–481. doi:10.1165/rcmb.2009-0018OC
67. Rossi G, Murer B, Cavazza et al (2004) Primary mucinous (so-called colloid) carcinomas of the lung: a clinicopathologic and immunohistochemical study with special reference to CDX-2 homeobox gene and MUC2 expression. Am J Surg Pathol 28(4):442–452.
68. Zenali MJ, Weissferdt A, Solis LM et al (2015) An update on clinicopathological, immunohistochemical, and molecular profiles of colloid carcinoma of the lung. Hum Pathol 46:836–842. doi:10.1016/j.humpath.2014.10.032
69. Nakatani Y, Kitamura H, Inayama Y et al (1998) Pulmonary adenocarcinomas of the fetal lung type: a clinicopathologic study indicating differences in histology, epidemiology, and natural history of low-grade and high-grade forms. Am J Surg Pathol 22:399–411
70. Nakatani Y, Masudo K, Miyagi Y et al (2002) Aberrant nuclear localization and gene mutation of beta-catenin in low-grade adenocarcinoma offetal lung type: up-regulation of the Wnt signaling pathway may be a common denominator for the development of tumors that form morules. Mod Pathol 15:617–624
71. Morita, AJSP, 2013
72. Suzuki M, Yazawa T, Ota S et al (2015) High-grade fetal adenocarcinoma of the lung is a tumour with a fetal phenotype that shows diverse differentiation, including high-grade neuroendocrine carcinoma: a clinicopathological, immunohistochemical mutational study of 20 cases. Histopathology 67:806–816
73. Sekine S, Shibata T, Matsuno Y et al (2003) Beta-catenin mutations in pulmonary blastomas: association with morule formation. J Pathol 200:214–221
74. Wu Y, Chen D, Li Y et al (2014) DICER1 mutations in a patient with an ovarian Sertoli-Leydig tumor, well-differentiated fetal adenocarcinoma of the lung, and familial multinodular goiter. Eur J Med Genet 57:621–625

75. Kock L, Bah I, Wu Y et al (2016) Germline and somatic DICER1 mutations in a well-differentiated fetal adenocarcinoma of the lung. J Thorac Oncol 11(3):e31–e33. doi:10.1016/j.jtho.2015.09.012
76. Inamura K, Satoh Y, Okumura S et al (2005) Pulmonary adenocarcinomas with enteric differentiation: histologic and immunohistochemical characteristics compared with metastatic colorectal cancers and usual pulmonary adenocarcinomas. Am J Surg Pathol 29(5):660–665
77. Yousem SA (2005) Pulmonary intestinal-type adenocarcinoma does not show enteric differentiation by immunohistochemical study. Mod Pathol 18:816–821
78. Maeda R, Isowa N, Onuma H et al (2008) Pulmonary intestinal-type adenocarcinoma. Interact Cardiovasc Thorac Surg 7:349–351. doi:10.1510/icvts.2007.168716
79. Li HC, Schmidt L, Greenson JK et al (2009) Primary pulmonary adenocarcinoma with intestinal differentiation mimicking metastatic colorectal carcinoma: case report and review of literature. Am J Clin Pathol 131:129–133. doi:10.1309/AJCPB04XWICTFERL
80. Hatanaka K, Tsuta K, Watanabe K et al (2011) Primary pulmonary adenocarcinoma with enteric differentiation resembling metastatic colorectal carcinoma: a report of the second case negative for cytokeratin 7. Pathol Res Pract 207:188–191. doi:10.1016/j.prp.2010.07.005
81. Lin D, Zhao Y, Li H et al (2013) Pulmonary enteric adenocarcinoma with villin brush border immunoreactivity: a case report and literature review. J Thorac Dis 5(1):E17–E20. doi:10.3978/j.issn.2072-1439.2012.06.06
82. Wang CX, Liu B, Wang YF et al (2014) Pulmonary enteric adenocarcinoma: a study of the clinicopathologic and molecular status of nine cases. Int J Clin Exp Pathol 7:1266–1274
83. László T, Lacza A, Tóth D et al (2014) Pulmonary enteric adenocarcinoma indistinguishable morphologically and immunohistologically from metastatic colorectal carcinoma. Histopathology 65:283–287. doi:10.1111/his.12403
84. Handa Y, Kai Y, Ikeda T et al (2015) Pulmonary enteric adenocarcinoma. Gen Thorac Cardiovasc Surg Jul 3. [Epub ahead of print]
85. Metro G, Valtorta E, Siggillino A et al (2015) Enteric-type adenocarcinoma of the lung harbouring a novel KRAS Q22K mutation with concomitant KRAS polysomy: a case report. Ecancermedicalscience 9:559. doi:10.3332/ecancer.2015.559
86. Garajová I, Funel N, Fiorentino M et al (2015) MicroRNA profiling of primary pulmonary enteric adenocarcinoma in members from the same family reveals some similarities to pancreatic adenocarcinoma-a step towards personalized therapy. Clin Epigenetics 7:129. doi:10.1186/s13148-015-0162-5

Chapter 2
Screening Lung Cancer with Low-Dose CT Combined with Molecular Markers

Yuichi Takiguchi

Abstract Early detection and surgical resection at an early stage is the bottom line for curing lung cancer. As attempts for reducing mortality from lung cancer with chest radiograph screening was unsuccessful as shown by Mayo Lung Project and the Prostate, Lung, Colorectal, and Ovarian Cancer Screening Study, screening with low-dose CT (LDCT) has been investigated for two decades. Despite ostensibly promising results from single-arm studies of LDCT in increased detections of early-stage lung cancer, early small-scale randomized studies for LDCT screening failed to show a reduction in lung cancer mortality compared to the control, suggesting an involvement of length bias or overdiagnosis bias. The National Lung Screening Trial was the first to demonstrate a reduced mortality of lung cancer by LDCT, whereas the results provided some issues relating to high false-positive rate, cost-effectiveness, and overdiagnosis. Combination of other diagnostic modalities such as computer-aided diagnosis of LDCT and serum/plasma molecular markers, together with tobacco control, may enhance the positive aspects and suppress the negative aspects of the LDCT screening.

Keywords Low-dose CT • CT screening • Biomarker • Early detection • Lung cancer

2.1 Introduction

Lung cancer is a leading cause of death in the developed countries and is estimated to account for 1.59 million deaths worldwide at 2012 [1] and 77,200 deaths in Japan at 2015 [2]. Despite the recent advancement of chemotherapy, molecular targeted therapy and immune-checkpoint therapy, lung cancer at an advanced stage is still incurable with a limited median overall survival time. Surgical resection with or without adjuvant chemotherapy at an early stage solely provides an opportunity to

Y. Takiguchi, M.D., Ph.D. (✉)
Department of Medical Oncology, Graduate School of Medicine,
Chiba University, Chiba, Japan
e-mail: takiguchi@faculty.chiba-u.jp

© Springer Science+Business Media Singapore 2017

Y. Takiguchi (ed.), *Molecular Targeted Therapy of Lung Cancer*,
DOI 10.1007/978-981-10-2002-5_2

cure the disease, except for a smaller chance of cure at a locally advanced stage when it was treated with a definitive chemoradiotherapy. Early detection and surgical resection of lung cancer, therefore, is the bottom line to reduce deaths from lung cancer.

The first randomized study for evaluating effectiveness of early detection of lung cancer, the Mayo Lung Project (MLP), compared a screening with chest radiography in every 4 months (the intervention group) with a screening with chest radiograph in every year (the control group) in high-risk group for the development of lung cancer, that is, in male heavy smokers [3, 4]. Among 4618 individuals assigned to the intervention group, 206 cases (4.5 %) with lung cancer were detected, whereas 160 cases (3.5 %) were detected among 4593 individuals assigned to the control group. Deaths from lung cancer were observed in 122 cases (2.6 %) in the former and in 115 (2.5 %) in the latter groups, respectively. Beside the disappointing result of the absent effect in reducing deaths from lung cancer, an overdiagnosis rate of as high as 29 % (206/160) in the intervention group over the control group was rather puzzling [3, 4]. To exclude a possibility of a length bias, an additional analysis with an extended follow-up period as long as 20 years was conducted, whereas this follow-up study also failed to resolve the puzzle [5]. A recent and the second randomized controlled trial evaluating the efficacy in reducing lung cancer mortality by means of screening with chest radiograph was a part of the Prostate, Lung, Colorectal, and Ovarian (PLCO) trial, and the results were published at 2011 [6]. Cumulative lung cancer incidence rates through 13-year follow-up period were comparable between the chest radiograph screening group (20.1 per 10,000 person-years) and the control group (19.2 per 10,000 person-years), and lung cancer deaths were also comparable between the two groups (1213 vs. 1230, resulting in a relative ratio of 0.99 with 95 % CI of 0.87–1.22). Again, screening lung cancer with chest radiograph failed to reduce mortality from lung cancer.

On the other hand, in Japan, screening of lung cancer that is conjunct with screening of pulmonary tuberculosis, mainly with a 10 × 10 cm miniature chest radiograph, has been carried out for decades as a government public health administrative policy. Although many retrospective and prospective cohort studies had suggested a reduction of deaths from lung cancer [7, 8], the degree of the effectiveness was judged small by a reconnaissance report by a government-funded evaluation team. The report concluded that further studies elucidating the efficacy of lung cancer screening taking advantage of more efficient modalities, such as CT, to detect early-stage lung cancer were required. The National Lung Screening Trial (NLST), at 2011, eventually disclosed a 20 % reduction in lung cancer mortality in a high-risk population by means of CT screening over chest radiograph screening. This chapter is overviewing the history of study on lung cancer screening until the NLST, major outcomes and limitations of NLST, and future study directions taking advantage of some new technologies such as computer-aided diagnosis and genome analysis.

2.2 History of Study on Early Detection of Lung Cancer with CT

2.2.1 *Single-Arm Retrospective or Prospective Studies*

The study on CT screening for lung cancer was initiated at the late 1190s followed by a number of single-arm retrospective or prospective studies [9–14] for individuals at high risk for lung cancer or for general residents. They showed promising results including the high detection rates (ranging 1–2.7 % in high-risk population [9, 11–14], and approximately 0.5 % in general resident [10, 15] and in a industry employee [16]), high rate (ranging from 80 to 90 %) of patients with Stage I in patients with detected lung cancer, good therapeutic outcomes [17], and that low-dose CT (LDCT) to minimize radiation exposure is sufficient to detect lung nodules suspicious for lung cancer [18]. For example, a Japanese study reported significantly higher detection rate of peripheral lung cancer by LDCT (0.43 %, 15/3457) than by chest radiograph (0.12 %, 4/3457) in the same study population [9]. An additional Japanese study conducted an LDCT screening for individuals whose chest radiograph screening results were turned out to be negative and found ten times more lung cancers in the LDCT screening (0.454 %, 15/3305) compared to the chest radiograph screening (0.044 %, 10/22,720 person-years) during a 4-year study period (Table 2.1) [15]. Therefore, its usefulness in terms of reducing mortality from lung cancer became the next major research topic. To minimize substantial bias, randomized studies elucidating a potential of LDCT screening to reduce mortality from lung cancer were warranted.

2.2.2 *Small-Scale Randomized Studies*

At least, three small-scale randomized studies on lung cancer screening with LDCT involving less than 5000 participants had reported the results (Table 2.2). Very disappointingly, none of them was successful in proving reduced mortality from lung

Table 2.1 Lung cancer detection rate in early study for LDCT

Ref #	Year published	Subject	Number[a]	Positive rate	Detected Lung cancer	Detection rate	Ratio of Stage I
[9]	1996	High risk	3457	17 %	15	0.43 %	93 %
[10]	1998	General resident	5483	2 %	19[b]	0.347 %[b]	84 %
[12]	2001	High risk	59,023	10 %	484	0.82 %	85 %
[15]	2008	General resident (LDCT)[c]	3305	10 %	15	0.454	100 %
[15]	2008	General resident (Radiography)[c]	22,720	1.4 %	10	0.044	60 %

[a]Cumulative number including initial and repeat screening
[b]Included two patients with atypical adenomatous hyperplasia
[c]Comparison between screenings with LDCT and chest radiography

Table 2.2 Major outcome from initial randomized studies on LDCT lung cancer screening

Study	Year published	Ref #	Control	Number	Follow-up period (median)	Mortality from lung cancer (%)		
						LDCT	Control	*P* value
DANTE	2009	[19]	Usual care	2472	34 M	20 (1.6)	10 (1.7)	0.83
DLCST	2012	[20]	Usual care	4104	58 M	15 (0.7)	11 (0.5)	0.43
MILD	2012	[21]	Usual care	4099	53 M	12/6 (1.0/0.5)[a]	7 (0.4)	0.21

[a]Annual/biennial screening

cancer by means of LDCT. The Detection and Screening of Early Lung Cancer by Novel Imaging Technology and Molecular Essays (DANTE) trial [19] conducted in Italy with a median follow-up period of 33.7 months screened 1276 individuals at high risk for lung cancer to detect 47 lung cancers. In addition, symptomatic 13 lung cancers were diagnosed outside screening, resulting in 60 lung cancers in total in the screening group (detection rate = 4,7 %; 60/1276). Among them, 33 patients (55 %; 33/60) were at Stage I. However, 20 patients among the 60 lung cancers were eventually fatal. On the other hand, in the control group consisting of 1196 individuals who were not screened as to lung cancer, 34 lung cancers were diagnosed (detection rate = 2.9 %; 34/1196) including 12 patients at Stage I (3.5 %; 12/34), and death occurred in 20 patients among them. The Danish Lung Cancer Screening Trial (DLCST) [20] with a median follow-up period of 58 months screened 2052 individuals at high risk for lung cancer and resulted in the diagnosis of lung cancer in 69 individuals including the diagnosis outside the screening (detection rate = 3,4 %; 69/2052). Among them, 47 patients (68.1 %; 47/69) were at Stage I. In the control group consisting of 2052 individuals who were not screened as to lung cancer, 24 lung cancers were diagnosed (detection rate = 1.2 %; 24/2052) throughout the follow-up period. In both studies, although the detection rates and ratios of early stage in the diagnosed lung cancer were significantly higher in the screening group than in the control group, mortalities from lung cancer were not reduced in the screening group. The Multicentric Italian Lung Detection (MILD) [21] also conducted in Italy for individuals at high risk. Again, detection rate with LDCT was higher (0.457 % with biennial LDCT screening and 0.620 % with annual LDCT screening) than that with chest radiograph (0.311 %). Among lung cancers detected by LDCT, 63 % were at Stage I. However, there was no reduction in lung cancer mortality with LDCT screening compared to chest radiograph screening of 0.109 % with biennial LDCT screening and 0.216 % with annual LDCT screening compared to 0.109 % with chest radiograph screening. These results clearly indicated the efficiency of LDCT screening in detecting lung cancers at high rate and at earlier stage, whereas the primary endpoint of efficacy in reducing mortality from lung cancer was not shown. However, these studies had common profound problems of small sample size and short follow-up period [22].

Another Italian study ITALUNG [23, 24] involving 3206 participants awaits us to review the results. A larger randomized study conducted in the Netherlands and Belgium (NELSON) [25, 26] involving 15,822 participants at high risk for lung cancer is also expected to provide the results in 2016.

2.3 Major Outcomes of the NLST

2.3.1 The First Positive Study of a Large-Scale Randomized Study

In contrast to the negative results in the previously mentioned three randomized studies involving relatively small study populations, the NLST is the first and only study at present that showed a statistically significant reduction in mortality from lung cancer by means of LDCT screening [27]. As the interim analysis disclosed a 20 % reduction in lung cancer mortality, the study was early terminated. This study is the largest in the scale, involving 53,454 participants with a median follow-up period of 6.5 years that is the longest one among the randomized studies mentioned above. Individuals at high risk for lung cancer defined by inclusion criteria of age ranging from 55 to 74 years, smoking history of more than 30 pack-years, and interval of smoking cessation of less than 15 years were enrolled to the study to be randomized either to the LDCT group (n=26,722) or to the chest radiograph (control) group (n=26,732). Participants in the LDCT or chest radiograph groups underwent annual screening with LDCT or chest radiograph, respectively, for 3 years with four screenings including the baseline screening. In LDCT group, noncalcified pulmonary nodules ≥4 mm in diameter were judged as positive. The primary study endpoint of reduced mortality from lung cancer was met, and the mortality from lung cancer was reduced by 20.0 % (95 % confidence interval, 6.8–26.7 %; $P = 0.004$) in the LDCT group compared to the chest radiograph group. In addition, mortality from all cause was reduced by 6.7 % (95 % CI, 1.2–13.6 %; $P = 0.02$) in the LDCT group compared to the chest radiograph group (Table 2.3). Lesser patients with Stage IV and more patients with Stage IA were diagnosed in the LDCT group than in the chest radiograph group (Table 2.4).

On the other hand, positive rates of screening were 24.2 and 6.9 % in the LDCT and chest radiograph groups, respectively. Because the frequencies of major complications relating to interventions after screening were similar in both groups (Table 2.5), absolute number of patients suffered from the major complication was significantly larger in the LDCT group. The high positive rate of 24.2 %, inevitably accompanied by more patients with complication, and a possible substantial extent of overdiagnosis seemed major concerns of the results. Incidence of lung cancer diagnosis (per 100,000 person-years) was 645 in the LDCT group and 572 in the chest radiograph group, respectively, suggesting an overdiagnosis in the LDCT

Table 2.3 Comparison of lung cancer or all-cause mortality between LDCT and chest radiography screening group in the NLST

Modality	Person-years (P-Y)	Death	Mortality per 100,000 P-Y	Reduction in mortality (%)
Lung cancer mortality				
CT	144,000	356	247	20.0 (p = 0.004)
Radiography	143,000	443	309	
All-cause mortality				
CT	167,000	1877	1123	6.7 (p = 0.02)
Radiography	166,000	2000	1205	

Table 2.4 Comparison of lung cancer stage distribution (in %) between LDCT and chest radiography screening group in the NLST

Modality	IA	IB	IIA	IIB	IIIA	IIIB	IV
LDCT	40.0	10.0	3.4	3.7	9.5	11.7	21.7
Radiography	21.1	10.0	3.4	4.5	11.7	13.1	36.1

Table 2.5 Comparison of frequencies of major complications with screening procedures between LDCT and chest radiography screening group in the NLST

		Frequency of major complications			
Modality	Positive rate	In positive screening	In positive result without lung cancer	In positive results with lung cancer	Deaths within 60 days after invasive procedures
LDCT	24.2 %	1.4 %	0.06 %	11.2 %	16 (10 with lung cancer)
Radiography	6.9 %	1.2 %	0.02 %	8.2 %	10 (all with lung cancer)

group over the chest radiograph group by as many as 73 per 100,000 person-years or 11.3 % (73/645) in the diagnosed lung cancer.

2.3.2 *Post Hoc Analyses of the NLST*

A series of post hoc analyses were performed and published after the publication of the major results. They included subset, cost-effectiveness, simulation, and other analyses. Stephanie et al. subcategorized the study population into five groups according to the risk for lung cancer. Reduction rates of lung cancer mortality were not different among the five subgroups, whereas lung cancer deaths in number prevented by LDCT were significantly different among them, and the number was larger in subgroups at higher risk than in subgroups at lower risk [28]. Extensive simulation analyses on the cost-effectiveness were also performed [29]. Excellent review articles comprehensively discuss on these issues [22, 30, 31]. Based on these studies and discussions, many US organizations published guidelines to recommend LDCT for individuals at high risk [32]. Among them, the US Preventive Service Task Force (USPSTF) proposed a guideline with a grade B recommendation for annual lung cancer screening with LDCT for individuals at age ranging from 55 to 80, with smoking history of more than 30 pack-years and with smoking cessation period of less than 15 years [33].

One of the most important post hoc analyses was on overdiagnosis caused by LDCT screening [34]. According to the study by Patz et al., the rates of overdiagnosis were 11.0 % (95 % CI, 3.2–18.2 %) overall, 14.4 % (6.1–21.8 %) in all non-small cell lung cancer, and as much as 67.6 % (95 % CI, 53.5–78.5 %) in bronchioloalveolar cell carcinoma (BAC) [35]. On the other hand, it would be also important to notice that the majority (86 %) of the overdiagnosis was caused by the initial screening with LDCT [36]. This fact may suggest that repeated screening in the same individuals might ameliorate the problem of overdiagnosis and may also suggest that the problem might be related to a lead-time bias but not to an overdiagnosis bias.

Recently, by analyzing the PLCO data, Pinsky and Kramer demonstrated a similar risk of developing lung cancer in individuals with smoking history of 20–29 pack-years compared to the NLST eligible individuals with smoking history ≥30 pack-years, warranting further evaluation of the LDCT in population with less smoking history [37, 38]. By the way, many studies linked chronic obstructive pulmonary disease (COPD) to an increase in the development of lung cancer [39]. A simulation study suggested a benefit from LDCT screening for individuals suffering from COPD with smoking history of less park-years than the NLST eligibility [40].

2.3.3 *Implementation of LDCT as a Public Health-Care Program in the USA*

Even after the recommendation by the USPSTF, debates existed whether to or not to implement LDCT screening as a public health-care program in the USA [41]. The International Association for the Study of Lung Cancer (IASLC), in response to the public comment recruitment to the USPSTF recommendation, raised several issues to be considered when implementing LDCT screening as a public health-care policy [42]. They included cost-effectiveness, radiological protocol including the value of volumetry for LDCT-detected pulmonary nodules, selection criteria for individuals to be screened in respect to age, smoking history, other risk factor assessment and comorbid conditions, and harms from LDCT screening. Based on the evidences from and extensive public discussions on the results of the NLST, the Center for Medicare and Medicaid Services (CMS) finally made a decision to approve the Medicare coverage for lung cancer screening with LDCT at 5 February in 2015 [37, 43]. Upon the same time of this decision, researchers of the CMS created major three challenges for translating research into policy and even into practice:

1. Eligible individuals must be accurately identified for screening as to age, smoking history, and willingness to adhere to a long-term screening program and undergo additional diagnostic procedures and treatment when necessary.
2. The screening must be performed as part of a cohesive screening program to enhance its likelihood of success, and the cohesive program includes an adherence to evidence-based careening with technical parameters for LDCT, criteria for radiologists and imaging center, the use of a standardized nodule-identification-and-reporting system, smoking cessation program, follow-up evaluation, and central registration.
3. Multidisciplinary involvement during the entire course from screening to treatment [43].

Then, they pointed out that population screening after implementing the LDCT screening in the public should confer similar benefits over time and minimizes risk as shown in the NLST study. Although it was solely a retrospective study, Nawa et al. disclosed a 24 % reduction (95 % CI of 14–33 %) in lung cancer mortality in a small district in Japan where lung cancer screening with LDCT had been implemented as a local government policy for 8 years [44]. Providing such data after implementing LDCT screening in the USA would further validate the usefulness of the screening.

2.4 Future Directions Beyond CT Screening

2.4.1 Computer-Aided Evaluation of CT Screening

One of the shortcomings of the NLST was a high positive rate and high false-positive rate throughout the screening. As they classified all noncalcified pulmonary nodules ≥4 mm as positive, the positive rate was as high as 24.2 % [45]. It would bring a substantial harm for individuals who were screened if the screening with a positive rate of 24.2 % were implemented as a health-care policy. Although the final results from the NELSON study have not been disclosed, results of a prespecified analysis assessing screening performance in a subset population (n = 7155) were published [46]. In this screening process where semiautomated volumetric evaluation of pulmonary nodules with software was utilized, screening results were initially classified as negative, indeterminate, or positive based on nodule presence and volume. Subsequently, individuals with an initial indeterminate result underwent follow-up screening to classify their final screening result as negative or positive based on volume doubling time calculated by the software. That is to say, noncalcified pulmonary nodules ≥500 mm^3 in volume, nodule volume doubling time <400 days plus increase in volume by ≥25 %, and newly appeared solid part in nonsolid nodules were judged as positive. Consequently, in contrast to the high positive rate (24.2 %) in the NLST, the positive rate was 2–3 %, and other parameters were excellent enough with sensitivity of 84.6 % (95% CI. 79.6–89.2), specificity of 98.6 %

Table 2.6 Comparison of major outcome parameters between the NLST and a subset of the NELSON

Parameters	NELSON	NLST
Positive rate	2.2 %	24.2 %
Positive predictive value	40.4 %	3.6 %[a]
Detected lung cancer	187 patients	356 patients
Detection rate	0.909 % (per person-years)	0.645 % (per person-years)
	2.6 % (per person)	1.3 % (per person)

[a]Calculated from the published data (Ref #27)

(95% CI, 98.5–98.8), positive predictive value of 40.4 % (95% CI, 35.9–44.7), and negative predictive value of 99.8 % (95% CI, 99.8–99.9). Although careful attention is needed because these data are still premature in contrast to the results from the NLST, comparison of the results between them seemed valuable as shown in Table 2.6. Computer-aided diagnosis (CAD) specific for lung cancer screening with LDCT has been developed to facilitate high sensitivity of LDCT screening without enforcing human labor burden [47, 48]. Although not fully established, a CAD equipped with automated volumetric software may further enhance sensitivity and suppress false-positive rate, resulting in excellent true positive predictive value.

2.4.2 *Molecular Marker-Assisted Lung Cancer Screening with LDCT*

Overdiagnosis is a distinct issue from the false-positive one. The issue is sometimes misunderstood because it is not recognized in clinical situation: it is solely recognized by statistical calculation based on randomized studies. It is an ultimate form of the length bias and is, in some extent, inevitable in any cancer screening [49–51]. Despite its direct harm to patients, patients with overdiagnosis are not distinguished, either with clinical or pathological findings, from patients with clinically meaningful lung cancer. Cancers of overdiagnosis would be at an early stage and have indolent nature in their biology. Molecular markers distinguishing biological aggressiveness from indolence of cancer, in theory, might help differentiate them.

On the other hand, extensive efforts are put in searching molecular markers that detect patients with early cancer or even individuals at high risk, although no clinically valid markers have been identified [52]. Although biomarkers in tumor or stromal tissue may identify indolent cancers in early-stage lung cancers [53, 54], an approach utilizing surgically resected samples cannot be used for identifying high-risk individuals suitable for enrolling into LDCT screening. A study population for large-scale LDCT screening is also an ideal study population for developing molecular markers for early detection of lung cancer. Individuals with or without lung cancer in the study population have unbiased comparable backgrounds including age range, gender ratio, comorbidities, and other conditions. Montani et al. developed

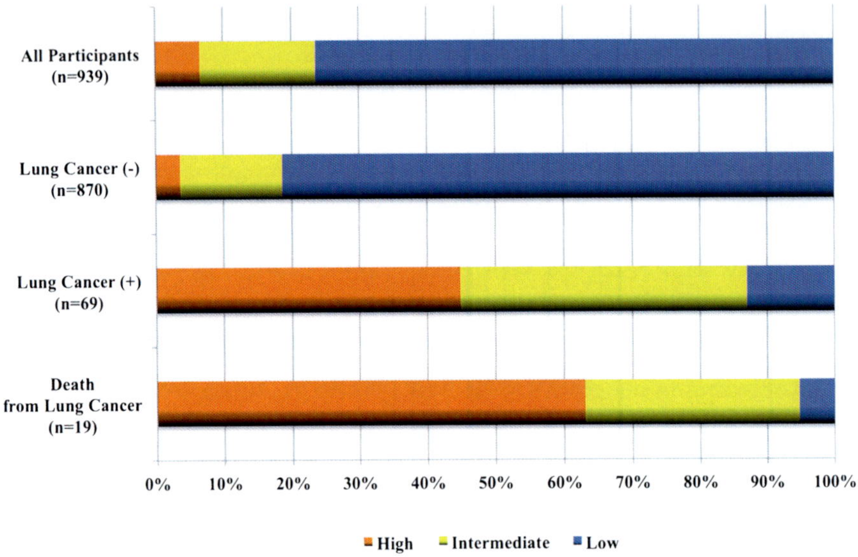

Fig. 2.1 Distribution of the risk grades accessed by the plasma miRNA signature classifiers (MSCs) according to demographics of participants in the MILD study. Frequency of participants with high or high+intermediate MSC grade increases according to the increasing risk of lung cancer

a serum microRNA (miRNA) signature, and performed a large-scale validation study of it in the study population of LDCT screening (the Continuous Observation of Smoking Subjects, COSMOS). The results were excellent with a sensitivity of 77.8 % (95 % CI, 64.2–91.4 %) and a specificity of 74.8 % (95% CI, 72.1–77.5 %) [55]. Boeri et al. also developed a plasma miRNA signature classifiers (MSCs) with 24 miRNAs [56] and validated its performance in the study population of the MILD trial [57]. They classified the test results of entire population into three categories: high risk, intermediate risk, and low risk, resulting in 6.7 % (63/939), 16.9 % (159/939), and 76.4 % (717/939) in each category, respectively. Individuals classified as high risk accounted for only 3.7 % in participants who were eventually proven to have no lung cancer (n = 870), whereas it accounted for 44.9 % in participants who were eventually proven to have lung cancer (n = 69) and 63.2 % in patients with lung cancer who were eventually dead from lung cancer (n = 19). Contrarily, ones classified as low risk accounted for 81.4, 13.0, and 5.3 % in the three populations, respectively (Fig. 2.1). These data suggest a high performance of this molecular test. Thereafter, they conducted a retrospective simulation analysis limited to participants assigned to the LDCT arm (n=652). They consisted of 594 participants without lung cancer and 58 participants who were eventually proven to have lung cancer. Here, the researchers redefined the screening positive as LDCT screening positive and/or MSC high/intermediate risk and redefined the screening negative as LDCT screening negative and MSC low risk. Consequently, compared to the diagnosis solely with LDCT, the complementary diagnosis of LDCT and MSC reduced false-positive rate from 19.4 % (115/594) to 3.7 % (22/594), with an accompanied reduction in sensitivity from 79 % (46/58) to 69 % (40/58). The most outstanding result of the study was survival data according to the MSC risk category. Among 63 individuals with high risk, 11 (17.5 %) were dead from lung cancer in 3 years, whereas 5 of 159 (3.1 %) with intermediate and none of 717 (0.0 %) with low risk were dead from lung cancer. Taking all things into consideration, it would be possible to hypothesize that serum or plasma biomarker tests might distinguish indolent cancers from aggressive cancers and that such biomarker tests complemented with LDCT might reduce the rate of overdiagnosis. As other studies for LDCT screening including the NLST and NELSON collected biomarker samples, further fruits from biomarker studies are awaited.

2.5 Conclusion

Lung cancer screening with LDCT for individuals at high risk was proven to reduce mortality from lung cancer and is recommended for the right population in the USA. The EU and Japan are awaiting results of the NELSON study to make their recommendation for their regions. Regardless of the results of the NELSON, lung cancer screening with LDCT has a potential to further improve its clinical relevance by incorporating the developing fruits of computer-aided diagnosis and molecular markers. Harmonization of these screening modalities might overcome some shortcomings of the current LDCT screening, including a high false-positive rate and overdiagnosis. As the technique involving the LDCT screening requires complex procedures, including pulmonary nodule detection, positivity judgment, follow-up study, and definitive diagnostic procedure and treatment, a sophisticated and established quality control is strictly required. More active intervention for smoking cessation in conjunction with the LDCT screening is another bottom line to reduce lung cancer mortality [58].

References

1. Lung Cancer Estimated Incidence, Mortality and Prevalence Worldwide in 2012. (2014) World Health Organization. http://globocan.iarc.fr/Pages/fact_sheets_cancer.aspx. Accessed 11 Nov 2015
2. 2015's estimated cancer statistics, mortality. (2015) National Cancer Center. http://gdb.ganjoho.jp/graph_db/gdb9?showData=&dataType=53&graphId=901&totalTarget=10&cardinalCd=1&smTypes=5&smTypes=8&smTypes=10&smTypes=12&smTypes=14&smTypes=16&smTypes=20&smTypes=67&sexType=0. Accessed 11 Nov 2015
3. Fontana RS, Sanderson DR, Woolner LB, Taylor WF, Eugene Miller W, Muhm JR, Bernatz PE, Spencer Payne W, Pairolero PC, Bergstralh EJ (1991) Screening for lung cancer. A critique of the mayo lung project. Cancer 67(S4):1155–1164. doi:10.1002/1097-0142(19910215)67:4+<1155::AID-CNCR2820671509>3.0.CO;2-0

4. Fontana RS (2000) The mayo lung project. Cancer 89(S11):2352–2355. doi:10.1002/1097-0142(20001201)89:11+<2352::AID-CNCR7>3.0.CO;2-5
5. Marcus PM, Bergstralh EJ, Fagerstrom RM, Williams DE, Fontana R, Taylor WF, Prorok PC (2000) Lung cancer mortality in the mayo lung project: impact of extended follow-up. J Natl Cancer Inst 92(16):1308–1316. doi:10.1093/jnci/92.16.1308
6. Oken MM, Hocking WG, Kvale PA, Andriole GL, Buys SS, Church TR, Crawford ED, Fouad MN, Isaacs C, Reding DJ, Weissfeld JL, Yokochi LA, O'Brien B, Ragard LR, Rathmell JM, Riley TL, Wright P, Caparaso N, Hu P, Izmirlian G, Pinsky PF, Prorok PC, Kramer BS, Miller AB, Gohagan JK, Berg CD (2011) Screening by chest radiograph and lung cancer mortality: the Prostate, Lung, Colorectal, and Ovarian (PLCO) randomized trial. JAMA 306(17):1865–1873. doi:10.1001/jama.2011.1591
7. Sobue T, Suzuki T, Naruke T (1992) A case-control study for evaluating lung-cancer screening in Japan. Japanese Lung-Cancer-Screening Research Group. Int J Cancer 50(2):230–237
8. Sagawa M, Nakayama T, Tsukada H, Nishii K, Baba T, Kurita Y, Saito Y, Kaneko M, Sakuma T, Suzuki T, Fujimura S (2003) The efficacy of lung cancer screening conducted in 1990s: four case-control studies in Japan. Lung Cancer 41(1):29–36
9. Kaneko M, Eguchi K, Ohmatsu H, Kakinuma R, Naruke T, Suemasu K, Moriyama N (1996) Peripheral lung cancer: screening and detection with low-dose spiral CT versus radiography. Radiology 201(3):798–802. doi:10.1148/radiology.201.3.8939234
10. Sone S, Takashima S, Li F, Yang Z, Honda T, Maruyama Y, Hasegawa M, Yamanda T, Kubo K, Hanamura K, Asakura K (1998) Mass screening for lung cancer with mobile spiral computed tomography scanner. Lancet 351(9111):1242–1245. doi:10.1016/s0140-6736(97)08229-9
11. Henschke CI, McCauley DI, Yankelevitz DF, Naidich DP, McGuinness G, Miettinen OS, Libby DM, Pasmantier MW, Koizumi J, Altorki NK, Smith JP (1999) Early lung cancer action project: overall design and findings from baseline screening. Lancet 354(9173):99–105. doi:10.1016/s0140-6736(99)06093-6
12. Henschke CI, Naidich DP, Yankelevitz DF, McGuinness G, McCauley DI, Smith JP, Libby D, Pasmantier M, Vazquez M, Koizumi J, Flieder D, Altorki N, Miettinen OS (2001) Early lung cancer action project: initial findings on repeat screenings. Cancer 92(1):153–159
13. Swensen SJ, Jett JR, Sloan JA, Midthun DE, Hartman TE, Sykes AM, Aughenbaugh GL, Zink FE, Hillman SL, Noetzel GR, Marks RS, Clayton AC, Pairolero PC (2002) Screening for lung cancer with low-dose spiral computed tomography. Am J Respir Crit Care Med 165(4):508–513. doi:10.1164/ajrccm.165.4.2107006
14. Diederich S, Wormanns D, Semik M, Thomas M, Lenzen H, Roos N, Heindel W (2002) Screening for early lung cancer with low-dose spiral CT: prevalence in 817 asymptomatic smokers. Radiology 222(3):773–781. doi:10.1148/radiol.2223010490
15. Fujikawa A, Takiguchi Y, Mizuno S, Uruma T, Suzuki K, Nagao K, Niijima M, Edo H, Hino M, Kuriyama T (2008) Lung cancer screening – comparison of computed tomography and X-ray. Lung Cancer 61(2):195–201
16. Nawa T, Nakagawa T, Kusano S, Kawasaki Y, Sugawara Y, Nakata H (2002) Lung cancer screening using low-dose spiral CT: results of baseline and 1-year follow-up studies. Chest 122(1):15–20
17. The International Early Lung Cancer Action Program I (2006) Survival of patients with stage I lung cancer detected on CT screening. N Engl J Med 355(17):1763–1771. doi:10.1056/NEJMoa060476
18. Zhu X, Yu J, Huang Z (2004) Low-dose chest CT: optimizing radiation protection for patients. Am J Roentgenol 183(3):809–816. doi:10.2214/ajr.183.3.1830809
19. Infante M, Cavuto S, Lutman FR, Brambilla G, Chiesa G, Ceresoli G, Passera E, Angeli E, Chiarenza M, Aranzulla G, Cariboni U, Errico V, Inzirillo F, Bottoni E, Voulaz E, Alloisio M, Destro A, Roncalli M, Santoro A, Ravasi G (2009) A randomized study of lung cancer screening with spiral computed tomography: three-year results from the DANTE trial. Am J Respir Crit Care Med 180(5):445–453. doi:10.1164/rccm.200901-0076OC

20. Saghir Z, Dirksen A, Ashraf H, Bach KS, Brodersen J, Clementsen PF, Dossing M, Hansen H, Kofoed KF, Larsen KR, Mortensen J, Rasmussen JF, Seersholm N, Skov BG, Thorsen H, Tonnesen P, Pedersen JH (2012) CT screening for lung cancer brings forward early disease. The randomised Danish Lung Cancer Screening Trial: status after five annual screening rounds with low-dose CT. Thorax 67(4):296–301. doi:10.1136/thoraxjnl-2011-200736
21. Pastorino U, Rossi M, Rosato V, Marchiano A, Sverzellati N, Morosi C, Fabbri A, Galeone C, Negri E, Sozzi G, Pelosi G, La Vecchia C (2012) Annual or biennial CT screening versus observation in heavy smokers: 5-year results of the MILD trial. Eur J Cancer Prev 21(3):308–315. doi:10.1097/CEJ.0b013e328351e1b6
22. Humphrey LL, Deffebach M, Pappas M, Baumann C, Artis K, Mitchell JP, Zakher B, Fu R, Slatore CG (2013) Screening for lung cancer with low-dose computed tomography: a systematic review to update the US Preventive services task force recommendation. Ann Intern Med 159(6):411–420. doi:10.7326/0003-4819-159-6-201309170-00690
23. Lopes Pegna A, Picozzi G, Mascalchi M, Maria Carozzi F, Carrozzi L, Comin C, Spinelli C, Falaschi F, Grazzini M, Innocenti F, Ronchi C, Paci E (2009) Design, recruitment and baseline results of the ITALUNG trial for lung cancer screening with low-dose CT. Lung Cancer 64(1):34–40. doi:10.1016/j.lungcan.2008.07.003
24. Pegna AL, Picozzi G, Falaschi F, Carrozzi L, Falchini M, Carozzi FM, Pistelli F, Comin C, Deliperi A, Grazzini M, Innocenti F, Maddau C, Vella A, Vaggelli L, Paci E, Mascalchi M, Group ftISR (2013) Four-Year Results of Low-Dose CT Screening and Nodule Management in the ITALUNG Trial. J Thorac Oncol 8(7):866–875. doi:10.1097/JTO.1090b1013e31828f31868d31826
25. Horeweg N, van der Aalst CM, Vliegenthart R, Zhao Y, Xie X, Scholten ET, Mali W, Thunnissen E, Weenink C, Groen HJ, Lammers JW, Nackaerts K, van Rosmalen J, Oudkerk M, de Koning HJ (2013) Volumetric computed tomography screening for lung cancer: three rounds of the NELSON trial. Eur Respir J 42(6):1659–1667. doi:10.1183/09031936.00197712
26. van Iersel CA, de Koning HJ, Draisma G, Mali WP, Scholten ET, Nackaerts K, Prokop M, Habbema JD, Oudkerk M, van Klaveren RJ (2007) Risk-based selection from the general population in a screening trial: selection criteria, recruitment and power for the Dutch-Belgian randomised lung cancer multi-slice CT screening trial (NELSON). Int J Cancer 120(4):868–874. doi:10.1002/ijc.22134
27. Aberle DR, Adams AM, Berg CD, Black WC, Clapp JD, Fagerstrom RM, Gareen IF, Gatsonis C, Marcus PM, Sicks JD (2011) Reduced lung-cancer mortality with low-dose computed tomographic screening. N Engl J Med 365(5):395–409. doi:10.1056/NEJMoa1102873
28. Kovalchik SA, Tammemagi M, Berg CD, Caporaso NE, Riley TL, Korch M, Silvestri GA, Chaturvedi AK, Katki HA (2013) Targeting of low-dose CT screening according to the risk of lung-cancer death. N Engl J Med 369(3):245–254. doi:10.1056/NEJMoa1301851
29. de Koning HJ, Meza R, Plevritis SK, ten Haaf K, Munshi VN, Jeon J, Erdogan SA, Kong CY, Han SS, van Rosmalen J, Choi SE, Pinsky PF, Berrington de Gonzalez A, Berg CD, Black WC, Tammemagi MC, Hazelton WD, Feuer EJ, McMahon PM (2014) Benefits and harms of computed tomography lung cancer screening strategies: a comparative modeling study for the U.S. Preventive Services Task Force. Ann Intern Med 160(5):311–320. doi:10.7326/m13-2316
30. Bach PB, Mirkin JN, Oliver TK, Azzoli CG, Berry DA, Brawley OW, Byers T, Colditz GA, Gould MK, Jett JR, Sabichi AL, Smith-Bindman R, Wood DE, Qaseem A, Detterbeck FC (2012) Benefits and harms of CT screening for lung cancer: a systematic review. JAMA 307(22):2418–2429. doi:10.1001/jama.2012.5521
31. Seigneurin A, Field JK, Gachet A, Duffy SW (2014) A systematic review of the characteristics associated with recall rates, detection rates and positive predictive values of computed tomography screening for lung cancer. Ann Oncol 25(4):781–791. doi:10.1093/annonc/mdt491
32. Field JK, Hansell DM, Duffy SW, Baldwin DR (2013) CT screening for lung cancer: countdown to implementation. Lancet Oncol 14(13):e591–e600. doi:10.1016/s1470-2045(13)70293-6

33. Moyer VA (2014) Screening for lung cancer: U.S. Preventive Services Task Force recommendation statement. Ann Intern Med 160(5):330–338. doi:10.7326/m13-2771
34. Harris RP (2015) Starting a new discussion about screening for lung cancer. JAMA 313(7):717–718. doi:10.1001/jama.2014.14769
35. Patz EF Jr, Pinsky P, Gatsonis C et al (2014) OVerdiagnosis in low-dose computed tomography screening for lung cancer. JAMA Intern Med 174(2):269–274. doi:10.1001/jamainternmed.2013.12738
36. Takiguchi Y, Sekine I, Iwasawa S (2013) Overdiagnosis in Lung Cancer Screening with Low-Dose Computed Tomography. J Thorac Oncol 8(11):e101–e102. doi:10.1097/JTO.1090b1013e3182a1476d1094
37. Jacobson FL (2015) Lung cancer screening with low-dose computed tomography beyond the national lung screening trial. J Natl Cancer Inst 107(11). doi:10.1093/jnci/djv286
38. Pinsky PF, Kramer BS (2015) Lung cancer risk and demographic characteristics of current 20–29 pack-year smokers: implications for screening. J Natl Cancer Inst 107(11). doi:10.1093/jnci/djv226
39. Takiguchi Y, Sekine I, Iwasawa S, Kurimoto R, Tatsumi K (2014) Chronic obstructive pulmonary disease as a risk factor for lung cancer. World J Clin Oncol 5(4):660–666. doi:10.5306/wjco.v5.i4.660
40. Lowry KP, Gazelle GS, Gilmore ME, Johanson C, Munshi V, Choi SE, Tramontano AC, Kong CY, McMahon PM (2015) Personalizing annual lung cancer screening for patients with chronic obstructive pulmonary disease: a decision analysis. Cancer 121(10):1556–1562. doi:10.1002/cncr.29225
41. Bindman A (2015) JAMA Forum: lung cancer screening and evidence-based policy. JAMA 313(1):17–18. doi:10.1001/jama.2014.16429
42. Field JK, Aberle DR, Altorki N, Baldwin DR, Dresler C, Duffy SW, Goldstraw P, Hirsch FR, Pedersen JH, de Koning HJ, Mulshine JL, Sullivan DC, Tsao MS, Travis WD (2014) The International Association Study Lung Cancer (IASLC) Strategic Screening Advisory Committee (SSAC) response to the USPSTF recommendations. J Thorac Oncol 9(2):141–143. doi:10.1097/jto.0000000000000060
43. Chin J, Syrek Jensen T, Ashby L, Hermansen J, Hutter JD, Conway PH (2015) Screening for Lung Cancer with Low-Dose CT — Translating Science into Medicare Coverage Policy. N Engl J Med 372(22):2083–2085. doi:10.1056/NEJMp1502598
44. Nawa T, Nakagawa T, Mizoue T, Kusano S, Chonan T, Hayashihara K, Suito T, Endo K (2012) A decrease in lung cancer mortality following the introduction of low-dose chest CT screening in Hitachi, Japan. Lung Cancer 78(3):225–228. doi:10.1016/j.lungcan.2012.09.012
45. The National Lung Screening Trial: Overview and Study Design (2011) Radiology 258(1):243–253. doi:10.1148/radiol.10091808
46. Horeweg N, Scholten ET, de Jong PA, van der Aalst CM, Weenink C, Lammers JW, Nackaerts K, Vliegenthart R, ten Haaf K, Yousaf-Khan UA, Heuvelmans MA, Thunnissen E, Oudkerk M, Mali W, de Koning HJ (2014) Detection of lung cancer through low-dose CT screening (NELSON): a prespecified analysis of screening test performance and interval cancers. Lancet Oncol 15(12):1342–1350. doi:10.1016/s1470-2045(14)70387-0
47. Mori K, Niki N, Kondo T, Kamiyama Y, Kodama T, Kawada Y, Moriyama N (2005) Development of a novel computer-aided diagnosis system for automatic discrimination of malignant from benign solitary pulmonary nodules on thin-section dynamic computed tomography. J Comput Assist Tomogr 29(2):215–222
48. Abe Y, Hanai K, Nakano M, Ohkubo Y, Hasizume T, Kakizaki T, Nakamura M, Niki N, Eguchi K, Fujino T, Moriyama N (2005) A computer-aided diagnosis (CAD) system in lung cancer screening with computed tomography. Anticancer Res 25(1b):483–488
49. Detterbeck FC (2012) Cancer, concepts, cohorts and complexity: avoiding oversimplification of overdiagnosis. Thorax 67(9):842–845. doi:10.1136/thoraxjnl-2012-201779
50. Esserman LJ, Thompson IM, Reid B, Nelson P, Ransohoff DF, Welch HG, Hwang S, Berry DA, Kinzler KW, Black WC, Bissell M, Parnes H, Srivastava S (2014) Addressing overdiag-

nosis and overtreatment in cancer: a prescription for change. Lancet Oncol 15(6):e234–e242. doi:10.1016/s1470-2045(13)70598-9

51. Esserman LJ, Thompson IM Jr, Reid B (2013) Overdiagnosis and overtreatment in cancer: an opportunity for improvement. JAMA 310(8):797–798. doi:10.1001/jama.2013.108415
52. Hassanein M, Callison JC, Callaway-Lane C, Aldrich MC, Grogan EL, Massion PP (2012) The state of molecular biomarkers for the early detection of lung cancer. Cancer Prev Res 5(8):992–1006. doi:10.1158/1940-6207.capr-11-0441
53. Gentles AJ, Bratman SV, Lee LJ, Harris JP, Feng W, Nair RV, Shultz DB, Nair VS, Hoang CD, West RB, Plevritis SK, Alizadeh AA, Diehn M (2015) Integrating tumor and stromal gene expression signatures with clinical indices for survival stratification of early-stage non-small cell lung cancer. J Natl Cancer Inst 107(10). doi:10.1093/jnci/djv211
54. Robles AI, Arai E, Mathe EA, Okayama H, Schetter AJ, Brown D, Petersen D, Bowman ED, Noro R, Welsh JA, Edelman DC, Stevenson HS, Wang Y, Tsuchiya N, Kohno T, Skaug V, Mollerup S, Haugen A, Meltzer PS, Yokota J, Kanai Y, Harris CC (2015) An integrated prognostic classifier for stage I lung adenocarcinoma based on mRNA, microRNA, and DNA Methylation Biomarkers. J Thorac Oncol 10(7):1037–1048. doi:10.1097/jto.0000000000000560
55. Montani F, Marzi MJ, Dezi F, Dama E, Carletti RM, Bonizzi G, Bertolotti R, Bellomi M, Rampinelli C, Maisonneuve P, Spaggiari L, Veronesi G, Nicassio F, Di Fiore PP, Bianchi F (2015) miR-Test: a blood test for lung cancer early detection. J Natl Cancer Inst 107(6):djv063. doi:10.1093/jnci/djv063
56. Boeri M, Verri C, Conte D, Roz L, Modena P, Facchinetti F, Calabro E, Croce CM, Pastorino U, Sozzi G (2011) MicroRNA signatures in tissues and plasma predict development and prognosis of computed tomography detected lung cancer. Proc Natl Acad Sci U S A 108(9):3713–3718. doi:10.1073/pnas.1100048108
57. Sozzi G, Boeri M, Rossi M, Verri C, Suatoni P, Bravi F, Roz L, Conte D, Grassi M, Sverzellati N, Marchiano A, Negri E, La Vecchia C, Pastorino U (2014) Clinical utility of a plasma-based miRNA signature classifier within computed tomography lung cancer screening: a correlative MILD trial study. J Clin Oncol 32(8):768–773. doi:10.1200/jco.2013.50.4357
58. Park ER, Gareen IF, Japuntich S, Lennes I, Hyland K, DeMello S, Sicks JD, Rigotti NA (2015) Primary care provider-delivered smoking cessation interventions and smoking cessation among participants in the national lung screening trial. JAMA Intern Med 175(9):1509–1516. doi:10.1001/jamainternmed.2015.2391

Chapter 3
PET-CT, Bio-imaging for Predicting Prognosis and Response to Chemotherapy in Patients with Lung Cancer

Kyoichi Kaira

Abstract Positron emission tomography (PET) with 2-[^{18}F]-fluoro-2-deoxy-D-glucose (^{18}F-FDG) is a clinically useful tool for the detection of malignant tumors. However, the uptake of ^{18}F-FDG is not tumor specific; thus, other PET tracers have been developed as imaging modalities for human neoplasms. PET tracers, including ^{18}F-FDG, are used as prognostic and monitoring tools after therapy in lung cancer. In our institution, we developed L-[3-^{18}F]-α-methyltyrosine (^{18}F-FAMT) as an amino acid PET tracer. ^{18}F-FAMT-PET is useful for differentiating malignant from benign lesions, as ^{18}F-FAMT is transported into tumor cells via L-type amino acid transporter 1 (LAT1). This review focuses on the prognostic and clinical significance of ^{18}F-FDG, ^{18}F-FAMT, and other forms of PET imaging after therapy in patients with lung cancer.

Keywords ^{18}F-FDG PET • ^{18}F-FAMT PET • Lung cancer • Amino acid transporter • Prognostic factor

3.1 Introduction

Lung cancer can be classified as non-small-cell lung cancer (NSCLC) and small-cell lung cancer (SCLC) and is a common malignancy with a poor prognosis after appropriate therapeutic treatment. Performance status (PS) and disease staging are thought to be significant predictors linked to poor outcome after treatment. However, no established biomarker has been discovered for improved outcome after appropriate treatment against lung cancer. Recently, positron emission tomography (PET) imaging has been reported to be a good modality for the detection of malignant lesions in various types of human neoplasm. Previous studies have shown that the accumulation of 2-[^{18}F]-fluoro-2-deoxy-D-glucose (^{18}F-FDG) within tumor cells is

---

K. Kaira, M.D., Ph.D. (✉)
Department of Oncology Clinical Development, Gunma University Graduate School of Medicine, 3-39-22 showa-machi, Maebashi, Gunma 371-8511, Japan
e-mail: kkaira1970@yahoo.co.jp

© Springer Science+Business Media Singapore 2017

Y. Takiguchi (ed.), *Molecular Targeted Therapy of Lung Cancer*,
DOI 10.1007/978-981-10-2002-5_3

predictive of the efficacy of systemic chemotherapy and prognosis after surgical resection [1–3]. Glucose metabolism, hypoxia, angiogenesis, and cell proliferation have been described as mechanisms of ^{18}F-FDG uptake within tumor cells in previous reports [4, 5]. Here, we review the clinical role of PET-computed tomography (CT) bio-imaging as a predictive marker in patients with lung cancer.

3.2 Prognostic Variables in Patients with Lung Cancer

Lung cancer is a leading cause of death worldwide. Surgical resection is the best treatment for patients with early-stage disease, whereas patients with advanced disease are treated with systemic chemotherapy. Although PS and disease stage are novel prognostic factors, there is no established biomarker for predicting prognosis after surgery or chemotherapy in patients with lung cancer. Several studies have found that progressive markers related to angiogenesis, cell proliferation, and metastases could be useful as predictive markers [4, 5]. Further investigation is warranted in order to discover a promising molecular marker for predicting outcome after therapy in patients with lung cancer.

3.3 Biological Significance of ^{18}F-FDG-PET

^{18}F-FDG-PET is useful as a noninvasive diagnostic modality during oncologic practice for various human cancers [6]. It was identified as a molecular imaging technique to measure glucose metabolism [7]. The increased expression of glucose transporter 1 (Glut1) has been proven to be closely associated with the accumulation of ^{18}F-FDG within tumor cells [4, 7, 8] (Fig. 3.1). The glucose phosphorylation enzyme hexokinase is also necessary for glucose metabolism in cancer cells, and glucose-6-phosphatase is reduced by enhanced concentrations of hexokinase [7, 8]. Glut1 is known to be a possible intrinsic marker of hypoxia, and the regulation of hypoxia via the hypoxia inducible factor (HIF)-1 pathway plays a crucial role in the increased expression of Glut1 [7, 8]. HIF-1α has been shown to aid tumor growth by the induction of angiogenesis via the expression of vascular endothelial growth factor (VEGF) in addition to anaerobic metabolic mechanisms [4, 5]. Therefore, the measurement of ^{18}F-FDG uptake is determined by the presence of glucose metabolism (Glut1), hypoxia (HIF-1α), and angiogenesis (VEGF) in human neoplasms [4, 5, 7, 8]. However, the underlying mechanisms of ^{18}F-FDG uptake within tumor cells are still a matter of debate in many human cancers, as the extent of ^{18}F-FDG accumulation is thought to be influenced by many factors.

Increased glucose uptake is necessary for the survival of cancer cells, and it is thought that glucose transport plays an important role in this process [4, 5, 7, 8]. Currently, there are 14 known glucose transporter protein subtypes, and Glut1 and Glut3 have been demonstrated to be highly expressed in several human cancers [9].

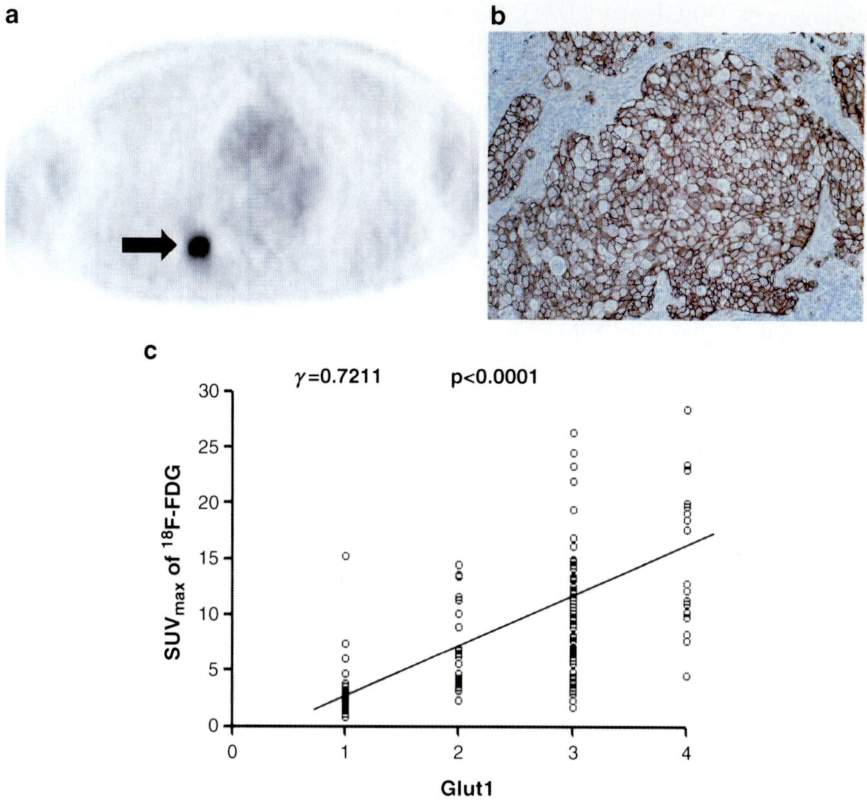

Fig. 3.1 ^{18}F-FDG-PET (transaxial section) showed increased uptake of tracers in a primary tumor of lung cancer (**a**) (*black arrow*). Immunohistochemical analysis of the resected tumor revealed that the immunohistostaining pattern of Glut1 was membranous (**b**). There was a statistically significant correlation between the expression of Glut1 and the SUV_{max} of ^{18}F-FDG uptake in lung cancer (**c**) (Modified from Ref. [5, 14])

In previous studies, it has been shown that expression levels of Glut1 are closely correlated with the accumulation of ^{18}F-FDG within tumor cells [4, 5]. However, we found no significant relationship between Glut 3 expression and ^{18}F-FDG uptake in cancer cells [4]. Recently, it has been reported that the amount of ^{18}F-FDG uptake is determined by not only glucose metabolism, angiogenesis, and cell proliferation, but also the mammalian target of rapamycin (mTOR) signaling pathway [10]. In some studies, the uptake of ^{18}F-FDG has been shown to be significantly linked with therapeutic response to mTOR inhibitors in cancer patients [11, 12]. Conversely, another report has documented that the amount of ^{18}F-FDG within tumor cells is not a potential predictive marker for the response to mTOR inhibitor therapy, but a pharmacodynamic measurement for Akt activation during mTOR inhibitor treatment [13]. In our previous study, we found that the amount of ^{18}F-FDG uptake was closely correlated with glucose metabolism (Glut1/hexokinase I), hypoxia (HIF-1α), angiogenesis (VEGF and CD34), and the PI3K/Akt/mTOR signaling pathway in patients with lung cancer [5]. Generally, the accumulation of ^{18}F-FDG within cancer cells is known to be markedly higher in non-adenocarcinoma (AC) than in AC, but the correlation with these molecular markers was proven to be stronger in AC compared to non-AC. It has also been reported that the amount of ^{18}F-FDG accumulation and glucose metabolism as determined by Glut1 expression has the potential to be significant prognostic markers for predicting outcome after surgical treatment in AC patients.

In experimental studies, the relationship between ^{18}F-FDG uptake and glucose metabolism has been investigated using tumor cell lines [4, 5]. Our in vitro studies depicted that uptake of ^{18}F-FDG was clearly decreased by inhibition of Glut1 and increased by Glut1 upregulation through the induction of HIF-1α in most cell lines [4, 5]. As it remains unclear whether ^{18}F-FDG uptake within tumor cells is controlled by the mTOR pathway, the relationship between inhibition of the mTOR pathway and ^{18}F-FDG uptake was investigated using lung cancer cell lines. Uptake of ^{18}F-FDG was reduced by mTOR inhibitors in a dose-dependent manner, and the inhibition of mTORC1 alone decreased ^{18}F-FDG accumulation in tumor cell lines. Therefore, these results suggest that the mTORC1 signaling pathway is associated with a biological mechanism for the uptake of ^{18}F-FDG in lung cancer cells.

3.4 Clinical Role of ^{18}F-FDG-PET as a Predictor of Outcome After Chemotherapy

^{18}F-FDG-PET is useful not only for the diagnosis and staging of lung cancer but also for the prediction of therapeutic outcome [1, 2]. A recent meta-analysis showed that the primary tumor-standardized uptake values (SUVs) measured on ^{18}F-FDG have prognostic value in lung cancer [3]. Many reports have investigated whether the value of maximal SUV (SUV_{max}) is best for use as a prognostic predictive marker after the treatment of lung cancer; however, SUV varies with several factors such as blood glucose levels and the interval between tracer injection and scanning time. Therefore, the amount of background ^{18}F-FDG uptake is thought to be affected by significant statistical error. Recent studies have shown that it is not the value of SUV_{max} but rather the ratio of the SUV_{max} of the tumor to the mean SUV of the mediastinum (T/M ratio) that is better for measurement of the accumulation of ^{18}F-FDG [14, 15]. However, we cannot currently conclude whether SUV_{max} or the T/M ratio is more appropriate to predict prognosis after chemotherapy against advanced lung cancer. Further investigation is warranted in order to discover new measurements of ^{18}F-FDG uptake that are useful for the prediction of outcome after treatment.

For the prediction of therapeutic outcome using ^{18}F-FDG-PET, appropriate timing of PET imaging after treatment is problematic. Many studies regarding PET in the assessment of chemotherapy focus on the early prediction of response by measuring differences between SUV_{max} before and after treatment. However, the relationship between initial SUV and subsequent response to chemotherapy remains unclear.

Previously, we have reported the prognostic significance of SUV_{max}, the T/M ratio, and the ratio of the SUV_{max} of the metastatic site to the SUV_{max} of the primary site (M/P ratio) measured on ^{18}F-FDG-PET in patients with advanced NSCLC who had received chemotherapy [16]. Comparison of the prognostic significance of SUV_{max}, T/M ratio, and M/P ratio showed that high M/P ratio was a significant independent factor for predicting a shorter prognosis after chemotherapy in patients with advanced NSCLC. Although SUV_{max} and T/M ratio were significantly related to a worse outcome by univariate analysis, the prognosis of non-AC was unaffected by the value of SUV_{max} and the T/M ratio. Additionally, a high M/P ratio indicated a poor response to initial chemotherapy, but this was not true for SUV_{max} and T/M ratio.

It has been shown that the accumulation of ^{18}F-FDG in preoperative primary tumors is a useful prognostic factor for predicting negative prognosis in NSCLC patients with postoperative recurrence who have received platinum-based chemotherapy [17]. The uptake of ^{18}F-FDG within tumor cells was significantly higher in the preoperative primary site than in the recurrence site, and low ^{18}F-FDG accumulation within the preoperative primary tumor correlated closely with the presence of epidermal growth factor receptor (*EGFR*) mutation. The amount of ^{18}F-FDG uptake in preoperative primary tumors may be important for predicting outcome after chemotherapy against recurrent NSCLC as compared to that in recurrent sites. Further investigation in a large prospective study is needed to confirm this result.

In order to evaluate the early prediction of therapeutic response to EGFR-tyrosine kinase inhibitor (TKI), the usefulness of ^{18}F-FDG-PET was investigated in advanced NSCLC harboring *EGFR* mutation [18]. In this preliminary study, 5 NSCLC patients with *EGFR* mutation underwent ^{18}F-FDG-PET to evaluate changes in ^{18}F-FDG accumulation at 2 days and 4 weeks after the initiation of gefitinib in comparison to ^{18}F-FDG-PET imaging prior to treatment. The SUV_{max} of ^{18}F-FDG uptake significantly decreased on day 2 after the initiation of gefitinib in patients with any response. On the other hand, the SUV_{max} of ^{18}F-FDG uptake markedly increased in patients with progressive disease. This preliminary data suggests that ^{18}F-FDG-PET could predict therapeutic response to EGFR-TKI in the early phase of therapy in patients with advanced NSCLC harboring *EGFR* mutation. Zander et al. compared changes in ^{18}F-FDG uptake after 1 (early) and 6 (late) weeks of erlotinib treatment in patients with advanced NSCLC [19]. Their study stated that patients with an early metabolic ^{18}F-FDG response showed significantly longer survival, but those with a late metabolic ^{18}F-FDG response did not. Therefore, early response monitoring using ^{18}F-FDG-PET may be predictive for outcome after EGFR-TKI treatment.

^{18}F-FDG-PET has also been shown to be useful for NSCLC patients who have received concurrent platinum-based chemoradiotherapy [20]. One hundred seventy-three subjects underwent posttreatment SUV analyses, and posttreatment ^{18}F-FDG-PET was performed at about 14 weeks after radiotherapy. High posttreatment SUV within tumor cells was related to negative survival in stage III NSCLC. Therefore, the appropriate timing of ^{18}F-FDG-PET monitoring remains controversial, even though there are several reports on this subject.

Currently, we cannot conclude whether SUV_{max} as a measurement of ^{18}F-FDG uptake is appropriate for predicting outcome after any treatment in patients with lung cancer. A recent meta-analysis reported the prognostic value of metabolic tumor volume (MTV) and total lesion glycolysis (TLG) in patients with lung cancer [21]. In 13 studies including 1581 patients, patients with high MTV or high TLG had a worse prognosis for adverse events and death, and MTV and TLG were considered to be significant prognostic factors in patients with early-stage (stage I/II) and advanced-stage (stage III/IV) NSCLC.

3.5 False-Positive Findings in ^{18}F-FDG-PET

^{18}F-FDG-PET can differentiate malignant tumors from benign lesions. Generally, physicians utilize ^{18}F-FDG-PET scans to diagnose malignant disease and perform disease staging for human cancers. However, false-positive findings on ^{18}F-FDG-PET are observed when inflammatory and granulomatous lesions such as pneumonia, pulmonary tuberculosis, aspergillus, and sarcoidosis are present [22, 23]. To overcome the disadvantage of false-positive ^{18}F-FDG uptake in benign lesions, the optimal cutoff point of SUV has been discussed. Although an SUV_{max} of 2.5 has been identified as an optimal cutoff point, malignant disease cannot be diagnosed according to the value of SUV_{max} on ^{18}F-FDG-PET [24].

In our institution, we developed L-[3-^{18}F]-α-methyltyrosine (^{18}F-FAMT) as an amino acid PET tracer [25]. ^{18}F-FAMT-PET is useful for differentiating malignant from benign lesions, as ^{18}F-FAMT is transported into tumor cells via L-type amino acid transporter 1 (LAT1) [26]. The LAT family consists of four subtypes: LAT1, LAT2, LAT3, and LAT4 [27, 28]. In particular, LAT1 is expressed extensively in many human neoplasms. There is no evidence of LAT1 expression in normal tissue or benign lesions. Since the uptake of ^{18}F-FAMT is correlated closely with the expression of LAT1, ^{18}F-FAMT is considered to be a specific PET tracer for malignant disease (Fig. 3.2). In previous studies, ^{18}F-FAMT-PET has been described to be able to differentiate malignant nodules from sarcoid lesions [23].

3.6 Development of Tumor-Specific PET Tracers

Generally, ^{18}F-FDG-PET is used as a tool for oncologic diagnosis, but there is debate as to whether it is the best tumor-specific PET tracer since it can show false-positive findings in inflammatory and granulomatous diseases. Therefore, the development of tumor-specific PET tracers that do not show false-positive PET findings in nonmalignant lesions is important. Currently, several types of tumor PET tracers in addition to ^{18}F-FDG-PET exist, which utilize amino acid metabolism, hypoxia,

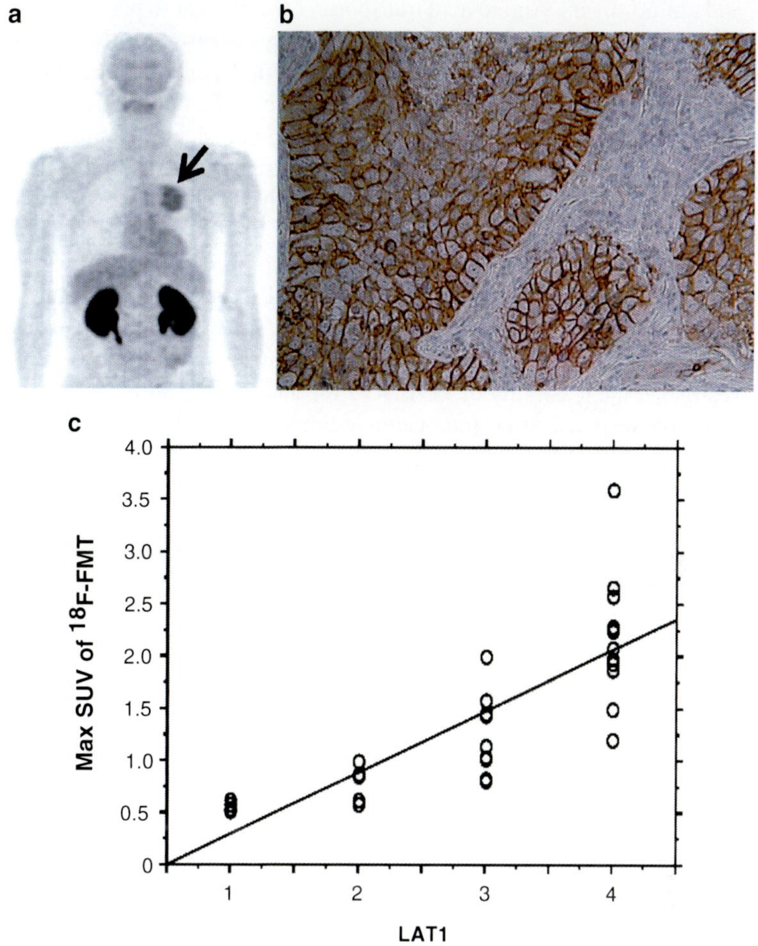

Fig. 3.2 ^{18}F-FAMT-PET (transaxial section) showed increased uptake of tracers in a primary tumor of lung cancer (**a**) (*black arrow*). Immunohistochemical analysis of the resected tumor revealed that the immunohistostaining pattern of LAT1 was membranous (**b**). A statistically significant correlation was observed between the expression of LAT1 and the SUV_{max} of ^{18}F-FAMT uptake in lung cancer (**c**) (Ref. [24, 26])

and DNA synthesis. In this section, we introduce the clinical significance of these new PET tracers.

1. Amino Acid Metabolism

Tumor cells require nutrients such as glucose, amino acids, fatty acids, and vitamins in order to grow and survive. This demand is met by the availability of nutrients achieved through vascular formation and upregulation of specific transporters. Amino acids are needed not only for protein synthesis but also as a source of nitrogen and carbon for the synthesis of purine and pyrimidine nucleotides, glutathione, and amino sugars. Amino acid transporters are thought to play a crucial role in tumor cell proliferation and development. Among the several different types of amino acid transporters that exist, system L is a Na^+-independent large and neutral amino acid transporter [27]. LAT1 transports large neutral amino acids such as leucine, isoleucine, valine, phenylalanine, tyrosine, tryptophan, methionine, and histidine [28]. As described above, ^{18}F-FAMT is transported into cells by LAT1, so we believe that ^{18}F-FAMT-PET represents an alternative method for the molecular imaging of LAT1.

Another amino acid PET tracer, ^{11}C-labeled methionine (MET), has been used in the imaging of various human cancers [29–31] and is better for the detection of malignant tumors than ^{18}F-FDG-PET because of its higher specificity, which improves the differentiation between malignant and benign processes. Ishimoto, et al. investigated the feasibility of ^{18}F-FDG-PET and ^{11}C-MET-PET for the evaluation of treatment response after stereotactic radiotherapy (SRT) in lung cancer [32]. ^{18}F-FDG-PET and ^{11}C-MET-PET were performed in 9 patients 1 week before and up to 8 months after SRT. The results showed that the SUV of ^{18}F-FDG and ^{11}C-MET uptake after SRT changed concordantly and the addition of ^{11}C-MET-PET provided no additional information over ^{18}F-FDG-PET. To our knowledge, there are no reports on prognosis and response after chemotherapy using ^{11}C-MET-PET in patients with lung cancer. Additionally, PET tracers using tyrosine derivatives such as 2-^{18}F-fluoro-L-tyrosine (^{18}F-FET) and 3-^{123}I-iodo-α-methyl-L-tyrosine (^{123}I-IMT) have also been investigated [33–35]. ^{18}F-FET-PET has been examined for its diagnostic performance in lung cancer, and the level of ^{18}F-FET accumulation was shown to be markedly different according to histology [33, 34]. It was found that all patients with squamous cell carcinoma (SQC) showed positive ^{18}F-FET findings, but negative ^{18}F-FET uptake was observed in most AC patients. Therefore, ^{18}F-FET-PET is thought to be useful in the differentiation between malignant and benign lesions in patients with SQC; however, it does not seem to be a meaningful method in AC patients. ^{123}I-IMT single-photon emission tomography (SPECT) shows a high sensitivity for detecting primary lung cancer lesions (94 %), but for small tumors (of less than 20 mm), its sensitivity is too low to be helpful [35]. In an experimental study, ^{123}I-IMT was found to be transported selectively via human LAT1 [36], while it has also been considered that ^{18}F-FET may be selectively transported via LAT1, which is expressed in normal cells [34]. In human cancers, however, it remains unclear whether ^{11}C-MET, ^{18}F-FET, and ^{123}I-IMT correlate closely with the expression of amino acid transporters such as LAT1. Currently, the uptake of ^{18}F-FAMT alone has been proven to be significantly correlated with the expression level of LAT1 in patients with lung cancer.

2. Hypoxic Imaging

Tumor hypoxia is associated with resistance to chemotherapy and radiotherapy and induces angiogenesis, metastasis, and tumor aggressiveness, resulting in poor prognosis. Therefore, it has a significant impact on the biological activity of various malignancies and acts as a target for tumor imaging [37, 38]. Recently, hypoxia-targeting radiopharmaceuticals have been developed for the diagnosis of malignancy and assessment of therapeutic response. ^{18}F-fluoromisonidazole (FMISO) and ^{60}Cu or ^{64}Cu-diacetyl-bis (4-(N)-methylthiosemicarbazone) (ASTM) are clinically available as the principal forms of such PET tracers [39, 40]. ^{60}Cu-ASTM-PET and ^{18}F-FDG-PET were compared to evaluate their usefulness for therapeutic monitoring in patients with NSCLC [41]. In 14 patients, both forms of PET imaging were performed to evaluate therapeutic response. The mean T/M ratio for ^{60}Cu-ASTM was significantly lower in responders than in nonresponders, but the mean SUV for ^{60}Cu-ASTM was not different between responders and nonresponders. In addition, the uptake of ^{60}Cu-ASTM did not correlate with that of ^{18}F-FDG. This preliminary study suggests that PET imaging with ^{60}Cu-ASTM may be effective for patients who are less likely to respond to any treatment. FMISO is a hypoxic PET tracer that is taken up selectively by hypoxic cells and shows a slower washout from normoxic cells than ^{60}Cu-ASTM. In eight patients with advanced NSCLC, FMISO-PET and ^{18}F-FDG-PET were assessed before and 2 weeks after chemotherapy [42]. This study suggested that changes in FMISO uptake within tumor cells measured the early response to chemotherapy and may be useful as a predictive marker related to survival. Cherk et al. described the evaluation of 17 patients with resectable NSCLC using FMISO-PET and ^{18}F-FDG-PET [43]. In this study, the accumulation level of FMISO was significantly lower than that of ^{18}F-FDG, and there was no correlation between either FMISO or ^{18}F-FDG uptake and microvessel density (MVD) and HIF-1α, VEGF, or Glut1 expression, but there was a weakly positive correlation between both FMISO and ^{18}F-FDG uptake and the proliferative marker Ki-67. However, this preliminary study had a small sample size that may have caused bias in the results. A recent meta-analysis showed that hypoxia modification could improve local control and survival after treatment [44]. Sachpekidis et al. assessed 13 patients with stage III NSCLC by PET imaging with FMISO and ^{18}F-FDG, and their results indicated that only one subject showed increased FMISO uptake within the tumor site. A lack of correlation between FMISO and ^{18}F-FDG implied a discordance between increased hypoxia and glycolysis in this type of cancer [45]. Currently, it seems to be difficult to utilize FMISO or ^{60}Cu-ASTM-PET imaging in order to assess therapeutic response and prognosis after chemotherapy in NSCLC patients.

3. DNA Synthesis

3′-Deoxy-3′-[^{18}F] fluorothymidine (^{18}F-FLT) is a thymidine analog that was developed to measure cell proliferation [46]. ^{18}F-FLT is phosphorylated by thymidine kinase and enters the salvage signaling pathway without incorporation into DNA molecules. According to recent studies, we know that the uptake level of ^{18}F-FLT clearly reflects tumor cell proliferation as determined by Ki-67 labeling index [47, 48]. Therefore, ^{18}F-FLT is considered to be a more tumor-specific PET imaging marker as compared to ^{18}F-FDG. However, during ^{18}F-FLT-PET, this tracer normally accumulates in the liver and bone marrow; thus, malignant and normal lesions in these organs cannot be differentiated using ^{18}F-FLT-PET. Recently, the relationship between ^{18}F-FLT uptake and Ki-67 labeling index was investigated in 18 NSCLC patients [47]. This study showed that there was a statistically significant correlation between ^{18}F-FLT uptake and Ki-67 labeling index ($\gamma = 0.77$; $p < 0.0002$) as well as between ^{18}F-FLT uptake and ^{18}F-FDG uptake ($\gamma = 0.81$; $p < 0.0001$). With regard to diagnostic performance, the sensitivity for detection of malignant lesions was 72 % for ^{18}F-FLT-PET and 89 % for ^{18}F-FDG-PET. Therefore, while ^{18}F-FLT-PET may be less sensitive for disease staging compared to ^{18}F-FDG-PET, the uptake of ^{18}F-FLT is better for measurement of tumor proliferative activity than ^{18}F-FDG. Other studies also support this evidence [49, 50]. Everitt et al. reported the usefulness of ^{18}F-FLT-PET for the assessment of therapeutic monitoring after concurrent radical chemoradiotherapy in 20 patients with NSCLC [51]. ^{18}F-FLT-PET and ^{18}F-FDG-PET were performed at baseline, week 2, and week 4, and the SUV_{max} of tumor PET uptake was measured. SUV_{max} on ^{18}F-FDG-PET was 14 at baseline, 10 at 2 weeks, and 10 at 4 weeks. SUV_{max} on ^{18}F-FLT-PET was 6 at baseline, 3 at 2 weeks, and 2 at 4 weeks. Therefore, this study suggests that ^{18}F-FLT-PET is a more sensitive tracer of early therapeutic response than ^{18}F-FDG-PET. Although ^{18}F-FLT-PET reflects different proliferative responses after treatment in NSCLC patients, it remains unclear whether it can predict outcome after therapy. Zander et al. also described that early ^{18}F-FLT response (1 week after treatment) could predict significantly longer progression-free survival after erlotinib therapy, but ^{18}F-FLT was not predictive for response after 6 weeks of therapy [19]. Based on these studies, ^{18}F-FLT-PET may be a better monitoring marker for predicting early response after therapy. Further investigation in patients with lung cancer is warranted for the confirmation of these results.

3.7 Clinical Significance of ^{18}F-FAMT-PET

Recently, it has been reported that ^{18}F-FAMT-PET is effective for the diagnosis of malignant diseases such as lung cancer, oral cancer, esophageal cancer, and multiple myeloma [26, 52–54]. In patients with NSCLC, a high uptake of ^{18}F-FAMT is closely related to a worse prognosis after therapy [24, 55]. It has been found that the accumulation of ^{18}F-FAMT is a significant predictive marker related to poor outcome in patients with AC, but not non-AC [24]. Although the amount of ^{18}F-FAMT uptake within tumor cells is significantly higher in patients with SQC than in those with AC, the prognostic significance of ^{18}F-FAMT seems to have more impact in patients with AC than in those with non-AC. Little is known regarding why the uptake of ^{18}F-FAMT plays a crucial role in the tumor aggressiveness of AC patients.

In our previous study, we compared ^{18}F-FAMT-PET and ^{18}F-FDG-PET for the diagnosis of malignant lesions in patients with NSCLC who had undergone surgical resection [26]. The sensitivity, specificity, and accuracy in malignant lymph nodes on ^{18}F-FAMT-PET and ^{18}F-FDG-PET were 57.8, 100, and 92.5 %, respectively, and 65.7, 91.0, and 86.5 %, respectively. Our study showed that the tumor specificity of ^{18}F-FAMT-PET is significantly superior to that of ^{18}F-FDG-PET, but the sensitivity of ^{18}F-FAMT-PET is lower compared to that of ^{18}F-FDG-PET. Moreover, the assessment of therapeutic monitoring with ^{18}F-FAMT-PET was performed in 18 patients with lung cancer who had received chemoradiotherapy [55]. The results of this study suggested that the accumulation of ^{18}F-FAMT, assessed by lymph node-to-primary tumor ratio after treatment, could predict the outcome of patients ($p = 0.014$), but the lymph node-to-primary tumor ratio of ^{18}F-FDG did not reveal a significant result. Tumor metabolic response also seemed to be more accurate on ^{18}F-FAMT-PET than on ^{18}F-FDG-PET. It has been stated that ^{18}F-FDG-PET can both underestimate and overestimate the response to treatment. ^{18}F-FAMT-PET is a tumor-specific form of PET and can differentiate lung cancer from sarcoidosis, which shows a false-positive finding on ^{18}F-FDG-PET (Fig. 3.3). Further investigation is warranted to improve the sensitivity for detection of malignant disease on ^{18}F-FAMT-PET.

3.8 Prognostic Significance of LAT1 Expression

LAT1 is highly expressed in various human neoplasms, and transports ^{18}F-FAMT into tumor cells. Therefore, it correlates closely with the uptake level of this tracer. In tumor tissues, the expression of LAT1 was shown to exhibit a close relationship with tumor cell proliferation, angiogenesis, and metastasis [14, 26]. Recent studies

a

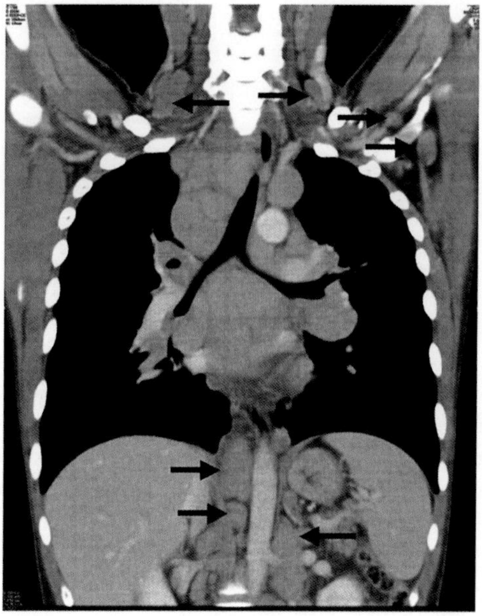

Fig. 3.3 A CT scan showed supraclavicular, bilateral hilar, mediastinal, cervical, and abdominal para-aortic lymphadenopathy (*black arrows*) (**a**). ^{18}F-FDG-PET revealed increased uptake in corresponding areas (**b**). ^{18}F-FAMT showed no increase in uptake corresponding to these lesions (**c**) (Ref. [23])

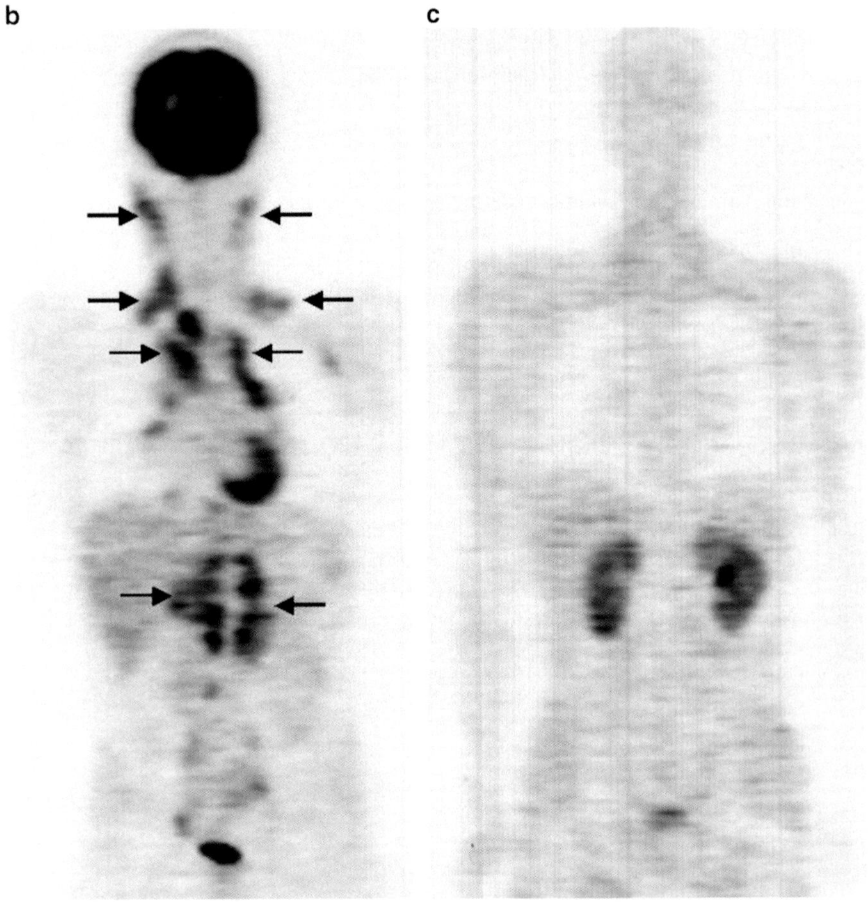

Fig. 3.3 (continued)

have demonstrated that increased expression of LAT1 is a significant marker for predicting poor outcome in various human cancers including lung cancer, breast cancer, gastrointestinal cancer, head and neck cancer, hematological malignancy, and malignant melanoma [52, 56–61]. Previously, we compared the prognostic significance of ^{18}F-FAMT accumulation in primary tumors with that of LAT1 expression in NSCLC patients [62]. Fifty-nine patients were analyzed in this study. Although high uptake of ^{18}F-FAMT and positive LAT1 expression were both significant predictors of poor outcome, we found that LAT1 expression was a stronger prognostic factor than ^{18}F-FAMT uptake in NSCLC patients. However, ^{18}F-FAMT-PET is considered to be an alternative method for molecular imaging of LAT1 expression in lung cancer.

3.9 Summary and Conclusion

In this review, we have discussed the clinical impact of ^{18}F-FDG-PET and ^{18}F-FAMT-PET in patients with lung cancer. The uptake of ^{18}F-FDG within tumor cells is determined by the presence of glucose metabolism (Glut1), cell proliferation (Ki-67), angiogenesis (VEGF), and hypoxia (HIF-1α). ^{18}F-FDG-PET could predict poor survival after chemotherapy; however, the optimal timing of ^{18}F-FDG-PET scanning remains unclear. Although ^{18}F-FDG-PET yields a high sensitivity for the detection of malignant tumors as compared to imaging using other PET tracers, it exhibits false-positive PET findings in inflammatory and granulomatous diseases such as sarcoidosis and pneumonia. To overcome this disadvantage of ^{18}F-FDG-PET scanning, tumor-specific PET tracers such as those involved in amino acid transport, hypoxia, and DNA synthesis have been developed in several institutions. In our institution, we developed ^{18}F-FAMT-PET to detect malignant tumors, and this PET tracer has been described to be useful for the monitoring of therapeutic response and prognosis after chemotherapy in patients with lung cancer [24, 55]. Moreover, ^{18}F-FAMT-PET is a tumor-specific form of PET and shows no false-positive findings in benign diseases such as sarcoidosis and pneumonia [14, 23]. LAT1 is highly expressed in various human cancers and is significantly correlated with the uptake of ^{18}F-FAMT; thus, we consider that ^{18}F-FAMT-PET could be an alternative method for the molecular imaging of LAT1. In this review, we have not discussed PET imaging of metabolic response with regards to monitoring after systemic chemotherapy in advanced lung cancer. Recently, we reported that ^{18}F-FAMT-PET could clarify the clinical significance of therapeutic response and outcome after systemic chemotherapy in patients with advanced lung cancer, as compared with ^{18}F-FDG-PET [63]. Ninety-five patients were eligible for analyses on both forms of PET scanning. Posttreatment SUV_{max} and metabolic response on ^{18}F-FAMT-PET were significantly correlated with tumor response. In all patients, univariate analysis showed that the posttreatment SUV_{max} of ^{18}F-FDG-PET and ^{18}F-FAMT-PET, as well as metabolic response on ^{18}F-FAMT-PET, was a significant prognostic marker for predicting poor outcome. Multivariate analysis confirmed that metabolic response on ^{18}F-FAMT-PET was a significant independent prognostic factor. This study suggests that metabolic response on ^{18}F-FAMT-PET could be a potential parameter for the prediction of prognosis after chemotherapy in patients with advanced lung cancer.

Several PET tracers including ^{18}F-FDG could be useful imaging modalities for representing underlying molecular biology in patients with lung cancer. Further investigation is needed to develop tumor-specific PET tracers in order to improve the monitoring quality of therapeutic response and outcome after treatment.

References

1. Sasaki R, Komaki R, Macapinlac H et al (2005) [18F] fluorodeoxyglucose uptake by positron emission tomography predicts outcome of non-small-cell lung cancer. J Clin Oncol 23:1136–1143
2. Vansteenkiste JF, Stroobants SG, Dupont PJ et al (1999) Prognostic importance of the standardized uptake value on ^{18}F-fluoro-2-deoxy-glucose-positron emission tomography in non-small cell lung cancer: an analysis of 125 cases. J Clin Oncol 17:3201–3206
3. Berghmans T, Dusart M, Paesmans M et al (2008) Primary tumor standardized uptake value (SUVmax) measured on fluorodeoxyglucose positron emission tomography (FDG-PET) is of prognostic value for survival in non-small cell lung cancer (NSCLC): a systematic review and meta-analysis (MA) by the European lung cancer working party for the IASLC Lung Cancer Staging Project. J Thorac Oncol 3:6–12
4. Kaira K, Endo M, Abe M et al (2010) Biologic correlation of 2-[^{18}F]-fluoro-2-deoxy-D-glucose uptake on positron emission tomography in thymic epithelial tumors. J Clin Oncol 28:3746–3753
5. Kaira K, Serizawa M, Koh Y et al (2014) Biological significance of 18F-FDG uptake on PET in patients with non-small-cell lung cancer. Lung Cancer 83:197–204
6. Vansteenkiste JF, Stroobants SG, De Leyn PR et al (1998) Lymph node staging in non-small-cell lung cancer with FDG-PET scan: a prospective study on 690 lymph node stations from 68 patients. J Clin Oncol 16:2142–2149
7. Mamede M, Higashi T, Kitaichi M et al (2005) [^{18}F] FDG uptake and PCNA, Glut-l, and Hexokinase-II expressions in cancers and inflammatory lesions of the lung. Neoplasia 7:369–379
8. Vleugel MM, Greijer AE, Shvarts A et al (2005) Differential prognostic impact of hypoxia induced and diffuse HIF-1alpha expression in invasive breast cancer. J Clin Pathol 58:172–177
9. Younes M, Brown RW, Stephenson M et al (1997) Overexpression of Glut1 and Glut3 in stage I nonsmall cell lung carcinoma is associated with poor survival. Cancer 80:1046–1051
10. Kaira K, Serizawa M, Koh Y et al (2012) Relationship between 18F-FDG uptake on positron emission tomography and molecular biology in malignant pleural mesothelioma. Eur J Cancer 48:1244–1254
11. Thomas GV, Tran C, Mellinghoff IK et al (2006) Hypoxia-inducible factor determines sensitivity to inhibitors of mTOR in kidney cancer. Nat Med 12:122–127
12. Novoga L, Bocellaard R, Kobe C et al (2009) Downregulation of ^{18}FDG uptake in PET as an early pharmacodynamic effect in treatment of non-small cell lung cancer with the mTOR inhibitor everlimus. J Nucl Med 50:1815–1819
13. Ma WW, Jacene H, Song D et al (2009) [18F]Fluorodeoxyglucose positron emission tomography correlates with Akt pathway activity but Is not predictive of clinical outcome during mTOR inhibitor therapy. J Clin Oncol 27:2697–2704
14. Kaira K, Oriuchi N, Shimizu K et al (2009) Evaluation of thoracic tumors with ^{18}F-FMT and ^{18}F-FDG PET-CT: a clinicopathological study. Int J Cancer 124:1152–1160
15. Endo M, Nakagawa K, Ohde Y et al (2008) Utility of ^{18}F-FDG-PET for differentiating the grade of malignancy in thymic epithelial tumors. Lung Cancer 61:350–355
16. Kaira K, Endo M, Asakura K et al (2010) Ratio of standardized uptake value on PET helps predict response and outcome after chemotherapy in advanced non-small cell lung cancer. Ann Nucl Med 24:697–705
17. Kaira K, Yamamoto N, Kenmotsu H et al (2014) Prognostic impact of F-FDG uptake on PET in non-small cell lung cancer patients with postoperative recurrence following platinum-based chemotherapy. Respir Investig 52:121–128
18. Sunaga N, Oriuchi N, Kaira K et al (2006) Usefulness of FDG-PET for early prediction of the response to gefitinib in non-small cell lung cancer. Lung Cancer 59:203–210

19. Zander T, Scheffler M, Nogova L et al (2011) Early prediction of nonprogression in advanced non-small-cell lung cancer treated with erlotinib by using [(18)F]fluorodeoxyglucose and [(18)F]fluorothymidine positron emission tomography. J Clin Oncol 29:1701–1708
20. Machtay M, Duan F, Siegel BA et al (2013) Prediction of survival by [18F]fluorodeoxyglucose positron emission tomography in patients with locally advanced non-small-cell lung cancer undergoing definitive chemoradiation therapy: results of the ACRIN 6668/RTOG 0235 trial. J Clin Oncol 31:3823–3830
21. Im HJ, Pak K, Cheon GJ et al (2015) Prognostic value of volumetric parameters of (18)F-FDG PET in non-small-cell lung cancer: a meta-analysis. Eur J Nucl Med Mol Imaging 42:241–251
22. Kaira K, Ishizuka T, Yanagitani N et al (2008) Laryngeal sarcoidosis detected by FDG positron emission tomography. Clin Nucl Med 33:878–879
23. Kaira K, Oriuchi N, Otani Y et al (2007) Diagnostic usefulness of fluorine-18-alpha-methyltyrosine positron emission tomography in combination with 18F-fluorodeoxyglucose in sarcoidosis patients. Chest 131:1019–1027
24. Kaira K, Oriuchi N, Shimizu K et al (2009) 18F-FMT uptake seen within primary cancer on PET helps predict outcome of non-small cell lung cancer. J Nucl Med 50:1770–1776
25. Tomiyoshi K, Amed K, Muhammad S et al (1997) Synthesis of isomers of 18F-labelled amino acid radiopharmaceutical: position 2- and 3-L-18F-alpha-methyltyrosine using a separation and purification system. Nucl Med Commun 18:169–175
26. Kaira K, Oriuchi N, Otani Y et al (2007) Fluorine-18-alpha-methyltyrosine positron emission tomography for diagnosis and staging of lung cancer: a clinicopathologic study. Clin Cancer Res 3:6369–6378
27. Kanai Y, Segawa H, Miyamoto K et al (1998) Expression cloning and characterization of a transporter for large neutral amino acids activated by the heavy chain of 4F2 antigen (CD98). J Biol Chem 273:23629–23632
28. Chairoungdua A, Segawa H, Kim JY et al (1999) Identification of an amino acid transporter associated with the cystinuria-related type II membrane glycoprotein. J Biol Chem 274:28845–28848
29. Yasukawa T, Yoshikawa K, Aoyagi H et al (2000) Usefulness of PET with 11C-methionine for the detection of hilar and mediastinal lymph node metastasis in lung cancer. J Nucl Med 41:283–290
30. Kubota K, Matsuzawa T, Fujiwara T et al (1990) Differential diagnosis of lung tumor with positron emission tomography: a prospective study. J Nucl Med 31:1927–1932
31. Tian M, Zhang H, Oriuchi N et al (2004) Comparison of 11C-choline PET and FDG PET for the differential diagnosis of malignant tumors. Eur J Nucl Med Mol Imaging 31:1064–1072
32. Ishimori T, Saga T, Nagata Y et al (2004) 18F-FDG and 11C-methionine PET for evaluation of treatment response of lung cancer after stereotactic radiotherapy. Ann Nucl Med 18:669–674
33. Pauleit D, Floeth F, Tellmann L et al (2004) Comparison of O-(2-18F-fluoroethyl)-L-tyrosine PET and 3-123I-iodo-alpha-methyl-L-tyrosine SPECT in brain tumors. J Nucl Med 45:374–381
34. Pauleit D, Stoffels G, Schaden W et al (2005) PET with O-(2-18F-Fluoroethyl)-L-Tyrosine in peripheral tumors: first clinical results. J Nucl Med 46:411–416
35. Jager PL, Groen HJ, van der Leest A et al (2001) L-3-[123I]iodo-alpha-methyl-tyrosine SPECT in non-small cell lung cancer: preliminary observations. J Nucl Med 42:579–585
36. Shikano N, Kanai Y, Kawai K et al (2003) Isoform selectivity of 3-125I-iodo-alpha-methyl-L-tyrosine membrane transport in human L-type amino acid transporters. J Nucl Med 44:244–246
37. Thomlinson RH, Gray LH (1955) The histological structure of some human lung cancers and the possible implications for radiotherapy. Br J Cancer 9:539–459
38. Evans SM, Koch CJ (2003) Prognostic significance of tumor oxygenation in humans. Cancer Lett 195:1–16

39. Rischin D, Hicks RJ, Fisher R, et al; Trans-Tasman Radiation Oncology Group Study 98.02 (2006) Prognostic significance of [18F]-misonidazole positron emission tomography-detected tumor hypoxia in patients with advanced head and neck cancer randomly assigned to chemoradiation with or without tirapazamine: a substudy of Trans-Tasman Radiation Oncology Group Study 98.02. J Clin Oncol 24:2098–2104
40. O'Donoghue JA, Zanzonico P, Pugachev A et al (2005) Assessment of regional tumor hypoxia using 18F-fluoromisonidazole and 64Cu(II)-diacetyl-bis(N4-methylthiosemicarbazone) positron emission tomography: comparative study featuring microPET imaging, Po2 probe measurement, autoradiography, and fluorescent microscopy in the R3327-AT and FaDu rat tumor models. Int J Radiat Oncol Biol Phys 61:1493–1502
41. Dehdashti F, Mintun MA, Lewis JS et al (2003) In vivo assessment of tumor hypoxia in lung cancer with 60Cu-ATSM. Eur J Nucl Med Mol Imaging 30:844–850
42. Gagel B, Reinartz P, Demirel C et al (2006) [18F] fluoromisonidazole and [18F] fluorodeoxyglucose positron emission tomography in response evaluation after chemo-/radiotherapy of non-small-cell lung cancer: a feasibility study. BMC Cancer 6:51
43. Cherk MH, Foo SS, Poon AM et al (2006) Lack of correlation of hypoxic cell fraction and angiogenesis with glucose metabolic rate in non-small cell lung cancer assessed by 18F-Fluoromisonidazole and 18F-FDG PET. J Nucl Med 47:1921–1926
44. Overgaard J, Horsman M (1996) Modification of hypoxia-induced radio resistance in tumours by the use of oxygen and sensitizers. Semin Radiat Oncol 6:10–21
45. Sachpekidis C, Thieke C, Askoxylakis V et al (2015) Combined use of (18)F-FDG and (18) F-FMISO in unresectable non-small cell lung cancer patients planned for radiotherapy: a dynamic PET/CT study. Am J Nucl Med Mol Imaging 5:127–142
46. Mankoff DA, Shields AF, Krohn KA (2005) PET imaging of cellular proliferation. Radiol Clin N Am 43:153–167
47. Yamamoto Y, Nishiyama Y, Kimura N et al (2008) Comparison of (18)F-FLT PET and (18) F-FDG PET for preoperative staging in non-small cell lung cancer. Eur J Nucl Med Mol Imaging 35:236–245
48. Leyton J, Latigo JR, Perumal M et al (2005) Early detection of tumor response to chemotherapy by 3′-deoxy-3′-[18F]fluorothymidine positron emission tomography: the effect of cisplatin on a fibrosarcoma tumor model in vivo. Cancer Res 65:4202–4210
49. Buck AK, Halter G, Schirrmeister H et al (2003) Imaging proliferation in lung tumors with PET: 18F-FLT versus 18F-FDG. J Nucl Med 44:1426–1431
50. Yap CS, Czernin J, Fishbein MC et al (2006) Evaluation of thoracic tumors with 18F-fluorothymidine and 18F-fluorodeoxyglucose-positron emission tomography. Chest 129:393–401
51. Everitt SJ, Ball DL, Hicks RJ et al (2014) Differential (18)F-FDG and (18)F-FLT Uptake on Serial PET/CT Imaging Before and During Definitive Chemoradiation for Non-Small Cell Lung Cancer. J Nucl Med 55:1069–1074
52. Suzuki S, Kaira K, Ohshima Y et al (2014) Biological significance of fluorine-18-α-methyltyrosine (FAMT) uptake on PET in patients with oesophageal cancer. Br J Cancer 110:1985–1991
53. Nobusawa A, Kim A, Kaira K et al (2013) Diagnostic Usefulness of ^{18}F-FAMT PET and L-Type Amino Acid Transporter 1 (LAT1) Expression in Oral Squamous Cell Carcinoma. Eur J Nucl Med Mol Imaging 40:1692–1700
54. Isoda A, Higuchi T, Nakano S et al (2012) ^{18}F-FAMT in patients with multiple myeloma: clinical utility compared to ^{18}F-FDG. Ann Nucl Med 26:811–816
55. Kaira K, Oriuchi N, Yanagitani N et al (2010) Assessment of therapy response in lung cancer with ^{18}F-α-methyl tyrosine PET. AJR Am J Roentgenol 195:1204–1211
56. Furuya M, Horiguchi J, Nakajima H et al (2012) Correlation of L-type amino acid transporter 1 and CD98 expression with triple negative breast cancer prognosis. Cancer Sci 103:382–389

57. Kaira K, Sunose Y, Ohshima Y et al (2013) Clinical significance of L-type amino acid transporter 1 expression as a prognostic marker and potential of new targeting therapy in biliary tract cancer. BMC Cancer 13:482
58. Kaira K, Oriuchi N, Imai H et al (2008) Prognostic significance of L-type amino acid transporter 1 expression in resectable stage I-III nonsmall cell lung cancer. Br J Cancer 98:742–748
59. Toyoda M, Kaira K, Ohshima Y et al (2014) Prognostic significance of amino-acid transporter expression (LAT1, ASCT2, and xCT) in surgically resected tongue cancer. Br J Cancer 110:2506–2513
60. Isoda A, Kaira K, Iwashina M et al (2014) Expression of L-type amino acid transporter 1 (LAT1) as a prognostic and therapeutic indicator in multiple myeloma. Cancer Sci 105:1496–1502
61. Shimizu A, Kaira K, Kato M et al (2015) Prognostic significance of L-type amino acid transporter 1 (LAT1) expression in cutaneous melanoma. Melanoma Res 25:399–405
62. Kaira K, Oriuchi N, Shimizu K et al (2010) Comparison of L-type amino acid transporter 1 expression and L-[3-18F]-α-methyl tyrosine uptake in outcome of non-small cell lung cancer. Nucl Med Biol 7:911–916
63. Kaira K, Higuchi T, Oriuchi N, et al. (2014) Usefulness of l-[3-^{18}f]-α-methyl tyrosine (^{18}F-FAMT) PET as therapeutic monitoring for patients with advanced lung cancer.. 2014 ASCO Annual Meeting, Chicago, Illinois, May 30-June 3.

Chapter 4
Methods in Molecular Diagnosis

Koichi Hagiwara

Abstract Identification of the epidermal growth factor receptor (*EGFR*) mutation dramatically changed the treatment of lung cancer. Much effort has been devoted in the research of this field, which have driven lung cancer as one of the leader in the era of the precision medicine. To further progress the understanding of lung cancer at the molecular level, and to better apply the accomplishment attained to the clinical practice, the use of proper molecular test is mandatory. This chapter describes characteristics of the molecular tests currently used and key points properly applying them in the clinical practice of lung cancer.

Keywords *EGFR* • Genetic test • PCR • Fish • IHC

4.1 Genomic Alterations in Lung Cancer [1]

Lung cancer develops from normal airway epithelium through a multistage process. Gain-of-function and loss-of-function mutations accumulate in the genes that have primary importance for keeping cellular homeostasis, and an overt cancer develops. Every cell is equipped with the DNA repairing system, and many of the DNA injuries caused by mutagens, ionizing radiations, or ultraviolet light are repaired. When the DNA damage is very severe and is unable to be reparied, the cell initiates apoptosis to eliminate itself from the body. Nonetheless, some cells survive without causing apoptosis, partly because the genes involving the apoptosis system are simultaneously damaged. Such cells accumulate genetic mutations, and gain an ability to autonomously proliferate. Cells with mutations occurring in the genes involved in invasion and metastasis gain a metastatic phenotype. Both non-small cell lung cancer (NSCLC) and small cell lung cancer (SCLC) are considered to develop through a similar process, although the sets of the genes mutated are different.

K. Hagiwara (✉)
Division of Pulmonary Medicine, Jichi Medical University,
3311-1 Yakushiji, Shimotsuke, Tochigi 329-0498, Japan
e-mail: hagiwark@me.com

© Springer Science+Business Media Singapore 2017
Y. Takiguchi (ed.), *Molecular Targeted Therapy of Lung Cancer*,
DOI 10.1007/978-981-10-2002-5_4

4.2 Concept of the Driver Mutations

Each gene mutated in the development of cancer has different importance. Inhibitors for the mutated epidermal growth factor receptor (*EGFR*) gene or the anaplastic lymphoma kinase (*ALK*)-fusion gene demonstrate a remarkable antitumor effect on the cancers harboring each mutation [2–4]. This indicates that the mutated *EGFR* gene or the *ALK*-fusion gene plays an essential role for maintaining the cancer phenotype. Such genes are called the driver genes. The mutated *EGFR* gene [5–7], the mutated Kirsten rat sarcoma viral oncogene homolog gene (*KRAS*) gene, the mutated *BRAF* gene [8], the *ALK*-fusion genes [9], the *ROS1*-fusion genes [10], and the *RET*-fusion genes [11] are considered to be driver genes. The driver genes exist in a mutually exclusive manner: only a single driver gene is found in one cancer cell [12]. It is considered that a single driver gene has sufficient power for a cell to acquire cancer phenotype, and thus no additional driver genes are required. This suggests that fewer mutational events may be required for cancers with a driver gene. This is consistent with the observation that cancers with a driver gene are frequently found in patients with a younger age.

4.3 Important Driver Genes

*4.3.1 The mutated Epidermal Growth Factor Receptor (*EGFR*) Gene*

The mutated *EGFR* is the driver gene most frequently found in lung cancer and found in up to 50 % of non-small cell lung cancer in Asians [13]. The mutated EGFR gene is more frequently seen in females and in nonsmokers. The mutations occur in the kinase domain and constitutively activate the kinase activity of the EGFR protein. Almost 90 % of the mutations are either an in-frame exon 19 deletion or a point mutation in exon 21 that changes leucine 858 to arginine (L858R) [14]. Mutant EGFR proteins with these mutations are sensitive to EGFR inhibitors including gefitinib, erlotinib, and afatinib. However, EGFR proteins with two mutations, one is the mutation observed before the treatment and the other is the T790M mutation that converts threonine 790 to methionine emerges in half of the patients during treatment. Such protein is resistant to the EGFR inhibitors [15]. A novel EGFR tyrosine kinase inhibitor, osimertinib, has been developed targeting the EGFR protein with the T790M mutation and has introduced into the market [16].

*4.3.2 The mutated v-Ki-ras2 Kirsten Rat Sarcoma Viral Oncogene Homolog (*KRAS*) Gene*

The *KRAS* gene is a member of Ras GTPase gene superfamily. Mutations occur in codons 12 or 13 and constitutively activate the protein. The frequency of mutation is higher in Caucasians and in smokers. Inhibitor of the mutant *KRAS* has been under an intense investigation without any remarkable successes. Accordingly, advanced lung cancer patients with *KRAS* mutations are treated by the cytotoxic chemotherapy or by the immune checkpoint inhibitors.

*4.3.3 The Anaplastic Lymphoma Receptor Tyrosine Kinase (*ALK*)-Fusion Gene*

In about 5 % of lung cancer, the carboxy-terminus of the ALK protein that contains kinase domain fuses to the amino-terminus of the echinoderm microtubule-associated protein-like 4 (EML4) protein or of the kinesin family member 5B (KIF5B) protein to form a constitutively active EML4-ALK or KIF5B-ALK-fusion protein. The *ALK* gene is not expressed in the airway, while *EML4* or *KIF5B* are expressed. When a fusion gene is formed, ALK kinase domain becomes under the control of the promoters for the EML4 or KIF5B gene, and is ectopically expressed in the airway. Moreover, the fusion proteins form dimers utilizing dimerization domain located in the amino-terminus of the EML4 or KIF5B proteins. This accelerates the self-phosphorylation of the fusion proteins, which transmits the constitutively activated signal into the cells. The ALK inhibitors exhibit remarkable effect on lung cancers harboring the *ALK*-fusion genes [4].

4.3.4 The mutated BRAF *Gene, the* ROS1*-Fusion Genes, and the* RET*-Fusion Genes*

Mutation in the *BRAF* gene, the *ROS1*-fusion genes, or *RET*-fusion genes is found in about 1 % of lung cancer. Although the frequency is not high, the molecular targeting drugs for each mutant gene have been identified, and their clinical efficacy has been demonstrated [17, 18]. Utilization of the mutations of these genes is anticipated.

Importance of molecular diagnosis in the clinical practice of lung cancer

The therapeutic procedures that may completely cure the disease have the primary importance in the treatment of lung cancer. Such procedures include surgical resection or chemoradiotherapy. However, about 2/3 of the lung cancer patients have an advanced disease where these procedures are mostly not applicable. Until two decades ago, cytotoxic chemotherapy is the only treatment regimen available. Nevertheless, the median overall survival of the systemic chemotherapies was less than a year [19]. After the emergence of the EGFR inhibitors, administration of them to the patients with the mutated *EGFR* gene has been established as the standard treatment. The overall survival has significantly elongated to more than 2 years, and it is expected to become more [2, 3, 20]. A similar result is obtained in cancers with the *ALK*-fusion gene by the use of ALK inhibitors [21]. These results indicate that the identification of a novel driver gene and the development of the corresponding molecular targeting drugs has a pivotal importance for improving the treatment of lung cancer.

4.4 Specimen Used for the Molecular Diagnosis of Lung Cancer

Both tissue and cytological specimens are used for the molecular diagnosis of lung cancer. The types of specimens heavily depend on the manner of clinical practice. In a country where bronchoscopy is applied to most of the patients, both types of specimens are employed, while in a country where needle biopsy is heavily applied, tissue specimens are mostly used. Figure 4.1 illustrates the types of specimen submitted to the *EGFR* mutation test in Japan [22]. Here, bronchoscopy is the primary procedure for establishing the diagnosis of lung cancer, about 1/3 of the specimens submitted to the *EGFR* mutation test are the cytological specimens [22]. Whichever types of specimens are employed, the presence of cancer cells in the specimens should be confirmed. Figure 4.2 shows the representative procedures.

Clinical samples contain both normal cells and cancer cells. Currently, pathological examination is the only way that is able to diagnose cancer. Then, how much is cancer cell content in pathologically cancer-positive samples? Figure 4.3a shows our result, suggesting that the cancer cell content in the cancer specimens are usually >1 % [23]. This indicates that the methods that are able to test samples with cancer cell content of 1 % is adequate for testing clinical samples [24].

The amount of DNA is very important for properly performing mutation test. In the test that identify mutations in a single-copy gene such as *EGFR*, the requirement for testing samples with a 1 % cancer cell content automatically determines the required amount of DNA. A single cell contains about 6.4 pg of genomic DNA. For detecting mutation occurring in 1 % of the cells, 640 pg of genomic DNA is required by a simple calculation. In actuality, the number of the cells sampled conforms to the binomial distribution, and thus more DNA is required (Fig. 4.3b). Two to 10 ng

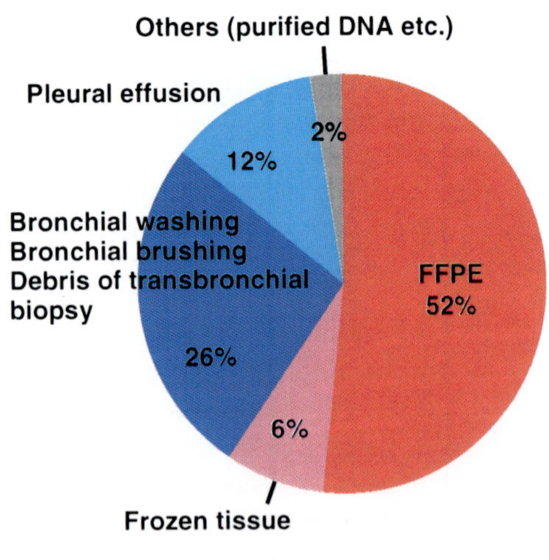

Fig. 4.1 Specimens used for the *EGFR* mutation test in Japan. The sample categories summarized from ~17,000 samples submitted to one of the laboratories in 2009 [38]. Tissue samples (i.e., FFPE and frozen tissue) are indicated in *warm colors*, while cytological samples (i.e., bronchoscopy specimens and pleural effusion) are indicated in *cold colors*

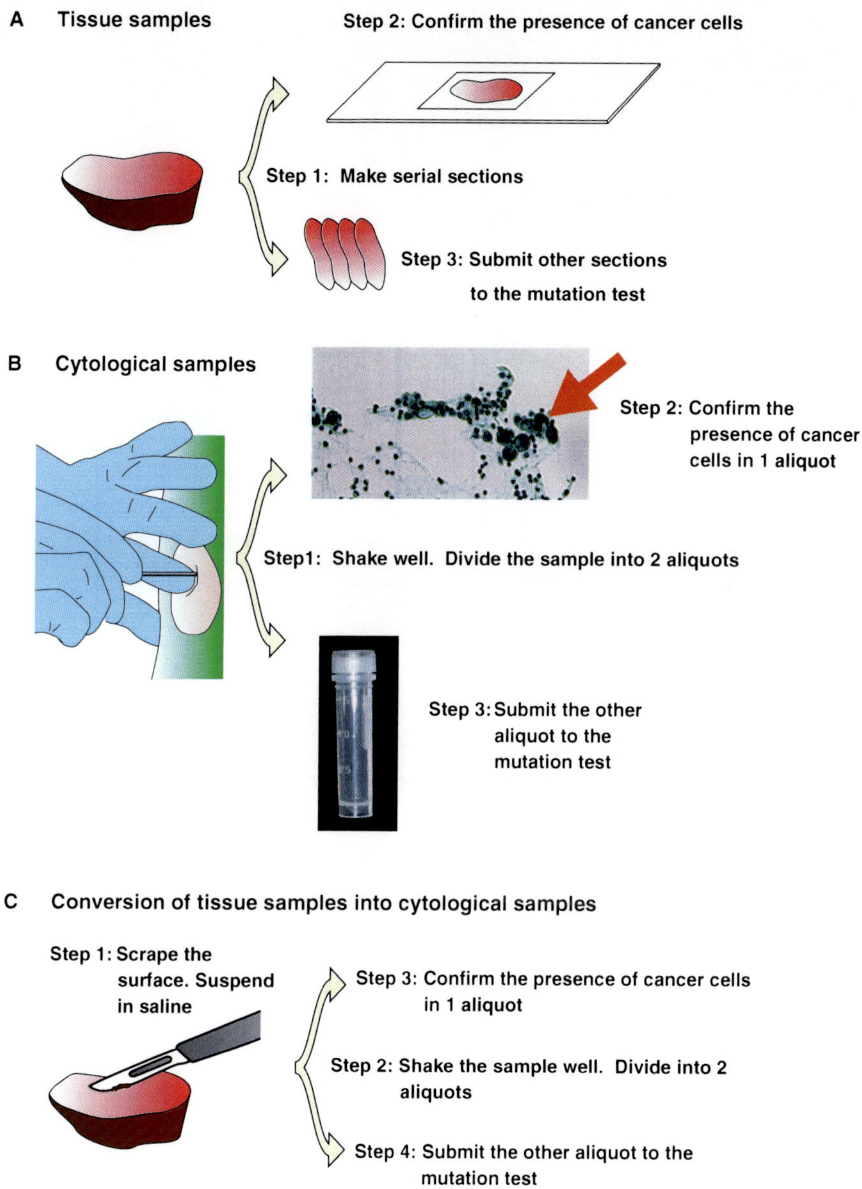

Fig. 4.2 Sample preparation procedures. (**a**) Tissue samples. Step 1: Serial sectioning. Step 2: The presence of cancer cells is confirmed in 1 section. Step 3: The *EGFR* mutation is investigated using other sections. Macro-dissection may be required to remove normal tissue before step 1. (**b**) Cytological samples. Step 1: Suspend the cells in saline. Divide the samples into 2 aliquots. Step 2: Confirm the presence of cancer cells in 1 aliquot. Step 3: Investigate the *EGFR* mutation using the other aliquot. (**c**) Preparation of cytological samples from tissue. Step 1: Scrape the surface of the tissue. Suspend the cells in saline. Step 2: Divide the samples into 2 aliquots. Step 3: Confirm the presence of cancer cells in 1 aliquot. Step 4: Investigate the *EGFR* mutation using the other aliquot

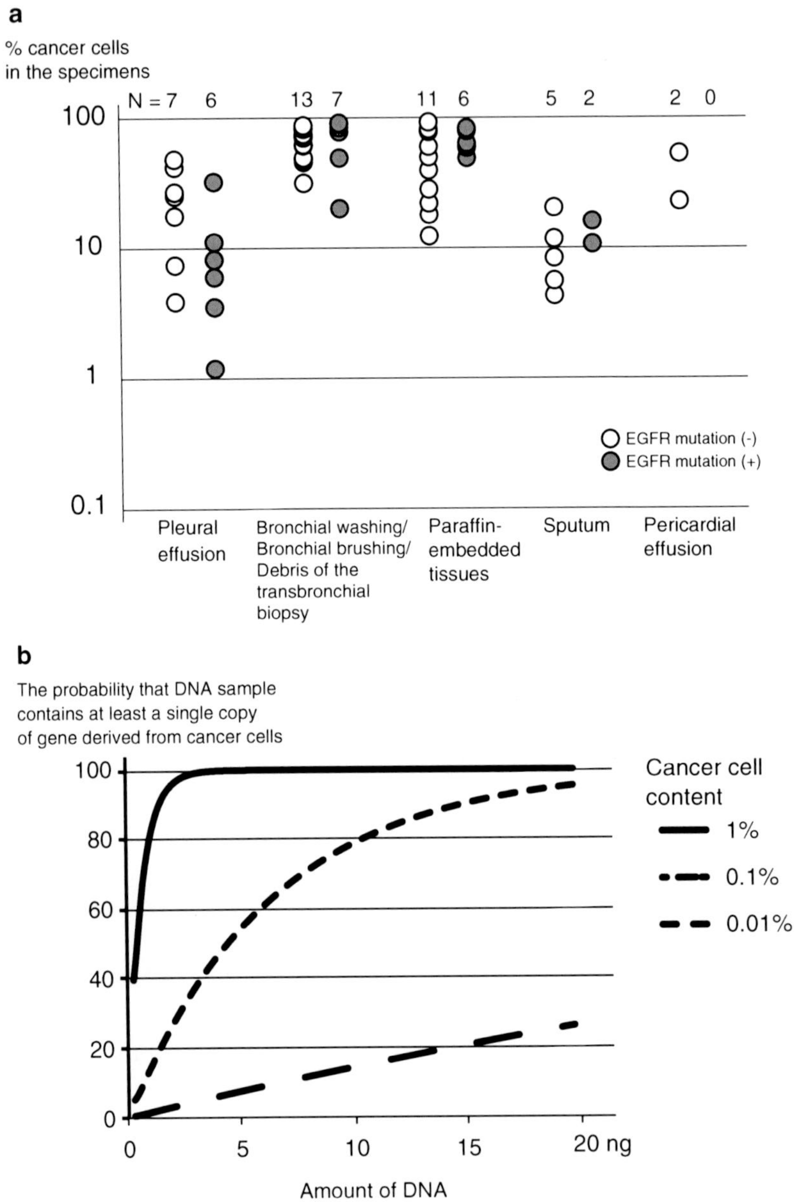

Fig. 4.3 Cancer cell content and the required amount of DNA. (**a**) The number of cancer cells and normal cells was counted in the microscopic slides randomly selected from those used for establishing the diagnosis of cancer. (**b**) The required amount of DNA. The probability that DNA sample contains at least a single copy of gene derived from cancer cells. When the cancer cell content is small, a huge amount of DNA is required to avoid false-negative result due to a sampling error

of DNA is recommended when testing samples with 1 % cancer cell content. If lesser amount of DNA is used, false-negative results may occur [22].

Formalin-fixed, paraffin-embedded (FFPE) tissue, snap-frozen tissue, and cytological samples are the types of samples most frequently used in the molecular diagnosis, and each has advantages and disadvantages. FFPE samples provide clear histological data and are best for fluorescent in situ hybridization (FISH) and immunohistological examination (IHC). However, FFPE samples are not the best for the examination of DNA and RNA. Procedures employed for fixation and embedding significantly degrade both DNA and RNA. DNA is often modified, providing artificially mutated sequences. RNA is usually severely damaged [25–28]. Snap-frozen tissue well preserves both DNA and RNA and allow histological examination. Although the quality of morphological information is not as good as that obtained from FFPE samples, they allow high-quality molecular biological analyses. Cytological samples do not provide information on the tissue architecture. However, DNA and RNA are well preserved in good quality, especially when cells are collected and stored in a preservative solution (e.g., RNAlater Solutions, ThermoFisher Scientific Inc.) just after sample isolation. The most prominent advantage of the cytological samples is that they can be isolated from almost all patients. It is therefore very important to utilize cytological samples for the molecular diagnosis for all patients benefit from molecular targeting drugs [22].

Cell-free DNA (cfDNA) is a source of DNA recently attracted much attention [29, 30]. It has been shown that the amount of DNA that is released from cancer cells into bloodstream is enough to allow molecular biological analyses in some patients with an advanced disease. DNA is isolated from plasma or urine. The problem is that, when mutation is not found, it is not easy to know whether cancer does not have mutation being investigated or the amount of DNA from cancer cell is not enough. Therefore, the interpretation of the result of a DNA test using cfDNA should be cautiously done, when a negative result is obtained.

4.5 Methods

The optimal methods for the molecular testing differ depending on the type of samples (tissue or cytological samples) and on the type of materials (DNA, RNA or protein). Most of the methods that are used for examining DNA or RNA at the nucleotide level utilize the polymerase chain reaction (PCR) or reverse-transcriptase polymerase chain reaction (RT-PCR). Gross changes of the chromosome structure are detected by the fluorescent in situ hybridization (FISH). The localization of a specific protein in the tissue is detected by the immunohistochemistry (IHC). Currently, the *EGFR*, *KRAS*, or *BRAF* mutations are detected by the PCR-based technique, while the presence of the *ALK*-, *ROS1*-, or *RET*-fusion genes is detected by RT-PCR or FISH [4]. The *ALK*-fusion genes are also detected by the IHC [31].

4.5.1 PCR-Based Methods

DNA is very stable and easy to isolate. Therefore, a DNA-based test is easy to perform. Many of the PCR-based methods amplify the mutation hotspots by PCR. Then the presence of mutated sequence is investigated by nucleotide sequencing or by a fluorescence-based technique like the TaqMan assay [32]. The methods include scorpion ARMS [33], the PNA-LNA clamp [34], the cycleave PCR [35], the PCR-invader [36], and the cobas EGFR mutation test [37]. The former four methods have confirmed to detect mutations from samples with a cancer cell content of 1 % [25], while the latter is able to detect from samples with 5 % cancer cells according to the package insert.

The most prominent characteristics of these methods are the implementation of special techniques that increase sensitivity, because the mutant sequences should be detected in the presence of large amount of backgrounds. For example, if the cancer cell content is 1 %, the wild-type *EGFR* sequence exists 200-fold more over the mutant *EGFR* sequence. Figure 4.4 illustrates how the PNA-LNA PCR clamp method [34] increase the sensitivity, which enables to detect mutant sequences from almost all clinical samples that are pathologically positive for cancer cells.

A weak point common to most of the PCR-based methods is that they are able to test only a limited number of locations on the genome. Therefore, PCR-based methods are not applicable when locations of mutations disperse in a wide range on a genome. A representative example is the *EML4-ALK*-fusion gene. The breakpoint that is the junction of the *EML4* gene and the *ALK* gene is not fixed and is found in a wide range in the introns. Therefore, it is hard to identify the presence of the *EML4-ALK*-fusion gene by identifying the breakpoint sequence. Accordingly, the *EML4-ALK*-fusion gene is identified directly using RT-PCR or indirectly using FISH or IHC.

4.5.2 RT-PCR-Based Methods

RNA is very labile and is easily degraded. Therefore, when handling RNA, a special precaution should be made. Moreover, RNA is not directly amplified by the DNA polymerases. It should be converted to DNA (complementary DNA: cDNA) by the reverse transcriptase (RT) and then amplified by PCR. The set of reactions is collectively called as RT-PCR. The mRNA of a fusion gene is a hybrid having the 5′ side of one gene and the 3′ side of the other gene. If a DNA fragment is amplified by RT-PCR using the primers specific for each side, it indicates that a fusion gene is present. Therefore, RT-PCR is the method of choice when good quality of RNA is available, as isolated from snap-frozen tissue or cytological sample immediately stored in a RNA preservation solution.

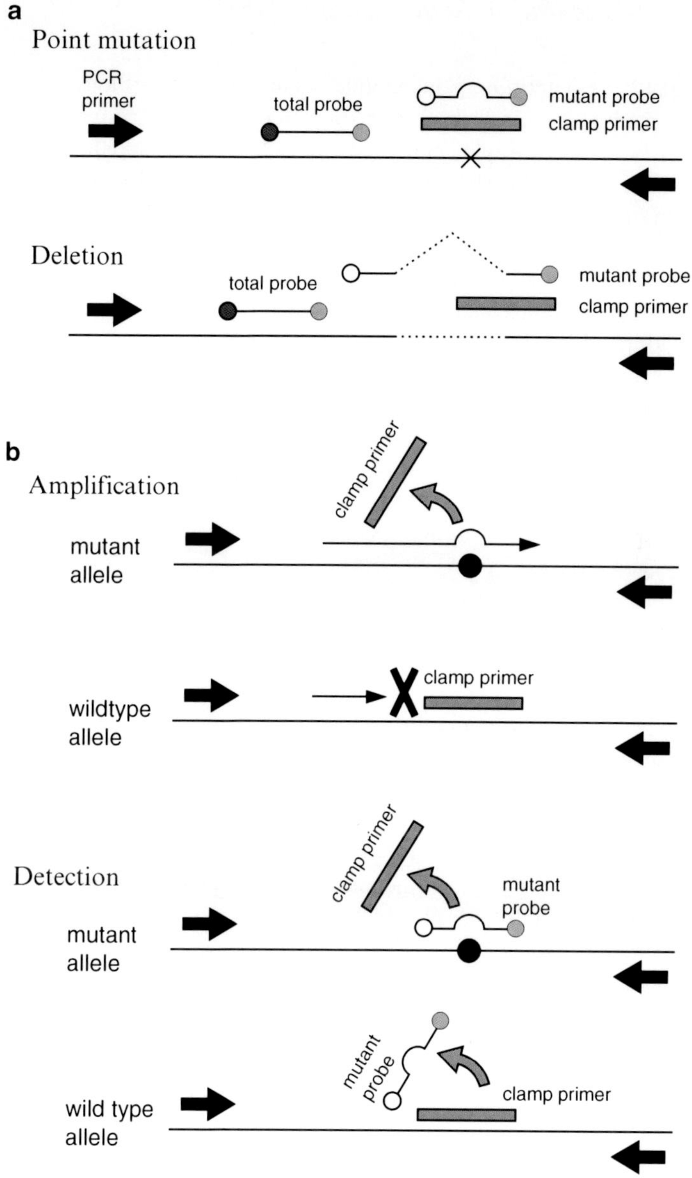

Fig. 4.4 PNA-LNA PCR clamp system [34]. (**a**) Primer and probes positions. Point mutation: Both the clamp primer and the mutant probe are located on the mutation site. The "total probe" to detect both mutant and wild-type fragments is on the adjoining sequence. X indicates the mutation. Deletion: The clamp primer partly covers the deletion. Sequences located on both sides of the deletion are joined to make the sequence of the mutation probe. A *dashed line* indicates a deletion. *Black arrows* are the PCR primers. (**b**) Schematic presentation of the reaction to detect point mutations. In the amplification process, the clamp primers fail to bind to the mutation sequences but binds to the wild type, resulting in the preferential amplification of mutant sequences. In the detection process, the mutant probes bind to the mutant sequences but fail to bind to the wild type partly because of mismatches and partly because of the competitive displacement by the clamp primers. The signals are detected by the 5′ nuclease assay. A *black circle* indicates a mutation

4.5.3 Next-Generation Sequencing (NGS)

NGS is capable of determining nucleotide sequence from millions of nucleotide fragments simultaneously. Therefore, using NGS is an attractive approach for simultaneously testing multiple genes from multiple samples, and the tests using the NGS is actively developed. The challenges are that the NGS were originally developed for determining a huge number of DNA fragments at the cost of correctness. Meanwhile, the mutation tests require correctness because the results are used in clinical practice. The NGS often misread normal sequence as mutant sequence, which causes serious consequences in the clinical practice. Procedures that reduce the false reading are vital for the development of NGS-based tests. Germ-line polymorphisms such as single nucleotide polymorphisms (SNPs) are not hard to detect. However, somatic mutations as found in the clinical samples with a cancer cell content of 1 % are not easy to correctly identify. Moreover, the turnaround time from sample submission to the report of the result takes weeks for NGS-based tests, while it should be less than a week for utilizing the result in clinical practice. More efforts are required for overcoming these problems in NGS.

4.5.4 Utilization of Liquid Biopsy Samples

Recently it has been reported that DNA fragments from cancer cells may shed into the bloodstream and can be isolated from serum or urine (cell-free DNA: cfDNA). Genetic mutations in cancer cells may be detected using cfDNA. Because the ratio of cancer cell-derived cfDNA to normal cell-derived DNA (endothelial cells, leukocytes, etc.) is hard to know, detection of mutation from cfDNA is performed using very sensitive methods. Therefore, the methods employed are digital PCR or highly sensitive PNA-LNA PCR clamp, which may detect mutations from samples of which cancer cell content is less than 1 %. However, when the cancer-derived cfDNA occupies less than 0.1 % of DNA detection is almost impossible, because the mount of input DNA automatically determines the limit of the detection (Fig. 4.5). Furthermore, use of more than 10 ng DNA is difficult, because the yield of cfDNA is not large. When the cancer-derived cfDNA occupies from 0.1 to 1 %, the effect of DNA polymerase error becomes evident. DNA polymerase sometimes misincorporates nucleotides during PCR. This produces mutant sequence from normal sequence, and normal DNA is misinterpreted as mutant DNA. When the ratio is more than 1 %, the mutation test itself is not difficult. The use of cfDNA is a very attractive approach, and if appropriately used, it enables to decide treatment without performing biopsy.

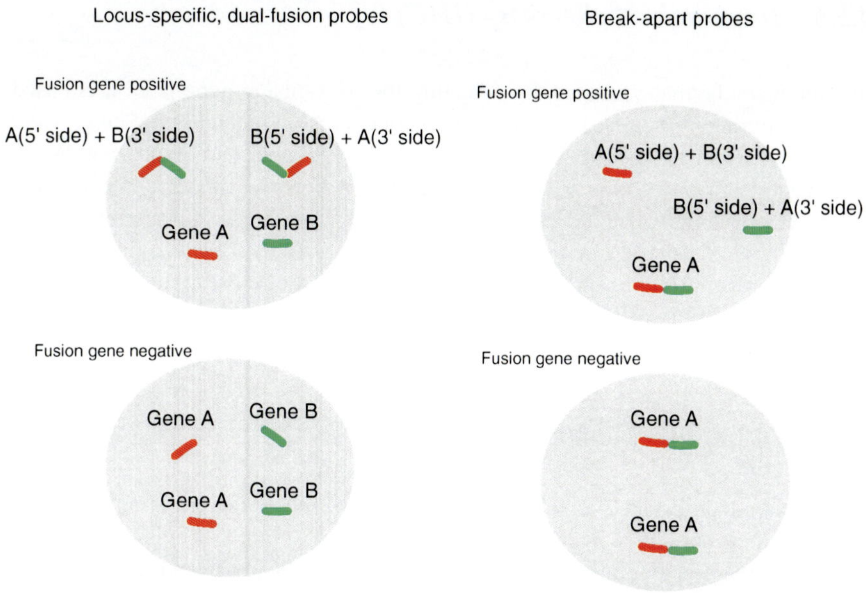

Fig. 4.5 FISH. (**a**) The test using the locus-specific, dual-fusion probes. Color probes are located on different genes, such as the *EML4* gene and the *ALK* gene. When these 2 genes form a fusion gene, both colors are observed at the same position. (**b**) The test using the break-apart probes. Color probes are located on the 5′ side and the 3′ side of a single gene. When one gene forms a fusion gene with the a fusion partner gene, two colors are separately observed

4.5.5 FISH (Fluorescent In Situ Hybridization)

FISH has been used in the test of hematological malignancies, in which chromosomal translocations or deletions are commonly found. Moreover, the *ALK* gene is one of the genes frequently found at the breakpoint of a chromosomal rearrangement and forms a fusion gene. As a result, FISH test for the ALK-fusion genes had already been established in the field of hematology when the *EML4-ALK*-fusion gene was found in lung cancer, and thus the test was directly applied to it.

There are two different FISH tests: tests using the locus-specific, dual-fusion probes and test using the break-apart probes (Fig. 4.5). For detecting *ALK*-fusion genes, *ROS1*-fusion genes, and *RET*-fusion genes, test using the break-apart probes is employed [4]. The test using break-apart probes is applicable to a gene that has multiple fusion partners. The weak points of FISH are that the procedure is complicated and the interpretation of the result under microscope is often difficult.

4.5.6 *Immunohistochemistry (IHC)*

Immunohistochemistry is used for detecting the *ALK*-fusion gene [31]. It is based on a very unique principle. The *ALK* gene is not expressed in the normal lung, while its fusion partner genes (the *EML4* gene or the *KIF5B* gene) are expressed. When *ALK* gene is fused to one of the partner genes, it becomes under control of the promoter of the partner gene and thus becomes expressed. The expression level is not high but is detectable by IHC with increased sensitivity. Because IHC is much easier to perform than FISH, it is a preferred method for detecting the *ALK*-fusion gene from the FFPE samples.

4.6 Companion Diagnostics

Companion diagnostics is devised by an idea that the molecular targeting drugs and diagnostic test should be available at the same time. At first, it seems to be a good idea. A companion diagnostics is a test with a certified sensitivity and specificity. By using the companion diagnostics, the drug can be appropriately used anywhere in the world. However, the companion diagnostics has many drawbacks. First, they are developed in the country the drug is developed. The clinical situation of the country may be very different from country to country. For example, most of the companion diagnostics developed in the USA are based on tissue samples. However, 1/3 of the clinical samples used for genetic testing are cytological samples in Japan. Therefore, companion diagnostics cannot be used for 1/3 of the lung cancer patients in Japan. Second, the companion diagnostics are not always the most reliable tests. Sometimes, they do not fulfill the specifications proposed by the academic committee [24]. The concept of the companion diagnostics should be carefully applied so that it helps the clinical practice rather than disturbing it.

4.7 Last Words

In this chapter, characteristics of a variety of genetic tests are described. With the increase of the number of genes that are worth investigating for the treatment of lung cancer, the test system increasingly becomes complicated. However, the turn-around time from the submission of the sample to the return of the result is vital for the information to be properly used in the clinical practice. The best balance between the complexity of the tests and the time required for obtaining the results should be continuously sought.

References

1. Vincent T, DeVita J, Lawrence TS, Rosenberg SA (2011) Cancer: principles & practice of oncology: primer of the molecular biology of cancer. Wolters Kluwer/Lippincott Williams & Wilkins, Philadelphia
2. Maemondo M, Inoue A, Kobayashi K, Sugawara S, Oizumi S, Isobe H, Gemma A, Harada M, Yoshizawa H, Kinoshita I, Fujita Y, Okinaga S, Hirano H, Yoshimori K, Harada T, Ogura T, Ando M, Miyazawa H, Tanaka T, Saijo Y, Hagiwara K, Morita S, Nukiwa T (2010) Gefitinib or chemotherapy for non-small-cell lung cancer with mutated EGFR. N Engl J Med 362(25):2380–2388. doi:10.1056/NEJMoa0909530
3. Mitsudomi T, Morita S, Yatabe Y, Negoro S, Okamoto I, Tsurutani J, Seto T, Satouchi M, Tada H, Hirashima T, Asami K, Katakami N, Takada M, Yoshioka H, Shibata K, Kudoh S, Shimizu E, Saito H, Toyooka S, Nakagawa K, Fukuoka M, West Japan Oncology Group (2010) Gefitinib versus cisplatin plus docetaxel in patients with non-small-cell lung cancer harbouring mutations of the epidermal growth factor receptor (WJTOG3405): an open label, randomised phase 3 trial. Lancet Oncol 11(2):121–128. doi:10.1016/S1470-2045(09)70364-X
4. Kwak EL, Bang YJ, Camidge DR, Shaw AT, Solomon B, Maki RG, Ou SH, Dezube BJ, Janne PA, Costa DB, Varella-Garcia M, Kim WH, Lynch TJ, Fidias P, Stubbs H, Engelman JA, Sequist LV, Tan W, Gandhi L, Mino-Kenudson M, Wei GC, Shreeve SM, Ratain MJ, Settleman J, Christensen JG, Haber DA, Wilner K, Salgia R, Shapiro GI, Clark JW, Iafrate AJ (2010) Anaplastic lymphoma kinase inhibition in non-small-cell lung cancer. N Engl J Med 363(18):1693–1703. doi:10.1056/NEJMoa1006448
5. Lynch TJ, Bell DW, Sordella R, Gurubhagavatula S, Okimoto RA, Brannigan BW, Harris PL, Haserlat SM, Supko JG, Haluska FG, Louis DN, Christiani DC, Settleman J, Haber DA (2004) Activating mutations in the epidermal growth factor receptor underlying responsiveness of non-small-cell lung cancer to gefitinib. N Engl J Med 350(21):2129–2139. doi:10.1056/NEJMoa040938
6. Paez JG, Janne PA, Lee JC, Tracy S, Greulich H, Gabriel S, Herman P, Kaye FJ, Lindeman N, Boggon TJ, Naoki K, Sasaki H, Fujii Y, Eck MJ, Sellers WR, Johnson BE, Meyerson M (2004) EGFR mutations in lung cancer: correlation with clinical response to gefitinib therapy. Science 304(5676):1497–1500. doi:10.1126/science.1099314
7. Pao W, Miller V, Zakowski M, Doherty J, Politi K, Sarkaria I, Singh B, Heelan R, Rusch V, Fulton L, Mardis E, Kupfer D, Wilson R, Kris M, Varmus H (2004) EGF receptor gene mutations are common in lung cancers from "never smokers" and are associated with sensitivity of tumors to gefitinib and erlotinib. Proc Natl Acad Sci U S A 101(36):13306–13311. doi:10.1073/pnas.0405220101
8. Shigematsu H, Takahashi T, Nomura M, Majmudar K, Suzuki M, Lee H, Wistuba II, Fong KM, Toyooka S, Shimizu N, Fujisawa T, Minna JD, Gazdar AF (2005) Somatic mutations of the HER2 kinase domain in lung adenocarcinomas. Cancer Res 65(5):1642–1646. doi:10.1158/0008-5472.CAN-04-4235
9. Soda M, Choi YL, Enomoto M, Takada S, Yamashita Y, Ishikawa S, Fujiwara S, Watanabe H, Kurashina K, Hatanaka H, Bando M, Ohno S, Ishikawa Y, Aburatani H, Niki T, Sohara Y, Sugiyama Y, Mano H (2007) Identification of the transforming EML4-ALK fusion gene in non-small-cell lung cancer. Nature 448(7153):561–566. doi:10.1038/nature05945
10. Takeuchi K, Soda M, Togashi Y, Suzuki R, Sakata S, Hatano S, Asaka R, Hamanaka W, Ninomiya H, Uehara H, Lim Choi Y, Satoh Y, Okumura S, Nakagawa K, Mano H, Ishikawa Y (2012) RET, ROS1 and ALK fusions in lung cancer. Nat Med 18(3):378–381. doi:10.1038/nm.2658
11. Kohno T, Ichikawa H, Totoki Y, Yasuda K, Hiramoto M, Nammo T, Sakamoto H, Tsuta K, Furuta K, Shimada Y, Iwakawa R, Ogiwara H, Oike T, Enari M, Schetter AJ, Okayama H, Haugen A, Skaug V, Chiku S, Yamanaka I, Arai Y, Watanabe S, Sekine I, Ogawa S, Harris CC, Tsuda H, Yoshida T, Yokota J, Shibata T (2012) KIF5B-RET fusions in lung adenocarcinoma. Nat Med 18(3):375–377. doi:10.1038/nm.2644

12. Saito M, Shiraishi K, Kunitoh H, Takenoshita S, Yokota J, Kohno T (2016) Gene aberrations for precision medicine against lung adenocarcinoma. Cancer Sci 107(6):713–720. doi:10.1111/cas.12941
13. Mitsudomi T (2014) Molecular epidemiology of lung cancer and geographic variations with special reference to EGFR mutations. Transl Lung Cancer Res 3(4):205–211. doi:10.3978/j.issn.2218-6751.2014.08.04
14. Tanaka T, Matsuoka M, Sutani A, Gemma A, Maemondo M, Inoue A, Okinaga S, Nagashima M, Oizumi S, Uematsu K, Nagai Y, Moriyama G, Miyazawa H, Ikebuchi K, Morita S, Kobayashi K, Hagiwara K (2010) Frequency of and variables associated with the EGFR mutation and its subtypes. Int J Cancer 126(3):651–655. doi:10.1002/ijc.24746
15. Kobayashi S, Boggon TJ, Dayaram T, Janne PA, Kocher O, Meyerson M, Johnson BE, Eck MJ, Tenen DG, Halmos B (2005) EGFR mutation and resistance of non-small-cell lung cancer to gefitinib. N Engl J Med 352(8):786–792. doi:10.1056/NEJMoa044238
16. Janne PA, Yang JC, Kim DW, Planchard D, Ohe Y, Ramalingam SS, Ahn MJ, Kim SW, Su WC, Horn L, Haggstrom D, Felip E, Kim JH, Frewer P, Cantarini M, Brown KH, Dickinson PA, Ghiorghiu S, Ranson M (2015) AZD9291 in EGFR inhibitor-resistant non-small-cell lung cancer. N Engl J Med 372(18):1689–1699. doi:10.1056/NEJMoa1411817
17. Mazieres J, Zalcman G, Crino L, Biondani P, Barlesi F, Filleron T, Dingemans AM, Lena H, Monnet I, Rothschild SI, Cappuzzo F, Besse B, Thiberville L, Rouviere D, Dziadziuszko R, Smit EF, Wolf J, Spirig C, Pecuchet N, Leenders F, Heuckmann JM, Diebold J, Milia JD, Thomas RK, Gautschi O (2015) Crizotinib therapy for advanced lung adenocarcinoma and a ROS1 rearrangement: results from the EUROS1 cohort. J Clin Oncol 33(9):992–999. doi:10.1200/JCO.2014.58.3302
18. Drilon A, Wang L, Hasanovic A, Suehara Y, Lipson D, Stephens P, Ross J, Miller V, Ginsberg M, Zakowski MF, Kris MG, Ladanyi M, Rizvi N (2013) Response to Cabozantinib in patients with RET fusion-positive lung adenocarcinomas. Cancer Discov 3(6):630–635. doi:10.1158/2159-8290.CD-13-0035
19. Schiller JH, Harrington D, Belani CP, Langer C, Sandler A, Krook J, Zhu J, Johnson DH, Eastern Cooperative Oncology Group (2002) Comparison of four chemotherapy regimens for advanced non-small-cell lung cancer. N Engl J Med 346(2):92–98. doi:10.1056/NEJMoa011954
20. Sequist LV, Yang JC, Yamamoto N, O'Byrne K, Hirsh V, Mok T, Geater SL, Orlov S, Tsai CM, Boyer M, Su WC, Bennouna J, Kato T, Gorbunova V, Lee KH, Shah R, Massey D, Zazulina V, Shahidi M, Schuler M (2013) Phase III study of afatinib or cisplatin plus pemetrexed in patients with metastatic lung adenocarcinoma with EGFR mutations. J Clin Oncol 31(27):3327–3334. doi:10.1200/JCO.2012.44.2806
21. Solomon BJ, Mok T, Kim DW, Wu YL, Nakagawa K, Mekhail T, Felip E, Cappuzzo F, Paolini J, Usari T, Iyer S, Reisman A, Wilner KD, Tursi J, Blackhall F, Investigators P (2014) First-line crizotinib versus chemotherapy in ALK-positive lung cancer. N Engl J Med 371(23):2167–2177. doi:10.1056/NEJMoa1408440
22. Hagiwara K, Kobayashi K (2013) Importance of the cytological samples for the epidermal growth factor receptor gene mutation test for non-small cell lung cancer. Cancer Sci 104(3):291–297. doi:10.1111/cas.12081
23. Tanaka T, Nagai Y, Miyazawa H, Koyama N, Matsuoka S, Sutani A, Huqun, Udagawa K, Murayama Y, Nagata M, Shimizu Y, Ikebuchi K, Kanazawa M, Kobayashi K, Hagiwara K (2007) Reliability of the peptide nucleic acid-locked nucleic acid polymerase chain reaction clamp-based test for epidermal growth factor receptor mutations integrated into the clinical practice for non-small cell lung cancers. Cancer Sci 98(2):246–252. doi:10.1111/j.1349-7006.2006.00377.x
24. Lindeman NI, Cagle PT, Beasley MB, Chitale DA, Dacic S, Giaccone G, Jenkins RB, Kwiatkowski DJ, Saldivar JS, Squire J, Thunnissen E, Ladanyi M (2013) Molecular testing guideline for selection of lung cancer patients for EGFR and ALK tyrosine kinase inhibitors: guideline from the college of American pathologists, international association for the study of lung cancer, and association for molecular pathology. J Thorac Oncol 8(7):823–859. doi:10.1097/JTO.0b013e318290868f

25. Goto K, Satouchi M, Ishii G, Nishio K, Hagiwara K, Mitsudomi T, Whiteley J, Donald E, McCormack R, Todo T (2012) An evaluation study of EGFR mutation tests utilized for non-small-cell lung cancer in the diagnostic setting. Ann Oncol 23:2914–2919. doi:10.1093/annonc/mds121
26. Greer CE, Wheeler CM, Manos MM (1994) Sample preparation and PCR amplification from paraffin-embedded tissues. PCR Methods Appl 3(6):S113–S122
27. Liu D, Nakano J, Ueno M, Masuya D, Nakashima T, Yokomise H, Yube K, Huang CL (2006) A useful protocol for analyses of mutations of the epidermal growth factor receptor gene. Oncol Rep 15(6):1503–1505
28. Marchetti A, Felicioni L, Buttitta F (2006) Assessing EGFR mutations. N Engl J Med 354(5):526–528. doi:10.1056/NEJMc052564; author reply 526–528
29. Cargnin S, Canonico PL, Genazzani AA, Terrazzino S (2016) Quantitative analysis of circulating cell-free DNA for correlation with lung cancer survival: a systematic review and meta-analysis. J Thorac Oncol. doi:10.1016/j.jtho.2016.08.002
30. Reck M, Hagiwara K, Han B, Tjulandin S, Grohe C, Yokoi T, Morabito A, Novello S, Arriola E, Molinier O, McCormack R, Ratcliffe M, Normanno N (2016) ctDNA determination of EGFR mutation status in European and Japanese patients with advanced NSCLC: the ASSESS study. J Thorac Oncol 11(10):1682–1689. doi:10.1016/j.jtho.2016.05.036
31. Takeuchi K, Choi YL, Togashi Y, Soda M, Hatano S, Inamura K, Takada S, Ueno T, Yamashita Y, Satoh Y, Okumura S, Nakagawa K, Ishikawa Y, Mano H (2009) KIF5B-ALK, a novel fusion oncokinase identified by an immunohistochemistry-based diagnostic system for ALK-positive lung cancer. Clin Cancer Res 15(9):3143–3149. doi:10.1158/1078-0432.CCR-08-3248
32. Holland PM, Abramson RD, Watson R, Gelfand DH (1991) Detection of specific polymerase chain reaction product by utilizing the 5′----3′ exonuclease activity of Thermus aquaticus DNA polymerase. Proc Natl Acad Sci U S A 88(16):7276–7280
33. Kimura H, Kasahara K, Kawaishi M, Kunitoh H, Tamura T, Holloway B, Nishio K (2006) Detection of epidermal growth factor receptor mutations in serum as a predictor of the response to gefitinib in patients with non-small-cell lung cancer. Clin Cancer Res 12(13):3915–3921. doi:10.1158/1078-0432.CCR-05-2324
34. Nagai Y, Miyazawa H, Huqun, Hagiwara K (2005) Genetic heterogeneity of the epidermal growth factor receptor in non-small cell lung cancer cell lines revealed by a rapid and sensitive detection system, the peptide nucleic acid-locked nucleic acid PCR clamp. Cancer Res 65(16):7276–7282. doi:10.1158/0008-5472.CAN-05-0331
35. Yatabe Y, Hida T, Horio Y, Kosaka T, Takahashi T, Mitsudomi T (2006) A rapid, sensitive assay to detect EGFR mutation in small biopsy specimens from lung cancer. J Mol Diagn 8(3):335–341. doi:10.2353/jmoldx.2006.050104
36. Hall JG, Eis PS, Law SM, Reynaldo LP, Prudent JR, Marshall DJ, Allawi HT, Mast AL, Dahlberg JE, Kwiatkowski RW, de Arruda M, Neri BP, Lyamichev VI (2000) Sensitive detection of DNA polymorphisms by the serial invasive signal amplification reaction. Proc Natl Acad Sci U S A 97(15):8272–8277. doi:10.1073/pnas.140225597
37. Kimura H, Ohira T, Uchida O, Matsubayashi J, Shimizu S, Nagao T, Ikeda N, Nishio K (2014) Analytical performance of the cobas EGFR mutation assay for Japanese non-small-cell lung cancer. Lung Cancer 83(3):329–333. doi:10.1016/j.lungcan.2013.12.012
38. Fujimoto H, Furumoto N, Takeda M, Matsuda K, Ogawa Y, Narisawa T, Hosogai N (2010) EGFR mutation test using the PNA-LNA PCR clamp performed in our company. Byori Gijutsu 73:67–69

Chapter 5
Accurate Nodal Staging and Biomarker Testing with Endobronchial Ultrasound-Guided Transbronchial Needle Aspiration

Takahiro Nakajima and Ichiro Yoshino

Abstract Endobronchial ultrasound-guided transbronchial needle aspiration (EBUS-TBNA) is a minimally invasive modality for sampling mediastinal and hilar lymph nodes. The guidelines for lung cancer staging suggest that EBUS-TBNA should be considered as the best first test of nodal staging of radiologically abnormal lymph nodes that are accessible by this approach.

EBUS-TBNA is increasingly being used in thoracic oncology, because it can be used to obtain specimens that can be assessed for molecular biomarkers. Informative immunohistochemistry and fluorescence in situ hybridization can be performed on adequate biopsy specimens. Oncogenic drivers such as mutations in the epidermal growth factor gene and the presence of the anaplastic lymphoma kinase fusion gene are now being routinely identified because of the high-quality samples obtained by EBUS-TBNA.

One of the advantages of EBUS-TBNA is that it can be performed repeatedly as a minimally invasive procedure. Restaging of the mediastinum after induction therapy can be easily performed, in contrast to repeated surgical procedures. Repeat biopsies performed for cases showing resistance to targeted therapy should become an important novel application for EBUS-TBNA. Because of improvements in the technology of molecular analysis, comprehensive gene expression analysis has become important for the management of patients with lung cancer. The repeatability of EBUS-TBNA means that it should be a powerful tool in this era of "precision medicine."

Keywords Nodal staging in lung cancer • Endobronchial ultrasound-guided transbronchial needle aspiration • Biomarker testing • Restaging • Rebiopsy

---

T. Nakajima, M.D., Ph.D. (✉) • I. Yoshino, M.D., Ph.D.
Department of General Thoracic Surgery, Chiba University Graduate School of Medicine, 1-8-1 Inohana, Chuo-ku, Chiba 260-8670, Japan
e-mail: takahiro_nakajima@med.miyazaki-u.ac.jp

© Springer Science+Business Media Singapore 2017

Y. Takiguchi (ed.), *Molecular Targeted Therapy of Lung Cancer*,
DOI 10.1007/978-981-10-2002-5_5

5.1 Introduction

Since the 2004 introduction of endobronchial ultrasound-guided transbronchial needle aspiration (EBUS-TBNA) for nodal staging in patients with lung cancer [1], several retrospective as well as prospective studies have been published. The convex-probe (CP)-EBUS bronchoscope consists of a linear ultrasound probe and instrument channel on the tip of a hybrid bronchoscope and is capable of performing a needle biopsy under real-time ultrasound guidance. EBUS-TBNA has opened the door to the image-guided minimally invasive histopathological assessment of mediastinal and hilar lymph nodes, as well as the lesions adjacent to the airway. EBUS-TBNA has been rapidly adopted worldwide [2].

EBUS-TBNA is now being used for the assessment of several respiratory disorders, including benign and infectious diseases. It has obtained significantly higher diagnostic yields compared to conventional diagnostic modalities, especially for early-stage sarcoidosis, which manifests lymphadenopathy with minimal changes in the pulmonary parenchyma [3]. Depending on the methods used to process the specimens obtained by EBUS-TBNA, they can be used for both cytological evaluations and histological evaluations, thus enabling molecular analysis, which is mandatory for treatments using molecularly targeted therapeutic agents [2].

One of the advantages of EBUS-TBNA is that it is a minimally invasive procedure. EBUS-TBNA can be performed for patients with advanced disease as well as for patients at high risk for surgery. Furthermore, EBUS-TBNA is a repeatable procedure; hence, it can be performed following a course of treatment to reassess the same targeted lesion. It allows identification of the initial oncogenic drivers and can then also monitor for genetic resistance during molecularly targeted therapy. EBUS-TBNA may be one of the most important technical advances in bronchology since the introduction of the flexible bronchoscope.

5.2 Accurate Nodal Staging by EBUS-TBNA

5.2.1 Endoscopic Nodal Staging for Patients with Lung Cancer

Several centers around the world have reported on the diagnostic yield of EBUS-TBNA for mediastinal staging in patients with lung cancer [1, 4, 5]. EBUS-TBNA has shown higher sensitivity and specificity than computed tomography (CT), positron emission tomography (PET), or integrated PET/CT [6, 7]. A previous meta-analysis of EBUS-TBNA for nodal staging in patients with lung cancer reported a pooled sensitivity of 88–93 % and pooled specificity of 100 % [8, 9]. There have been two important prospective studies that compared endoscopic staging to surgical staging that included mediastinoscopy. Yasufuku et al. reported a direct comparison between EBUS-TBNA and mediastinoscopy [10]. They enrolled surgically resectable lung cancer patients and performed both EBUS-TBNA and mediastinoscopy in the operating room, with the patients under general anesthesia. EBUS-TBNA was first performed by a thoracic surgeon, who then performed mediastinoscopy without knowing the results of EBUS-TBNA. In the patients who subsequently underwent surgery, surgical pathology served as the standard for assessing the diagnostic yield. The investigators concluded that there was no difference in the diagnostic yields obtained by mediastinoscopy vs EBUS-TBNA [10]. In the ASTER trial, Annema et al. compared the diagnostic yields of surgical staging and endoscopic staging (EBUS plus endoscopic ultrasound [EUS]) [11]. The patients were randomly allocated into two groups, as follows: (A) surgical staging and (B) endoscopic staging followed by surgical staging if endoscopic staging did not identify metastatic nodes. The study found that endoscopic staging had higher sensitivity and resulted in fewer unnecessary thoracotomies [11]. A combination of EBUS-TBNA and EUS-guided fine needle aspiration (EUS-FNA) can evaluate most of the mediastinal nodes. Some areas of the left mediastinum and the dorsal side of subcarinal lymph node can occasionally be easily assessed by EUS-FNA (Fig. 5.1). Actually, the combined technique is reportedly more sensitive than EBUS-TBNA or EUS-FNA alone [12, 13].

Studies have demonstrated the cost benefits of EBUS-TBNA over mediastinoscopy for the diagnosis of patients with isolated mediastinal lymphadenopathy [14]. The ASTER trial made a similar observation [15]. In 2013, the third edition of the American College of Chest Physician (ACCP) guidelines for the management of patients with lung cancer was published [12]. The new guidelines for staging the mediastinum now regard endoscopic ultrasound staging procedures, including EBUS-TBNA and EUS-FNA, the best first-line tests, better than surgical staging for radiologically suspect lymph nodes that are accessible by endoscopy [12]. However, the guidelines also mentioned concerns about the quality of endoscopic staging and the necessity of surgical staging for patients whose lymph nodes might have a high probability of metastatic disease but were found to be negative on EBUS and/or EUS [12]. Similar guidelines were also published by European societies, which recommend ultrasound-guided nodal staging as a combination of EBUS-TBNA and EUS-FNA [16]. They recommend that endoscopists should be trained for both EBUS and EUS, so that complete endoscopic nodal staging can be performed in a single session [16]. In addition, continuous education and training is mandatory for maintaining a high diagnostic yield and maximum safety during mediastinal staging by EBUS-TBNA and/or EUS-FNA [17]. Recently, guidelines that focus on the technical aspects of EBUS-TBNA were published, and they will greatly facilitate the standardization of EBUS-TBNA procedures [18].

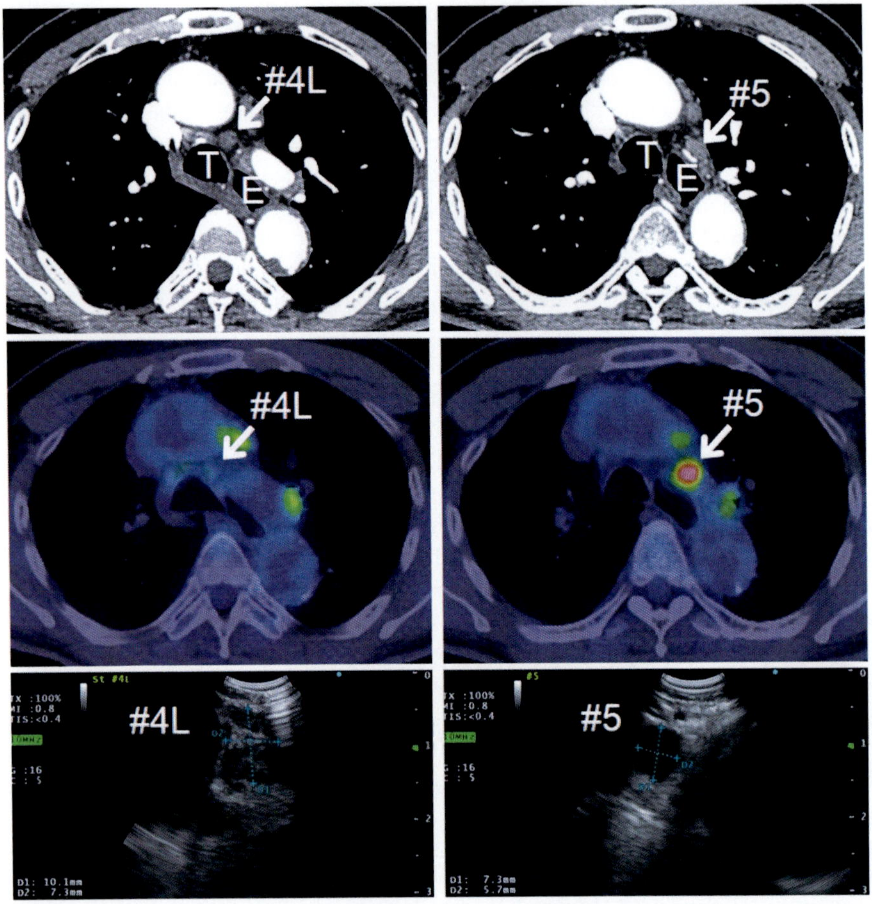

Fig. 5.1 Nodal staging using EBUS-TBNA combined with EUS-(B)-FNA. A CP-EBUS was used for EBUS-TBNA, which evaluated station 4L. The same bronchoscope was then introduced into the esophagus, and station 5 was evaluated by EUS-(B)-FNA. There were no malignant cells in the mediastinal lymph nodes, and the patient underwent radical surgery

5.2.2 *N1 Staging Using EBUS-TBNA*

The several advantages of EBUS-TBNA include minimal invasiveness, repeatability, and ability to easily perform hilar lymph node sampling. EBUS-TBNA can assess the lymph nodes and lesions adjacent to the airway and within the reach of the EBUS scope. EBUS-TBNA can assess the N1 nodes in addition to the N2/N3 nodes and accurately differentiate between N0 and N1 stages [19]. The diagnosis of N1 disease may sometimes affect surgical planning prior to surgery (Fig. 5.2). Another application of EBUS-TBNA being increasingly used is the staging of mediastinal and hilar lymph nodes for patients who will undergo stereotactic radiation therapy (SBRT) [20]. Patients who are being considered for SBRT often have

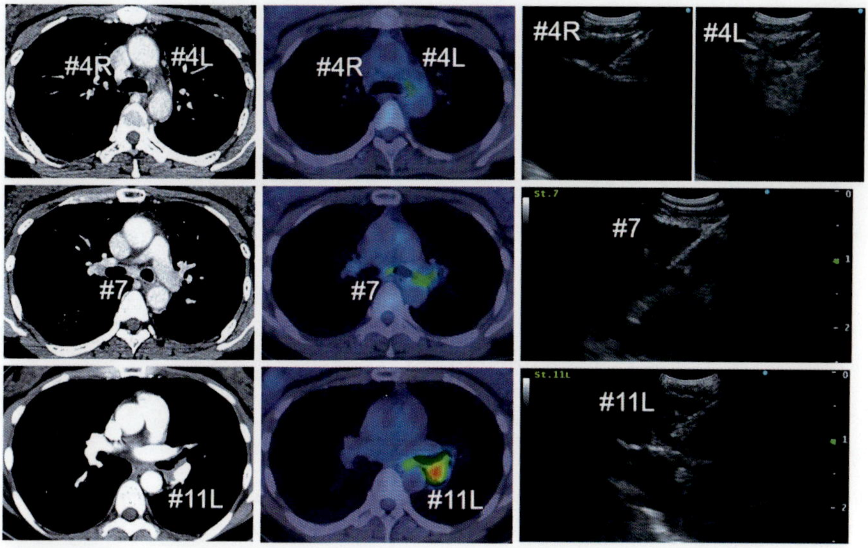

Fig. 5.2 Preoperative N1 staging by EBUS-TBNA. The patient had adenocarcinoma in the *left lower* pulmonary lobe. Nodal staging by EBUS-TBNA revealed malignant cells only in the station 11L lymph node (N1 disease). The patient underwent *left lower* lobectomy with lingular segmentectomy, and complete resection (R0) was achieved

comorbidities that preclude them from undergoing surgical staging. However, in contrast to surgery, the existence of N1 disease is a contraindication for SBRT. A clinical trial looking at the role of EBUS-TBNA in mediastinal and hilar staging for patients scheduled for SBRT is under way (NCT01786590).

N1 staging is also important for the treatment of small cell lung cancer. Surgery is only considered for stage I disease; therefore, the hilar lymph nodes should be carefully evaluated. EBUS-TBNA staging of small cell lung cancer has obtained a high diagnostic yield, with a sensitivity, specificity, and diagnostic accuracy of 96.4 %, 100 %, and 97.2 %, respectively [21]. Patients who have undergone surgery for N0 small cell lung cancer after nodal staging by EBUS-TBNA have had a favorable outcome [21]. Currently, the accessibility of interlobar/lobar lymph nodes to CP-EBUS is limited because of the size and angle of the scope. A novel thin EBUS scope (prototype thin convex-probe EBUS; BF-Y0046, Olympus, Japan: Fig. 5.3) is under development to improve the ability of the scope to access the distal lymph nodes [22].

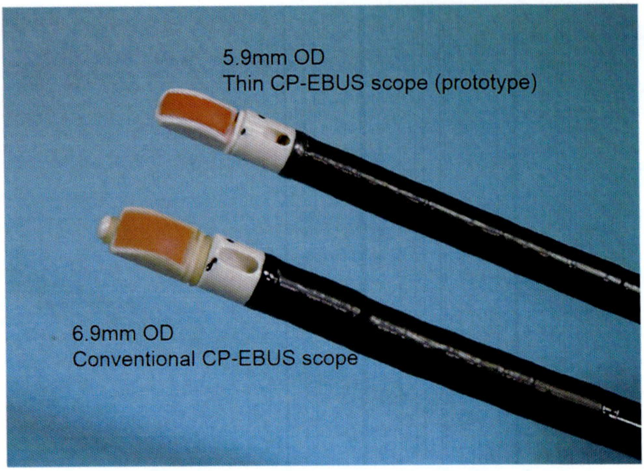

Fig. 5.3 Conventional CP-EBUS and a prototype thin CP-EBUS. The prototype thin CP-EBUS is 1 mm thinner than a conventional bronchoscope

5.2.3 EBUS-TBNA for the Restaging of Patients with Lung Cancer

Mediastinoscopy has been regarded as the "gold standard" for the mediastinal staging of patients with lung cancer, because of its high diagnostic yield and safety [23]. However, the restaging of the mediastinum by mediastinoscopy (remediastinoscopy) after induction therapy is technically difficult because of mediastinal adhesions [24]. Selected patients who undergo induction therapy (chemotherapy and/or radiotherapy) before undergoing surgery require accurate restaging of the mediastinum before the resection is performed. The diagnostic yield of remediastinoscopy is limited, compared with the initial mediastinoscopy [24]. By contrast, EBUS-TBNA is an easily repeatable procedure, and EBUS-TBNA for the restaging of the mediastinum has been reported to be useful [25]. Although the diagnostic yield of restaging EBUS-TBNA is also reduced compared to an initial evaluation; EBUS-TBNA is an acceptable minimally invasive modality [26, 27]. However, most of the data on the diagnostic yield of restaging EBUS-TBNA has been retrospective. Additional prospective studies are warranted.

EBUS-TBNA has been found to be useful for assessing mediastinal lymphadenopathy in patients with previously treated lung cancer. When CT and/or PET show abnormalities within the mediastinum or the hilum, i.e., enlarged lymph nodes or 18F-fluorodeoxyglucose (FDG)-avid lymph nodes, histological confirmation is often needed to rule out reactive adenopathy vs recurrence, since radiological findings alone are not reliable [28]. In our experience, EBUS-TBNA of radiologically abnormal mediastinal lymph nodes has confirmed histological features different from those of the nodes sampled before treatment. Those findings were helpful for deciding on subsequent therapy [29], especially for the patients who were treated by

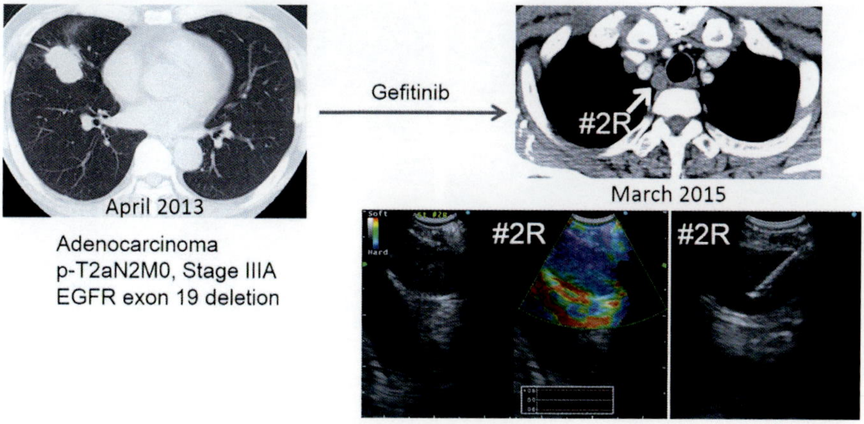

Fig. 5.4 Transformation to small cell lung cancer during EGFR-TKI treatment. The patient was administered gefitinib as adjuvant chemotherapy in a clinical trial setting. An enlarged station 2R lymph node was evaluated by EBUS-TBNA, and the lymph node was histologically diagnosed as small cell lung cancer with the use of immunostaining. The *EGFR* exon 19 deletion was also detected in the rebiopsied sample

epidermal growth factor receptor (EGFR) tyrosine kinase inhibitors and developed resistance that was sometimes associated with the transformation to small cell lung cancer (Fig. 5.4) [30]. Therefore, rebiopsy and tissue confirmation is important in the management of previously treated lung cancer.

5.2.4 EBUS Image Analysis for Distinguishing Between Benign and Malignant Lymph Nodes

The indication for invasive mediastinal staging in lung cancer depends on radiologic findings, which include CT and PET. In general, ultrasound can provide a more detailed high-resolution evaluation of nodal staging than CT or PET. Because of the resolution of ultrasound, there have been several important studies that have evaluated the EBUS image analysis of lymph nodes. The first study report was on the B-mode image classification of mediastinal and hilar lymph nodes using the first-generation EBUS ultrasound processor (EU-C2000, Olympus, Tokyo). B-mode images were classified according to six indicators, including size, shape (oval or round), margin (indistinct or distinct), echogenicity (homogeneous or heterogeneous), presence of central hilar structure, and presence of central necrosis sign [31]. Four predictors, round shape, distinct margin, heterogeneous echogenicity, and presence of coagulation necrosis sign, were identified as independent predictors for nodal metastasis [32]. A study that used a second-generation EBUS ultrasound processor (EU-ME1; Olympus, Tokyo) with the capability of Doppler mode evaluation evaluated vascular patterns within the lymph nodes as predictors of nodal

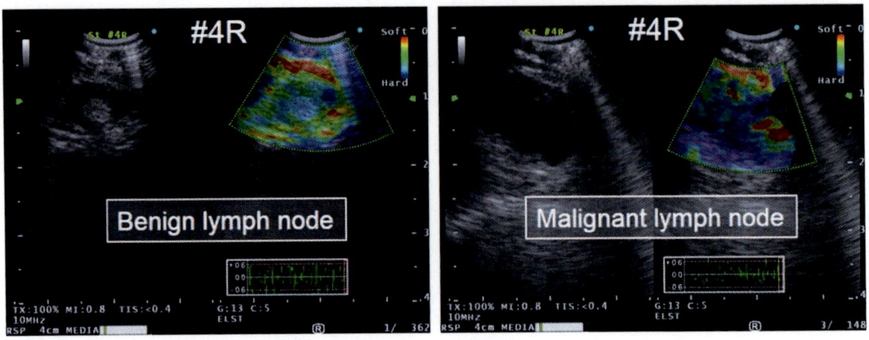

Fig. 5.5 Representative elastography images of benign and malignant lymph nodes. A benign lymph node appears colored *green* to *yellow/red* on the monitor. By contrast, a malignant lymph node appears *blue*, which indicates relative hardness compared with the surrounding tissue

disease [32]. This study categorized four vascular patterns and determined whether or not bronchial arterial flow was present. The use of vascular pattern classifications obtained a diagnostic yield for predicting positive or negative nodal metastasis of about 85 % [32]. Subsequent studies using B-mode classification found that EBUS was effective for differentiating metastatic vs normal lymph nodes [33–35].

The most recent ultrasound processor is equipped with an additional imaging feature, elastography, which is a strain imaging technique that assesses tissue stiffness and visualizes the distribution of stiffness in the region of interest. Malignant tissues tend to be stiffer than normal tissues because of the increased density of tumor cells and vascular structures (Fig. 5.5). Izumo et al. subjectively classified EBUS elastographic images into three categories: Type 1, predominantly non-blue; Type 2, partly blue; and Type 3, predominantly blue. They classified Type 3 as malignant and reported that 94.6 % of Type 3 lymph nodes were positive for metastasis [36]. However, for appropriate subjective categorization and classification of ultrasound image characteristics, bronchoscopists must have sufficient knowledge and experience with EBUS image analysis. We recently reported the utility of the "stiff area ratio" measured by EBUS elastography [37]. The "stiff area ratio" is a method of objective evaluation and should be more helpful in guiding bronchoscopists during selective sampling of a suspicious station or lymph node within the same station.

5.3 Multidirectional Analysis Using Samples Obtained by EBUS-TBNA

5.3.1 Acquisition and Preparation of EBUS-TBNA Samples

EBUS-TBNA samples undergoing immunohistochemical analysis in addition to conventional histomorphological evaluations have led to higher rates of identifying the histological subtypes of non-small cell lung cancer (NSCLC) specimens [38]. However, the quantity of material that can be obtained by needle biopsy is small, and therefore attempts have been made to improve the processing methods used for to EBUS-TBNA samples to obtain a pathological diagnosis [39]. A needle biopsy specimen is fundamentally cytological material, and "proper preconditioning" is important for maximizing the information derived from a very small sample. However, although the effectiveness of rapid on-site evaluation (ROSE) for improving the diagnostic yield of EBUS-TBNA is still controversial, ROSE may aid in deciding how to process the sample for additional evaluations [40]. In addition, ROSE was found to increase the chances of performing successful lung cancer genotyping from small numbers of needle biopsy samples that were obtained by EBUS-TBNA [41]. The recent guidelines on the techniques of EBUS-TBNA recommended obtaining samples for histopathological diagnosis and additional samples for molecular testing [18]. The cell block method [42] and the "tissue coagulation clot" method [43] have allowed histological evaluations and have been reported to improve the diagnostic yield of EBUS-TBNA. These "core" building techniques may facilitate biomarker testing in lung cancer [44]. The World Association for Bronchology and Interventional Pulmonology recently published guidelines on the acquisition and preparation of needle aspiration samples. The guidelines also encourage bronchoscopists to have discussions with their pathologist colleagues about the most suitable methods for processing specimens [45] (Fig. 5.6).

5.3.2 Detection of Oncogenic Drivers Using EBUS-TBNA Samples

Specimens obtained by EBUS-TBNA can be processed for multidirectional analysis [39]. Gefitinib, the first epidermal growth factor receptor-tyrosine kinase inhibitor (EGFR-TKI), was introduced for the treatment of lung cancer in 2002 [46]. It marked the beginning of molecularly targeted therapy for lung cancer. The association between *EGFR* gene mutations and sensitivity to gefitinib was reported in 2004 [47, 48]. Since then, clinical molecular testing for the detection of molecular targets has increasingly been required. Because the majority of the lung cancer patients have advanced disease at the time of diagnosis, the development of molecular testing techniques on a very small biopsy specimen was warranted. The initial attempt

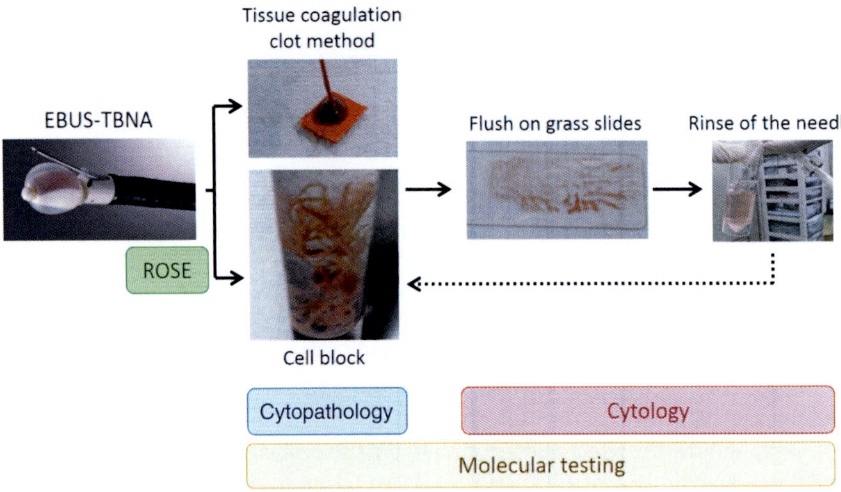

Fig. 5.6 Specimen handling at Chiba University Hospital. First, the sample obtained by EBUS-TBNA was pushed out of the needle by a stylet, and the "core" material was processed by the tissue coagulum clot cell block method or another cell block method. Then the needle was flushed with air, and the material in the expelled air was sprayed onto glass slides. And last, the needle was rinsed with normal saline. Molecular testing of samples at each step could be performed

to detect *EGFR* mutations in specimens provided by EBUS-TBNA was reported in 2007 [49]. Following improvements in molecular analysis, the detection sensitivity improved, and other investigators reported similar attempts at detecting *EGFR* mutations in EBUS-TBNA specimens [50, 51]. Multiplex mutation testing [52, 53] and mutation testing of the solution used to rinse the TBNA needle are used currently to increase the sensitivity of mutation detection [53]. In addition to *EGFR* gene mutations, the anaplastic lymphoma kinase (ALK) fusion gene was found to be a very strong oncogenic driver in 2007 [54]. The detection of the aberrant fusion gene will initially require immunohistochemical techniques and reverse transcription polymerase chain reaction (RT-PCR) or FISH. Sakairi et al. was able to detect the EML4-ALK fusion gene in an EBUS-TBNA sample using RT-PCR, with the result confirmed by FISH [55]. Biomarker testing of biopsy material is important for identifying which molecularly targeted therapeutic agents are useful for the patient with NSCLC. Testing should be performed prior to the start of treatment [56]. Patients with lung cancer who received the appropriate targeted therapy were reported to obtain improved survival compared with patients not receiving targeted therapy or with no oncogenic driver targets [57]. The use of cell block preparations of EBUS-TBNA specimens for molecular testing has obtained excellent results for lung adenocarcinoma; EBUS-TBNA specimens from 93 % of patients were sufficient for at least one round of molecular testing for *EGFR* mutations, the ALK fusion gene, and *KRAS* mutations [58].

5.3.3 Rebiopsy and Repeat Molecular Testing by EBUS-TBNA

The repeatable nature of EBUS-TBNA should be a powerful tool for identifying the appropriate molecularly targeted therapeutic agents. Gefitinib initially showed dramatic effects in patients with lung cancers harboring sensitive *EGFR* mutations; however, patients develop resistance to EGFR-TKIs. The well-known secondary changes which result in resistance to EGFR-TKIs include the substitution of methionine for threonine at position 790 (T790 M) in exon 20, which was reported in 2005 [59] and focal amplification of the MET proto-oncogene, which was reported in 2007 [60]. The secondary changes leading to resistance to ALK-TKIs include a mutation in the ATP-binding domain (L1196 M) and a mutation in the non-ATP-binding domain (C1156Y), which were reported in 2010 [61]. In addition, many other mechanisms have been reported to be involved in the development of ALK-TKI resistance [62]. Second- and third-generation TKIs have been developed to overcome various types of resistance [62]. Actually, the new-generation drugs have shown dramatic benefits for the patients with acquired resistance to initial TKI treatment [66, 64]. A previous analysis of tumor specimens taken at the time of acquired resistance to EGFR-TKI found secondary mutations, aberrant gene amplification, and transformation to small cell lung cancer [65]. The European Society for Medical Oncology (ESMO) guidelines have stated that the emergence of molecular resistance suggests that a repeat biopsy should be performed at the time of tumor progression [66]. Rebiopsy by EBUS-TBNA has been thought to be feasible [67] (Fig. 5.7); however, rebiopsy itself has so far not been a standard clinical practice because of patient factors (tolerability), physician preference, and limited resources [68].

5.3.4 Application of EBUS-TBNA Samples to Molecular Analysis

Multidirectional analysis of samples obtained by EBUS-TBNA has allowed assessment of oncogenic drivers in clinics, as well as the identification of genetic signatures for lung cancer research. We can extract DNA, RNA, and protein from EBUS-TBNA samples, and these materials can be used for transcriptome and proteome analysis. Analysis of specimens with aberrant DNA methylation can be used for assessment of chemosensitivity [69] and also used for making higher sensitivity for the detection of nodal metastasis by EBUS-TBNA [70]. Expression of the unique vascular endothelial growth factor-c (*VEGF-C*) mRNA was reported from EBUS-TBNA samples of metastatic lymph nodes. The higher expression of *VEGF-C* was observed in metastatic lymph nodes with high vascularity features using Doppler mode image [71]. High-quality EBUS-TBNA samples can be used for comprehensive mRNA and microRNA expression analysis using microarray technology [72]. Primary xenograft technology using EBUS-TBNA samples has been evaluated for overcoming the limitations in sample size [73, 74].

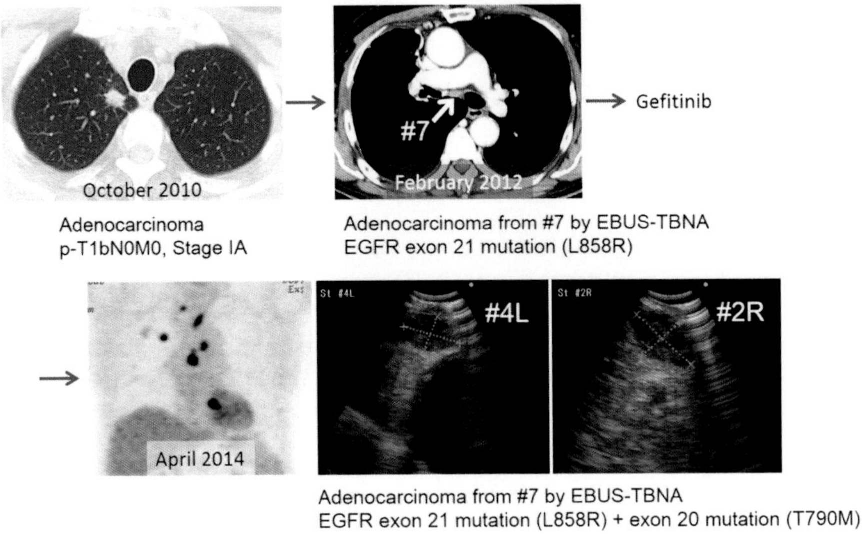

Fig. 5.7 Rebiopsy and repeat molecular testing. The patient first underwent EBUS-TBNA for the diagnosis of nodal recurrence in station 7. An *EGFR* mutation in exon 21 mutation was detected, and gefitinib was prescribed. Progression of disease was observed in the mediastinum, and a rebiopsy of station 4L and 2R was performed. An *EGFR* mutation in exon 20 mutation was detected, and the patient received cytotoxic chemotherapy

We often encounter difficulties obtaining tissue samples, since many patients have advanced disease at the time of first presentation and are not eligible for surgery. EBUS-TBNA can solve that problem, because it can obtain tumor samples from patients with advanced disease who are not candidates for surgery. Because it can easily and safely obtain samples that are evaluable by today's technology, EBUS-TBNA may greatly expand the knowledge base that supports lung cancer research.

5.4 Conclusions

EBUS-TBNA is now a necessary diagnostic modality for the staging of lung cancer and for providing specimens for biomarker testing. EBUS-TBNA is a repeatable procedure that can be performed to monitor the patient after treatment. Further advances in EBUS technology as well as in the needles used for tissue sampling will likely help both bronchoscopists and lung cancer investigators acquire ideal tissue samples for analysis.

Acknowledgments Funding: This work was supported by JSPS KAKENHI (Grant-in-Aid for Scientific Research (C)), Grant Number 26462122 (T.N.).

Conflict of Interest Takahiro Nakajima received honoraria and lecture fees from Olympus Medical Systems for EBUS-TBNA training courses.

References

1. Yasufuku K, Chiyo M, Koh E, Moriya Y, Iyoda A, Sekine Y, Shibuya K, Iizasa T, Fujisawa T (2005) Endobronchial ultrasound guided transbronchial needle aspiration for staging of lung cancer. Lung Cancer 50:347–354
2. Nakajima T, Yasufuku K, Yoshino I (2013) Current status and perspective of EBUS-TBNA. Gen Thorac Cardiovasc Surg 61:390–396
3. von Bartheld MB, Dekkers OM, Szlubowski A, Eberhardt R, Herth FJ, In't Veen JC, de Jong YP, van der Heijden EH, Tournoy KG, Claussen M, van den Blink B, Shah PL, Zoumot Z, Clementsen P, Porsbjerg C, Mauad T, Bernardi FD, van Zwet EW, Rabe KF, Annema JT (2013) Endosonography vs conventional bronchoscopy for the diagnosis of sarcoidosis: the GRANULOMA randomized clinical trial. JAMA 309:2457–2464
4. Rintoul RC, Skwarski KM, Murchison JT, Wallace WA, Walker WS, Penman ID (2005) Endobronchial and endoscopic ultrasound-guided real-time fine-needle aspiration for mediastinal staging. Eur Respir J 25:416–421
5. Herth FJ, Eberhardt R, Vilmann P, Krasnik M, Ernst A (2006) Real-time endobronchial ultrasound guided transbronchial needle aspiration for sampling mediastinal lymph nodes. Thorax 61:795–798
6. Yasufuku K, Nakajima T, Motoori K, Sekine Y, Shibuya K, Hiroshima K, Fujisawa T (2006) Comparison of endobronchial ultrasound, positron emission tomography, and CT for lymph node staging of lung cancer. Chest 130:710–718
7. Hwangbo B, Kim SK, Lee HS, Lee HS, Kim MS, Lee JM, Kim HY, Lee GK, Nam BH, Zo JI (2009) Application of endobronchial ultrasound-guided transbronchial needle aspiration following integrated PET/CT in mediastinal staging of potentially operable non-small cell lung cancer. Chest 135:1280–1287
8. Gu P, Zhao YZ, Jiang LY, Zhang W, Xin Y, Han BH (2009) Endobronchial ultrasound-guided transbronchial needle aspiration for staging of lung cancer: a systematic review and meta-analysis. Eur J Cancer 45:1389–1396
9. Adams K, Shah PL, Edmonds L, Lim E (2009) Test performance of endobronchial ultrasound and transbronchial needle aspiration biopsy for mediastinal staging in patients with lung cancer: systematic review and meta-analysis. Thorax 64:757–762
10. Yasufuku K, Pierre A, Darling G, de Perrot M, Waddell T, Johnston M, da Cunha SG, Geddie W, Boerner S, Le LW, Keshavjee S (2011) A prospective controlled trial of endobronchial ultrasound-guided transbronchial needle aspiration compared with mediastinoscopy for mediastinal lymph node staging of lung cancer. J Thorac Cardiovasc Surg 142:1393–1400
11. Annema JT, van Meerbeeck JP, Rintoul RC, Dooms C, Deschepper E, Dekkers OM, De Leyn P, Braun J, Carroll NR, Praet M, de Ryck F, Vansteenkiste J, Vermassen F, Versteegh MI, Veseliç M, Nicholson AG, Rabe KF, Tournoy KG (2010) Mediastinoscopy vs endosonography for mediastinal nodal staging of lung cancer: a randomized trial. JAMA 304:2245–2252
12. Silvestri GA, Gonzalez AV, Jantz MA, Margolis ML, Gould MK, Tanoue LT, Harris LJ, Detterbeck FC (2013) Methods for staging non-small cell lung cancer: diagnosis and management of lung cancer, 3rd ed: American College of Chest Physicians evidence-based clinical practice guidelines. Chest 143:e211S–e250S
13. Zhang R, Ying K, Shi L, Zhang L, Zhou L (2013) Combined endobronchial and endoscopic ultrasound-guided fine needle aspiration for mediastinal lymph node staging of lung cancer: a meta-analysis. Eur J Cancer 49:1860–1867

14. Navani N, Lawrence DR, Kolvekar S, Hayward M, McAsey D, Kocjan G, Falzon M, Capitanio A, Shaw P, Morris S, Omar RZ, SM J, REMEDY Trial Investigators (2012) Endobronchial ultrasound-guided transbronchial needle aspiration prevents mediastinoscopies in the diagnosis of isolated mediastinal lymphadenopathy: a prospective trial. Am J Respir Crit Care Med 186:255–260
15. Sharples LD, Jackson C, Wheaton E, Griffith G, Annema JT, Dooms C, Tournoy KG, Deschepper E, Hughes V, Magee L, Buxton M, Rintoul RC (2012) Clinical effectiveness and cost-effectiveness of endobronchial and endoscopic ultrasound relative to surgical staging in potentially resectable lung cancer: results from the ASTER randomised controlled trial. Health Technol Assess 16:1–75
16. Vilmann P, Clementsen PF, Colella S, Siemsen M, De Leyn P, Dumonceau JM, Herth FJ, Larghi A, Vazquez-Sequeiros E, Hassan C, Crombag L, Korevaar DA, Konge L, Annema JT (2015) Combined endobronchial and esophageal endosonography for the diagnosis and staging of lung cancer: European Society of Gastrointestinal Endoscopy (ESGE) Guideline, in cooperation with the European Respiratory Society (ERS) and the European Society of Thoracic Surgeons (ESTS). Endoscopy 47:545–559
17. Sakairi Y, Saegusa F, Yoshida S, Takiguchi Y, Tatsumi K, Yoshino I (2012) Evaluation of a learning system for endobronchial ultrasound-guided transbronchial needle aspiration. Respir Inves 50:46–53
18. Wahidi MM, Herth F, Yasufuku K, Shepherd RW, Yarmus L, Chawla M, Lamb C, Casey KR, Patel S, Silvestri GA, Feller-Kopman DJ (2016) Technical aspects of endobronchial ultrasound guided transbronchial needle aspiration: CHEST Guideline and Expert Panel Report. Chest 149:813–835
19. Yasufuku K, Nakajima T, Waddell T, Keshavjee S, Yoshino I (2013) Endobronchial ultrasound-guided transbronchial needle aspiration for differentiating N0 versus N1 lung cancer. Ann Thorac Surg 96:1756–1760
20. Nakajima T, Yasufuku K, Nakajima M, Baba M, Yoshikawa K, Kamada T, Hiroshima K, Nakatani Y, Fujisawa T, Yoshino I (2010) Endobronchial ultrasound-guided transbronchial needle aspiration for lymph node staging in patients with non-small cell lung cancer in non-operable patients pursuing radiotherapy as a primary treatment. J Thorac Oncol 5:606–611
21. Wada H, Nakajima T, Yasufuku K, Fujiwara T, Yoshida S, Suzuki M, Shibuya K, Hiroshima K, Nakatani Y, Yoshino I (2010) Lymph node staging by endobronchial ultrasound-guided transbronchial needle aspiration in patients with small cell lung cancer. Ann Thorac Surg 90:229–234
22. Wada H, Hirohashi K, Nakajima T, Anayama T, Kato T, Grindlay A, McConnell J, Yoshino I, Yasufuku K (2015) Assessment of the new thin convex probe endobronchial ultrasound bronchoscope and the dedicated aspiration needle: a preliminary study in the porcine lung. J Bronchology Interv Pulmonol. 22:20–27
23. Detterbeck FC, Jantz MA, Wallace M, Vansteenkiste J, Silvestri GA; American College of Chest Physicians (2007) Invasive mediastinal staging of lung cancer: ACCP evidence-based clinical practice guidelines (2nd edition). Chest 132:202S–220S
24. De Leyn P, Stroobants S, De Wever W, Lerut T, Coosemans W, Decker G, Nafteux P, Van Raemdonck D, Mortelmans L, Nackaerts K, Vansteenkiste J (2006) Prospective comparative study of integrated positron emission tomography-computed tomography scan compared with remediastinoscopy in the assessment of residual mediastinal lymph node disease after induction chemotherapy for mediastinoscopy-proven stage IIIA-N2 Non-small-cell lung cancer: a Leuven Lung Cancer Group Study. J Clin Oncol 24:3333–3339
25. Herth FJ, Annema JT, Eberhardt R, Yasufuku K, Ernst A, Krasnik M, Rintoul RC (2008) Endobronchial ultrasound with transbronchial needle aspiration for restaging the mediastinum in lung cancer. J Clin Oncol 26:3346–3350
26. Szlubowski A, Herth FJ, Soja J, Kołodziej M, Figura J, Cmiel A, Obrochta A, Pankowski J (2010) Endobronchial ultrasound-guided needle aspiration in non-small-cell lung cancer

restaging verified by the transcervical bilateral extended mediastinal lymphadenectomy–a prospective study. Eur J Cardiothorac Surg 37:1180–1184

27. Nasir BS, Bryant AS, Minnich DJ, Wei B, Dransfield MT, Cerfolio RJ (2014) The efficacy of restaging endobronchial ultrasound in patients with non-small cell lung cancer after preoperative therapy. Ann Thorac Surg 98:1008–1012
28. Yamamoto T, Sakairi Y, Nakajima T, Suzuki H, Tagawa T, Iwata T, Mizobuchi T, Yoshida S, Nakatani Y, Yoshino I (2015) Comparison between endobronchial ultrasound-guided transbronchial needle aspiration and 18F-fluorodeoxyglucose positron emission tomography in the diagnosis of postoperative nodal recurrence in patients with lung cancer. Eur J Cardiothorac Surg 47:234–238
29. Anraku M, Pierre AF, Nakajima T, de Perrot M, Darling GE, Waddell TK, Keshavjee S, Yasufuku K (2011) Endobronchial ultrasound-guided transbronchial needle aspiration in the management of previously treated lung cancer. Ann Thorac Surg 92:251–255
30. Yu HA, Arcila ME, Rekhtman N, Sima CS, Zakowski MF, Pao W, Kris MG, Miller VA, Ladanyi M, Riely GJ (2013) Analysis of tumor specimens at the time of acquired resistance to EGFR-TKI therapy in 155 patients with EGFR-mutant lung cancers. Clin Cancer Res 19:2240–2247
31. Fujiwara T, Yasufuku K, Nakajima T, Chiyo M, Yoshida S, Suzuki M, Shibuya K, Hiroshima K, Nakatani Y, Yoshino I (2010) The utility of sonographic features during endobronchial ultrasound-guided transbronchial needle aspiration for lymph node staging in patients with lung cancer: a standard endobronchial ultrasound image classification system. Chest 138:641–647
32. Nakajima T, Anayama T, Shingyoji M, Kimura H, Yoshino I, Yasufuku K (2012) Vascular image patterns of lymph nodes for the prediction of metastatic disease during EBUS-TBNA for mediastinal staging of lung cancer. J Thorac Oncol 7:1009–1014
33. Memoli JS, El-Bayoumi E, Pastis NJ, Tanner NT, Gomez M, Huggins JT, Onicescu G, Garrett-Mayer E, Armeson K, Taylor KK, Silvestri GA (2011) Using endobronchial ultrasound features to predict lymph node metastasis in patients with lung cancer. Chest 140:1550–1556
34. Wang L, Wu W, Hu Y, Teng J, Zhong R, Han B, Sun J (2015) Sonographic features of endobronchial ultrasonography predict intrathoracic lymph node metastasis in lung cancer patients. Ann Thorac Surg 100:1203–1209
35. Alici IO, Yılmaz Demirci N, Yılmaz A, Karakaya J, Özaydın E (2016) The sonographic features of malignant mediastinal lymph nodes and a proposal for an algorithmic approach for sampling during endobronchial ultrasound. Clin Respir J 10:606–613
36. Izumo T, Sasada S, Chavez C, Matsumoto Y, Tsuchida T (2014) Endobronchial ultrasound elastography in the diagnosis of mediastinal and hilar lymph nodes. Jpn J Clin Oncol 44:956–962
37. Nakajima T, Inage T, Sata Y, Morimoto J, Tagawa T, Suzuki H, Iwata T, Yoshida S, Nakatani Y, Yoshino I (2015) Elastography for predicting and localizing nodal metastases during endobronchial ultrasound. Respiration 90:499–506
38. Navani N, Brown JM, Nankivell M, Woolhouse I, Harrison RN, Jeebun V, Munavvar M, Ng BJ, Rassl DM, Falzon M, Kocjan G, Rintoul RC, Nicholson AG, Janes SM (2012) Suitability of endobronchial ultrasound-guided transbronchial needle aspiration specimens for subtyping and genotyping of non-small cell lung cancer: a multicenter study of 774 patients. Am J Respir Crit Care Med 185:1316–1322
39. Nakajima T, Yasufuku K (2011) How I do it–optimal methodology for multidirectional analysis of endobronchial ultrasound-guided transbronchial needle aspiration samples. J Thorac Oncol 6:203–206
40. Ko HM, da Cunha SG, Darling G, Pierre A, Yasufuku K, Boerner SL, Geddie WR (2013) Diagnosis and subclassification of lymphomas and non-neoplastic lesions involving mediastinal lymph nodes using endobronchial ultrasound-guided transbronchial needle aspiration. Diagn Cytopathol 41:1023–1030

41. Trisolini R, Cancellieri A, Tinelli C, de Biase D, Valentini I, Casadei G, Paioli D, Ferrari F, Gordini G, Patelli M, Tallini G (2015) Randomized trial of endobronchial ultrasound-guided transbronchial needle aspiration with and without rapid on-site evaluation for lung cancer genotyping. Chest 148:1430–1437
42. Sanz-Santos J, Serra P, Andreo F, Llatjós M, Castellà E, Monsó E (2012) Contribution of cell blocks obtained through endobronchial ultrasound-guided transbronchial needle aspiration to the diagnosis of lung cancer. BMC Cancer 12:34
43. Yung RC, Otell S, Illei P, Clark DP, Feller-Kopman D, Yarmus L, Askin F, Gabrielson E, Li QK (2012) Improvement of cellularity on cell block preparations using the so-called tissue coagulum clot method during endobronchial ultrasound-guided transbronchial fine-needle aspiration. Cancer Cytopathol 120:185–195
44. Jurado J, Saqi A, Maxfield R, Newmark A, Lavelle M, Bacchetta M, Gorenstein L, Dovidio F, Ginsburg ME, Sonett J, Bulman W (2013) The efficacy of EBUS-guided transbronchial needle aspiration for molecular testing in lung adenocarcinoma. Ann Thorac Surg 96:1196–1202
45. der Heijden EH v, RF C, Trisolini R, DP S, Hwangbo B, Nakajima T, Guldhammer-Skov B, Rossi G, Ferretti M, FF H, Yung R, Krasnik M, World Association for Bronchology and Interventional Pulmonology, Task Force on Specimen Guidelines (2014) Guideline for the acquisition and preparation of conventional and endobronchial ultrasound-guided transbronchial needle aspiration specimens for the diagnosis and molecular testing of patients with known or suspected lung cancer. Respiration 88:500–517
46. Ranson M, Hammond LA, Ferry D, Kris M, Tullo A, Murray PI, Miller V, Averbuch S, Ochs J, Morris C, Feyereislova A, Swaisland H, Rowinsky EK (2002) ZD1839, a selective oral epidermal growth factor receptor-tyrosine kinase inhibitor, is well tolerated and active in patients with solid, malignant tumors: results of a phase I trial. J Clin Oncol 20:2240–2250
47. Lynch TJ, Bell DW, Sordella R, Gurubhagavatula S, Okimoto RA, Brannigan BW, Harris PL, Haserlat SM, Supko JG, Haluska FG, Louis DN, Christiani DC, Settleman J, Haber DA (2004) Activating mutations in the epidermal growth factor receptor underlying responsiveness of non-small-cell lung cancer to gefitinib. N Engl J Med 350:2129–2139
48. Paez JG, Jänne PA, Lee JC, Tracy S, Greulich H, Gabriel S, Herman P, Kaye FJ, Lindeman N, Boggon TJ, Naoki K, Sasaki H, Fujii Y, Eck MJ, Sellers WR, Johnson BE, Meyerson M (2004) EGFR mutations in lung cancer: correlation with clinical response to gefitinib therapy. Science 304:1497–1500
49. Nakajima T, Yasufuku K, Suzuki M, Hiroshima K, Kubo R, Mohammed S, Miyagi Y, Matsukuma S, Sekine Y, Fujisawa T (2007) Assessment of epidermal growth factor receptor mutation by endobronchial ultrasound-guided transbronchial needle aspiration. Chest 132:597–602
50. Garcia-Olivé I, Monsó E, Andreo F, Sanz-Santos J, Taron M, Molina-Vila MA, Llatjós M, Castellà E, Moran T, Bertran-Alamillo J, Mayo-de-Las-Casas C, Queralt C, Rosell R (2010) Endobronchial ultrasound-guided transbronchial needle aspiration for identifying EGFR mutations. Eur Respir J. 35:391–395
51. Schuurbiers OC, Looijen-Salamon MG, Ligtenberg MJ, van der Heijden HF (2010) A brief retrospective report on the feasibility of epidermal growth factor receptor and KRAS mutation analysis in transesophageal ultrasound- and endobronchial ultrasound-guided fine needle cytological aspirates. J Thorac Oncol 5:1664–1667
52. Nakajima T, Yasufuku K, Nakagawara A, Kimura H, Yoshino I (2011) Multigene mutation analysis of metastatic lymph nodes in non-small cell lung cancer diagnosed by endobronchial ultrasound-guided transbronchial needle aspiration. Chest 140:1319–1324
53. Sakairi Y, Sato K, Itoga S, Saegusa F, Matsushita K, Nakajima T, Yoshida S, Takiguchi Y, Nomura F, Yoshino I (2014) Transbronchial biopsy needle rinse solution used for comprehensive biomarker testing in patients with lung cancer. J Thorac Oncol 9:26–32
54. Soda M, Choi YL, Enomoto M, Takada S, Yamashita Y, Ishikawa S, Fujiwara S, Watanabe H, Kurashina K, Hatanaka H, Bando M, Ohno S, Ishikawa Y, Aburatani H, Niki T, Sohara Y,

Sugiyama Y, Mano H (2007) Identification of the transforming EML4-ALK fusion gene in non-small-cell lung cancer. Nature 448:561–566

55. Sakairi Y, Nakajima T, Yasufuku K, Ikebe D, Kageyama H, Soda M, Takeuchi K, Itami M, Iizasa T, Yoshino I, Mano H, Kimura H (2010) EML4-ALK fusion gene assessment using metastatic lymph node samples obtained by endobronchial ultrasound-guided transbronchial needle aspiration. Clin Cancer Res 16:4938–4945
56. Lindeman NI, Cagle PT, Beasley MB, Chitale DA, Dacic S, Giaccone G, Jenkins RB, Kwiatkowski DJ, Saldivar JS, Squire J, Thunnissen E, Ladanyi M (2013) Molecular testing guideline for selection of lung cancer patients for EGFR and ALK tyrosine kinase inhibitors: guideline from the College of American Pathologists, International Association for the Study of Lung Cancer, and Association for Molecular Pathology. J Thorac Oncol 8:823–859
57. Kris MG, Johnson BE, Berry LD, Kwiatkowski DJ, Iafrate AJ, Wistuba II, Varella-Garcia M, Franklin WA, Aronson SL, Su PF, Shyr Y, Camidge DR, Sequist LV, Glisson BS, Khuri FR, Garon EB, Pao W, Rudin C, Schiller J, Haura EB, Socinski M, Shirai K, Chen H, Giaccone G, Ladanyi M, Kugler K, Minna JD, Bunn PA (2014) Using multiplexed assays of oncogenic drivers in lung cancers to select targeted drugs. JAMA 311:1998–2006
58. Pao W, Miller VA, Politi KA, Riely GJ, Somwar R, Zakowski MF, Kris MG, Varmus H (2005) Acquired resistance of lung adenocarcinomas to gefitinib or erlotinib is associated with a second mutation in the EGFR kinase domain. PLoS Med 2:e73
59. Engelman JA, Zejnullahu K, Mitsudomi T, Song Y, Hyland C, Park JO, Lindeman N, Gale CM, Zhao X, Christensen J, Kosaka T, Holmes AJ, Rogers AM, Cappuzzo F, Mok T, Lee C, Johnson BE, Cantley LC, Jänne PA (2007) MET amplification leads to gefitinib resistance in lung cancer by activating ERBB3 signaling. Science 316:1039–1043
60. Choi YL, Soda M, Yamashita Y, Ueno T, Takashima J, Nakajima T, Yatabe Y, Takeuchi K, Hamada T, Haruta H, Ishikawa Y, Kimura H, Mitsudomi T, Tanio Y, Mano H, ALK Lung Cancer Study Group (2010) EML4-ALK mutations in lung cancer that confer resistance to ALK inhibitors. N Engl J Med 363:1734–1739
61. Steuer CE, Ramalingam SS (2014) ALK-positive non-small cell lung cancer: mechanisms of resistance and emerging treatment options. Cancer 120:2392–2402
62. Jänne PA, Yang JC, Kim DW, Planchard D, Ohe Y, Ramalingam SS, Ahn MJ, Kim SW, Su WC, Horn L, Haggstrom D, Felip E, Kim JH, Frewer P, Cantarini M, Brown KH, Dickinson PA, Ghiorghiu S, Ranson M (2015) AZD9291 in EGFR inhibitor-resistant non-small-cell lung cancer. N Engl J Med 372:1689–1699
63. Gadgeel SM, Gandhi L, Riely GJ, Chiappori AA, West HL, Azada MC, Morcos PN, Lee RM, Garcia L, Yu L, Boisserie F, Di Laurenzio L, Golding S, Sato J, Yokoyama S, Tanaka T, Ou SH (2014) Safety and activity of alectinib against systemic disease and brain metastases in patients with crizotinib-resistant ALK-rearranged non-small-cell lung cancer (AF-002JG): results from the dose-finding portion of a phase 1/2 study. Lancet Oncol 15:1119–1128
64. Kerr KM, Bubendorf L, Edelman MJ, Marchetti A, Mok T, Novello S, O'Byrne K, Stahel R, Peters S, Felip E, Panel Members (2014) Second ESMO consensus conference on lung cancer: pathology and molecular biomarkers for non-small-cell lung cancer. Ann Oncol 25:1681–1690
65. Chouaid C, Dujon C, Do P, Monnet I, Madroszyk A, Le Caer H, Auliac JB, Berard H, Thomas P, Lena H, Robinet G, Baize N, Bizieux-Thaminy A, Fraboulet G, Locher C, Le Treut J, Hominal S, Vergnenegre A (2014) Feasibility and clinical impact of re-biopsy in advanced non small-cell lung cancer: a prospective multicenter study in a real-world setting (GFPC study 12-01). Lung Cancer 86:170–173
66. Tan CS, Gilligan D, Pacey S (2015) Treatment approaches for EGFR-inhibitor-resistant patients with non-small-cell lung cancer. Lancet Oncol 16:e447–e459
67. Nakajima T, Yasufuku K, Suzuki M, Fujiwara T, Shibuya K, Takiguchi Y, Hiroshima K, Kimura H, Yoshino I (2009) Assessment of Chemosensitivity-related Aberrant Methylation of Nonsmall Cell Lung Cancer by EBUS-TBNA. J Bronchology Interv Pulmonol 16:10–14

68. Darwiche K, Zarogoulidis P, Baehner K, Welter S, Tetzner R, Wohlschlaeger J, Theegarten D, Nakajima T, Freitag L (2013) Assessment of SHOX2 methylation in EBUS-TBNA specimen improves accuracy in lung cancer staging. Ann Oncol 24:2866–2870
69. Nakajima T, Anayama T, Koike T, Shingyoji M, Castle L, Kimura H, Yoshino I, Yasufuku K (2012) Endobronchial ultrasound doppler image features correlate with mRNA expression of HIF1-α and VEGF-C in patients with non-small-cell lung cancer. J Thorac Oncol 7:1661–1667
70. Nakajima T, Zamel R, Anayama T, Kimura H, Yoshino I, Keshavjee S, Yasufuku K (2012) Ribonucleic acid microarray analysis from lymph node samples obtained by endobronchial ultrasonography-guided transbronchial needle aspiration. Ann Thorac Surg 94:2097–2101
71. Leong TL, Marini KD, Rossello FJ, Jayasekara SN, Russell PA, Prodanovic Z, Kumar B, Ganju V, Alamgeer M, Irving LB, Steinfort DP, Peacock CD, Cain JE, Szczepny A, Watkins DN (2014) Genomic characterisation of small cell lung cancer patient-derived xenografts generated from endobronchial ultrasound-guided transbronchial needle aspiration specimens. PLoS One 9:e106862
72. Nakajima T, Geddie W, Anayama T, Ko HM, da Cunha SG, Boerner S, Wang T, Wang YH, Li M, Pham NA, Tsao MS, Yasufuku K (2015) Patient-derived tumor xenograft models established from samples obtained by endobronchial ultrasound-guided transbronchial needle aspiration. Lung Cancer 89:110–114

Chapter 6
Next-Generation Sequencing and Bioinformatics

Reika Kawabata-Iwakawa, Hidemasa Bono, and Masahiko Nishiyama

Abstract Improvement of next-generation sequencing (NGS) has revealed novel driver genes and new therapeutic targets for cancer patients. We summarize here the history of NGS development, current situation of typical analysis pipeline utilizing NGS – especially whole exome sequencing (WES) – whole transcriptome sequencing (WTS/RNA-seq), and useful public databases and tools for NGS data analysis. Large-scale sequencing projects such as The Cancer Genome Atlas (TCGA) and the International Cancer Genome Consortium (ICGC) have sequenced hundreds of tumor samples of different subtypes to provide catalogues of cancer genome. Chromothripsis, kataegis, and chromoplexy were discovered from those studies using second-generation technologies. Recent novel findings, such as *RET* gene fusion in lung cancer using these NGS following clinical translation, are also noted here. Clinical sequencing to find personalized treatment strategies for cancer patients has also been done for lung cancer. Mutations of EGFR and ALK are widely applied to screening for administration of EGFR-TKI and crizotinib, respectively. WES – targeting >100s mutation sequencing using NGS – has rapidly become a common way of clinical sequencing. Actionable mutations are used not only as targets of molecular therapy but also as markers for better stratifications in clinical trials.

Keywords Next-generation sequencing • Bioinformatics • Lung cancer • Database • Clinical sequencing

---

R. Kawabata-Iwakawa
Department of Molecular Pharmacology and Oncology, Gunma University Graduate School of Medicine, 3-39-22 Showa-Machi, Maebashi, Gunma 371-8511, Japan

H. Bono
Database Center for Life Science (DBCLS), National Institute of Genetics, Research Organization of Information and Systems, Mishima, Shizuoka, Japan

M. Nishiyama (✉)
Department of Molecular Pharmacology and Oncology, Gunma University Graduate School of Medicine, 3-39-22 Showa-Machi, Maebashi, Gunma 371-8511, Japan

Research Program for Omics-Based Medical Science, Division of Integrated Oncology Research, Gunma University Initiative for Advanced Research, Maebashi, Gunma, Japan
e-mail: m.nishiyama@gunma-u.ac.jp

© Springer Science+Business Media Singapore 2017
Y. Takiguchi (ed.), *Molecular Targeted Therapy of Lung Cancer*,
DOI 10.1007/978-981-10-2002-5_6

6.1 History of NGS Technology

6.1.1 Second-Generation Technologies

After completion of the Human Genome Project, large-scale sequencing approaches and sequence data analysis inspired the development of next-generation sequencing (NGS) methods, which made it possible for the fragments of nucleotides to be sequenced in a massive parallel way. The number of reads obtained by NGS was some orders of magnitude higher than that obtained by traditional capillary electrophoresis-based Sanger sequencing. However, this was achieved at the cost of both read length and accuracy. The first available next-generation sequencer with second-generation technologies provided an approximately 100-fold increase in throughput compared with Sanger sequencers; however, read lengths were ~100 base pairs (bp) compared with the ~700 bp read lengths by Sanger sequencing [1]. NGS platforms from several manufacturers have been in relatively wide use since 2007 [2], and numerous improvements and the emergence of new technologies have also occurred over the past decade. Nevertheless, read lengths provided by second-generation sequencing (SGS) platforms in used today, which are based on sequence by synthesis, range from ~100 to 500 bp [3]. NGS technologies also present an exponentially higher computational power than previous technologies because of the massive increase in the number of reads [4]. Development of enrichment methods such as exonic region provided systematic analyses of a considerable number of samples. Large-scale sequencing projects such as The Cancer Genome Atlas (TCGA) and the International Cancer Genome Consortium (ICGC) have sequenced hundreds of tumor samples of different subtypes to provide catalogues of the cancer genome. It is notable that chromothripsis, kataegis, and chromoplexy were discovered as a result of those studies using SGS technologies [5–7].

6.1.2 Third-Generation Technologies

Third-generation sequencing (TGS) technologies in NGS, which entirely eliminate DNA amplification, have recently begun to be used. This newest generation of sequencing methods is generally composed of technologies that interrogate single DNA molecules instead of clusters of DNA templates, thereby offering several advantages over SGS approaches, such as the elimination of amplification biases that emerged from polymerase chain reaction (PCR) [8]. One of the most widely used TGS technologies is SMRT sequencing, first developed in 2009 [9]. One of the main advantages of SMRT sequencing lies in its ability to produce unusually long read lengths; average read lengths have reached 21 kilobases (kb), and they continue to improve with the development of new reagent kits [10, 11]. Advantages provided by SMRT sequencing, however, also result in higher error rates, most often due to insertions and deletions (indels); but, these errors are randomly introduced and generally not context specific. Algorithms and software that help reduce the defects have been developed (e.g., the Quiver consensus algorithm [12]). The technology is also biased toward the identification of long fragments: A recent study shows that novel transcript isoforms less than 300 bp in length identified by short-read sequencing were generally not validated by SMRT sequencing [13]. To overcome this, various researchers have shown that a combination of SMRT and short-read sequencing, termed hybrid sequencing, can provide highly accurate sequence results, especially for complex genomic regions [14–16] and transcript isoforms [17–20].

Although SMRT sequencing is often used for study involving assembly of small bacterial genomes [21], the long read lengths are also well suited for sequencing large human cancer-related loci, such as gene fusion products. Two recent studies confirmed the presence of kinase domain mutations within *FLT3* genes by activating internal tandem duplications (FLT3-ITD) [22, 23] that affect ~20 % of AML patients and are associated with poor prognosis. SMRT sequencing has also been used to sequence the entire *BCR-ABL1* fusion gene transcript, allowing for the detection of compound mutations and splice isoforms [24]. Such assays are expected to be beneficial in the clinical setting, where mutations that confer resistance to TKI-based therapy could possibly be readily identified. Though the pyrosequencing platform had previously been used to address this issue, read lengths are not sufficient to cover the entire transcript, introducing the possibility of amplification biases affecting the measurement of mutation frequencies [25, 26]. SMRT sequencing has also been applied to the detection of other structural variants, such as deletion and translocation breakpoint determination [27].

6.1.3 Nanopore Sequencing

More recently, nanopore-based sequencing technologies have emerged as possibly yielding a single-molecule sequencing method. Oxford Nanopore, which is leading the development of this technology, released its portable sequencer called MinION to a selected community of researchers for testing, as part of the MinION Access Program (MAP) [28]. The company promised that the device would cost only $1000 and would provide read lengths orders of magnitude longer than existing NGS technologies. The platform was released in 2014, and preliminary reports suggest that the technology, while promising, requires further improvement [29, 30]. The concept of a truly portable high-throughput sequencing platform is attractive in several applications, including fieldwork and point of care diagnostics [28–31].

6.2 Typical Analysis Pipeline Utilizing NGS

6.2.1 Exome Sequencing

Whole exome sequencing (WES) targeting the exon regions that are transcribed to mRNA (1 % of the whole human genome) is mainly used in cancer genomic study. WES data processing is summarized in Fig. 6.1a. BWA-MEN is often utilized for mapping reads (generally in FASTQ format) to the reference genome, where BAM-formatted mapping results are obtained. These BAM files are then trimmed to remove duplicate reads by using SAMtools or Picard (http://picard.sourceforge.net/) [32]. Reads mapped to exonic regions can be extracted using BEDTools [33]. SAMtools can make variant call format (VCF) files as follows: Identify the genomic position of single nucleotide variants (SNVs), insertion, or deletion, and add the annotation of their mutational format – synonymous, nonsynonymous, missense, or frameshift.

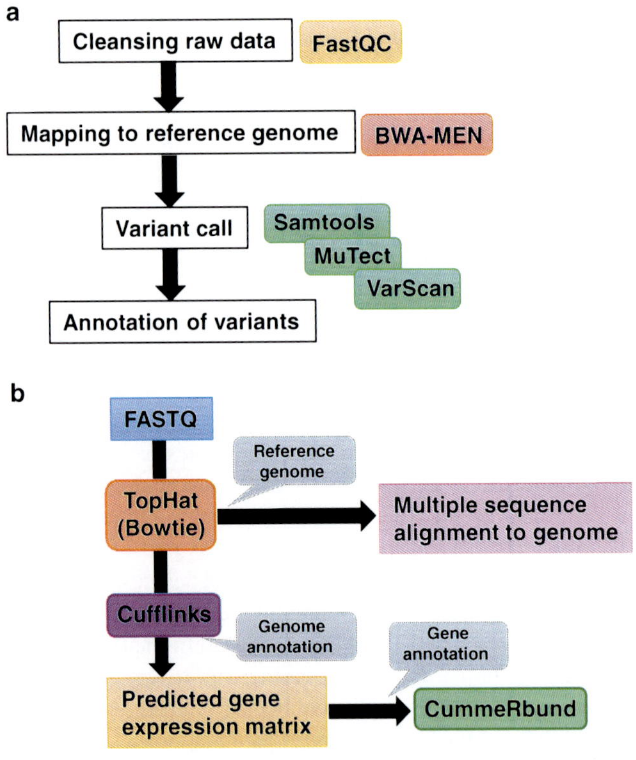

Fig. 6.1 Typical primary analysis pipeline of WES (**a**) and WTS (**b**). Briefly, data processing is done as follows: data cleansing, mapping to reference genome, and variant call/read counting and annotation. Various methods are used to detect somatic mutations or deferential expressed genes in the resulting sequence reads. Representative tools are described

Considering different patterns in base substitutions and mutant allele frequencies among cancer types and individual subjects, there are limitations in detecting mutations correctly by using the above software for detection of single nucleotide polymorphisms (SNPs). Tools developed exclusively for detection of somatic mutations, such as MuTect [34] or VarScan [35], are often used with somatic mutation calling for tumor/normal-paired sample. Briefly, MuTect has high sensitivity and is good at detecting low allele mutation, while VarScan has high specificity in detecting somatic mutations [36].

Usuyama et al. have recently developed a novel way of somatic mutation called HapMuC, using heterozygous germ line variants near candidate mutations. The algorithm had superior specificity and sensitivity compared with previous methods [37].

Gene polymorphism databases such as dbSNP and 1000 Genomes Project are usually used to remove SNPs from SNVs and/or short indel. Panel of normal samples is used to further filter and to remove false-positive somatic mutations caused by sequencing errors in normal samples.

Hundreds to thousands of somatic mutations are often detected in cancer genome. Among these, several tools, such as MutSigCV, OncodriveFM, and OncodriveCLUST, have been proposed to identify driver genes.

MutSigCV allows researchers to calculate the significance of genomic mutational status for cancer association by using not only background mutation rate but also DNA replication timing and transcriptional activity of the gene [38].

OncodriveFM provides a functional impact using three well-known methods (SIFT, PolyPhen2, and MutationAssessor) [39]: It is based on the assumption that any bias toward the accumulation of variants with high functional impact is an indication of positive selection and can thus be used to detect candidate driver genes or gene modules.

OncodriveCLUST (http://bg.upf.edu/oncodrive-clust) is a method to identify genes in which mutations accumulate within specific regions of the protein, because this denotes events selected by affecting the tumor [40]. It computes a score measuring the mutation clustering of a gene across the protein sequence and then compares it with a background model.

6.2.2 *Transcriptome Sequencing*

RNA expression profiling is possible by using whole transcriptome sequencing (WTS/RNA-seq) to replace the usage of microarray methods. WTS also allows scientists to look at alternative gene-spliced transcripts, posttranscriptional modifications, gene fusion, mutations/SNPs, and changes in gene expression [41]. RNA expression of each gene or transcript is often measured by RPKM/FPKM (reads/fragments per kilobase per million mapped reads) or TPM (transcripts per million). RPKM/FPKM is the normalized value of mapped read/fragment counts normalized by transcript length and total reads/fragments.

Bowtie or BWA is used for mapping of WTS reads. Alignment that can take account of splicing variants is needed to determine WTS data compared with exome analysis. So in WTS analysis, TopHat software in conjunction with Bowtie is used for mapping, to detect fusion gene and call SNVs (Fig. 6.1b) [42]. HISAT will be the core of the next version of TopHat (http://nextgenseek.com/2015/03/hisat-a-fast-and-memory-lean-rna-seq-aligner/). Cufflinks yield not only transcriptome assembly in conjunction with the splicing variants using genome annotation file (GTF format, as usual), FPKM, but also differentially expressed genes (DEGs) between two specified groups.

The STAR software package performs this task with higher levels of accuracy and speed. In addition to detecting annotated and novel splice junctions, STAR can discover more complex RNA sequence arrangements, such as chimeric and circular RNA [43].

Normalization of tag count data strategies is updated day by day using R packages: TCC (an acronym for Tag Count Comparison) is an R/Bioconductor package that provides a series of functions for differential expression analysis of tag count data. The package incorporates multistep normalization methods to remove potential DEGs before performing data normalization. TCC provides a simple unified interface that can perform such analyses with combinations of functions provided by edgeR, DESeq, and baySeq [44].

6.3 Public Databases and Tools for NGS Analysis

After identification of genomic/transcriptome alterations, several databases are available to extract the biomarker or therapeutic target for cancer (Table 6.1).

6.3.1 COSMIC

The Catalogue of Somatic Mutations in Cancer (COSMIC) is the most popular database: It includes SNV, insertion, deletion, gene fusion, genomic rearrangement, copy number, and differential expression data from over one million cancer genomes [45]. It can be confirmed whether detected alterations are known somatic mutations or not by comparing them with COSMIC database. Mutational frequency and mutational status of genes involved in tumors/cell lines from the dataset are also available.

Table 6.1 Public database and tools for NGS analysis

Database	Contents	URL
COSMIC	SNV, insertion, deletion, gene fusion, genomic rearrangement, copy number, and differential expression data	http://cancer.sanger.ac.uk/cosmic
cBioPortal	Mutations, CNV, RNA/protein expression, clinical data, and their correlations	http://www.cbioportal.org/index.do
DAVID	Characteristics of involved gene set using GSEA	https://david.ncifcrf.gov/
DGIdb	Drug-gene interactions	http://dgidb.genome.wustl.edu/
Genomics of Drug Sensitivity in Cancer	Cell line drug sensitivity data	http://www.cancerrxgene.org/
Mitelman Database	Fusion genes or chromosomal aberrations	http://cgap.nci.nih.gov/Chromosomes/Mitelman
RefEx	Expression profile of gene of interest in each normal organ and cell line	http://refex.dbcls.jp/
PrognoScan	Meta-analysis of the prognostic value of genes	http://www.prognoscan.org/

6.3.2 cBioPortal

The cBioPortal for Cancer Genomics makes possible visualization and analysis of large-scale cancer genomic dataset, not only of mutations, CNV, RNA/protein expression, and clinical data but also their correlations. The database contained data from 105 cancer genomic studies in October 2015 [46, 47].

6.3.3 DAVID

The Database for Annotation, Visualization and Integrated Discovery (DAVID) software can aid in knowing the characteristics of involved gene set using GSEA (Gene Set Enrichment Analysis) method [48]. DAVID software interprets annotated data with OMIM, gene ontology, and pathway.

6.3.4 DGIdb

The DGIdb is used to look at drug-gene interactions and potentially "druggable" genes [49]. Information on clinical trial status is also available.

6.3.5 Genomics of Drug Sensitivity in Cancer

This database provides cell line drug sensitivity data for 140 drugs representing >48,000 cell line-drug interactions [50]. Drug sensitivity data have been correlated with mutations in cancer genes in order to identify genetic factors associated with drug sensitivity or resistance.

6.3.6 Mitelman Database of Chromosome Aberrations and Gene Fusions in Cancer

This database can be used to confirm the frequency of detected fusion genes or chromosomal aberrations. In total, 10,026 gene fusions of 65,975 cases have been registered in 2015 (Mitelman F, Johansson B, and Mertens F (Eds.), Mitelman Database of Chromosome Aberrations and Gene Fusions in Cancer (2015). http://cgap.nci.nih.gov/Chromosomes/Mitelman).

6.3.7 RefEx

RefEx (Reference Expression dataset; http://refex.dbcls.jp/) attempts to achieve the reference of mammalian tissue gene expression data using various methods, such as expressed sequence tag (EST), microarray (GeneChip), and CAGE (cap analysis gene expression) and WTS/RNA-seq. This database is useful to find out the expression profiles of genes of interest in each normal organ, and the recent update of RefEx in collaboration with FANTOM5 project enables us to browse gene expression profiles in cell lines, primary cultures, and adult and fetal tissues from human and mouse [51].

6.3.8 PrognoScan

PrognoScan is an online biomarker validation tool for meta-analysis of the prognostic value of genes [52]. 8626 cases of 14 cancer types in 74 datasets are registered in this database.

6.4 Identification of Novel Therapeutic Targets for Lung Cancer by NGS

Recent studies using NGS platform have revealed novel driver genes affecting lung cancer carcinogenesis and have also formulated new therapeutic target repressing oncogenic addiction (Table 6.2).

Table 6.2 Recent studies using NGS in lung cancer

Author	Histology	No. of patients	Methods using NGS	Summary of novel findings	References
Kohno	ADC	30	WTS	*KIF5B-RET* fusion gene	[53]
Imielinski	ADC	183	WGS, WES	Mutations of *U2AF1*, *RBM10*, *ARID1A* genes. Structural variants of *EGFR*, *SIK2* genes	[58]
Seo	ADC	87	WES, WTS	Mutations of *LMTK2*, *ARID1A*, *NOTCH2*, *SMARCA4* genes. Fusion of *ALK*, *RET*, *ROS1*, *FGFR2*, *AXL*, *PDGFRA* genes	[59]
TCGA	ADC	230	WGS, WES, WTS	*NF1*, *RIT1* mutations. *MGA* mutations that occur mutually exclusive of MYC amplification	[60]
Fernandez-Cuesta	ADC	25	WTS	*CD74-NRG1* gene fusion in mucinous subtype	[61]
Jang	ADC	153	WTS	*SND1-BRAF* fusion gene	[62]
TCGA	SQC	178	WGS, WES	Significantly altered pathways included *NFE2L2* and *KEAP1* and/or deletion or mutation of *CUL3*. Amplification of *FGFR1* and *WHSC1L1*	[63]
Kim	SQC	104	WES	Similar spectrum of alterations between Korean and North American lung squamous cell carcinoma. *FGFR3-TACC3* fusion gene	[64]
Peifer	SCLC	29	WGS, WES, WTS	*TP53*, *RB1* inactivation in all cases. Frequent mutation of *CREBBP*, *EP300*, *MLL*, *PTEN*, *SLIT2*, *EPHA7* genes. *FGFR1* gene amplification	[67]
Rudin	SCLC	53	WGS, WES, WTS	22 significantly mutated genes. Frequent amplification of *SOX2* gene	[68]
George	SCLC	110	WGS, WTS	Frequent mutation in *TP73* and NOTCH family genes. Chromothripsis affecting	[69]
Govindan	NSCLC (16 ADC, 1 LCC)	17	WGS,WTS	Ten times higher mutation frequency among smokers than in nonsmokers. *EGFR* and *KRAS* mutations play initiation role in lung cancer carcinogenesis both in smokers and never-smokers	[70]

6.4.1 Lung Adenocarcinoma

KIF5-RET fusion gene was identified as a new driver gene by using WTS and/or whole-genome sequencing (WGS) of lung adenocarcinoma (LADC) patients in 2012 [53–56]. RET fusion was found in 1–2 % of LADC patients from both Asia and Europe, numbering approximately 12,000 lung cancer patients per year worldwide. The occurrence of RET fusion was found more frequently among the young and was specific among LADC patients [57].

Imielinski et al. identified *U2AF1*, *RBM10*, and *ARID1A* as novel driver genes of lung cancer using WES of surgically resected 183 lung adenocarcinomas. The genomic rearrangements in *EGFR* and *SIK2* genes were discovered by WGS analysis of 24 lung adenocarcinomas [58].

Seo et al. identified novel driver mutations in *LMTK2*, *ARID1A*, *NOTCH2*, and *SMARCA4* genes using WES of surgical specimens from 76 lung adenocarcinoma patients. WTS of 77 cases revealed fusion genes involving tyrosine kinase genes such as *FGFR2*, *AXL*, and *PDGFRA* in addition to *ALK*, *RET*, and *ROS1*, previously known fusion genes in lung adenocarcinomas [59].

TCGA researchers investigated omics landscape of 230 resected lung adenocarcinomas: They identified 18 genes as statistically significant mutated genes, including *RIT1* activating mutations and newly described loss-of-function *MGA* mutations, which are mutually exclusive with focal MYC amplification. Aberrations in *NF1*, *MET*, *ERBB2*, and *RIT1* occurred in 13 % of cases and were enriched in samples otherwise lacking an activated oncogene, suggesting a driver role for these events in some tumors. MAPK and PI(3)K pathway activity at the protein level was explained by known mutations in only a fraction of cases, suggesting additional, unexplained mechanisms of pathway activation [60].

Fernandez-Cuesta et al. discovered *CD74-NRG1* fusion gene using WTS of 25 lung adenocarcinomas of never-smokers. In addition to 102 pan-negative lung adenocarcinoma patients of never-smokers, five cases carried *CD74-NRG1* fusion gene. All positive cases were female of the invasive mucinous subtype [61].

Jang et al. identified *SND1-BRAF* fusion in 5/153 never-smoker lung adenocarcinoma patients by using WTS. Ectopic expression of *SND1-BRAF* in H1299 cells showed upregulated phosphorylation levels of MEK/ERK, cell proliferation, and spheroid formation compared with parental mock-transfected control [62].

6.4.2 Squamous Cell Lung Cancer

TCGA researchers revealed a comprehensive genomic landscape of squamous cell lung cancer in 2012. About 178 cases of squamous cell lung cancer patients were analyzed using WES and WTS. They detected novel loss-of-function mutation in the *HLA-A* class I major histocompatibility gene. Significantly altered pathways included *NFE2L2* and *KEAP1* and/or deletion or mutation of CUL3 in 34 % of tumors. They identified actionable alterations for therapeutic targets in most tumors [63].

Kim et al. clarified a similar spectrum of alterations between Korean and North American lung squamous cell carcinoma, in contrast to the differences seen in lung adenocarcinoma. They also identified recurrence of therapeutically actionable *FGFR3-TACC3* fusion in lung squamous cell carcinoma [64].

FGFR1 gene is amplified in up to ~20 % of squamous cell lung cancer patients. Clinical trials with FGFR inhibitors are currently underway [65, 66] (My Cancer Genome http://www.mycancergenome.org/content/disease/lung-cancer/fgfr1/58/ (Updated November 15)). Precision medicine targeting FGFR pathway will improve the prognoses of patients with lung squamous cell carcinoma.

6.4.3 Small-Cell Lung Cancer

Peifer et al. used 29 SCLCs for WES, WTS, and/or WGS analyses: All cases showed signatures of inactivation of p53 and Rb. In addition to frequent mutations in *CREBBP*, *EP300*, and *MLL*, which encode histone modification protein, frequent mutations in *PTEN*, *SLIT2*, and *EPHA7* genes and focal amplification of *FGFR1* gene were observed [67].

Rudin et al. identified 22 candidate driver genes by WES and WTS analysis of 36 primary SCLC and 17 SCLC cell lines. *SOX2* amplification was detected in ~27 % of SCLC tumors, and utility was confirmed as the therapeutic target for SCLC [68].

We identified frequent mutation in *TP73* and NOTCH family genes by WGS and WTS of 110 SCLCs; rearrangement of *TP73* gene induced oncogenic transcript TP73Δex2/3; and WGS revealed that chromothripsis on chromosomes 3 and 11 affects tumors with wild-type RB1 [69].

6.4.4 Tobacco Smoking and Lung Cancer Genome

Govindan et al. sequenced entire genome and transcriptome of 17 tumor-adjacent normal sample pairs from non-small cell lung cancer (NSCLC) patients: The samples revealed ten times higher mutation frequency among smokers than in never-smokers. *EGFR* and *KRAS* mutations were found in the foundation clones both among smokers and never-smokers by using deep digital sequencing, suggesting that *EGFR* and *KRAS* mutations play initiation roles in lung cancer carcinogenesis. In addition, 54 genes were identified as targetable mutations for therapy [70].

Gou et al. analyzed several datasets of NGS study in a total of 739 lung cancer tumors (390 adenocarcinomas, 282 squamous cell carcinomas, and 67 small cell carcinomas). They also demonstrated that smokers have many more somatic mutations than nonsmokers (nonsmokers ADC, 0.98/Mb; smokers ADC, 12.67/Mb; SQC, 8.75/Mb; SCLC, 15.87/Mb). The cancer genomes of smokers were more complicated when compared with nonsmokers [71].

6.5 NGS for Precision Medicine

6.5.1 Cancer Immunotherapy

Rizvi et al. used WES for 34 NSCLC cases treated with pembrolizumab, an antibody targeting programmed cell death-1 (PD-1) therapy, which revealed that a higher nonsynonymous mutation burden in tumors was associated with improved objective response, durable clinical benefit, and progression-free survival [72].

6.5.2 Clinical Sequencing

Clinical sequencing to make personalized treatment strategies for cancer patients has been used in lung cancer. Mutations of EGFR and ALK were widely used in screening for administration of EGFR-TKI and crizotinib, respectively. WES or targeting >100 s mutation sequencing using NGS rapidly is becoming a common method of clinical sequencing. Actionable mutations are used not only as targets of molecular therapy but also as markers to achieve better stratifications for clinical trials. In the USA, extensive clinical trials (master protocol) using >1000 squamous cell lung cancers have begun [73]. The "Foundation One" platform is a mutational screening using NGS for consulting adaptation in five clinical trials, including FGFR-TKI and anti-PDL1 antibody treatment.

Several clinical trials for NSCLC patients with KIF5B-RET rearrangements are currently ongoing using previously known RET inhibitors including cabozantinib, lenvatinib, vandetanib, sunitinib, and AUY922 (Table 6.3; ClinicalTrials.gov 2015; [74]).

Genomic analyses in lung cancer were mainly obtained using surgically resected tumors. To consider the availability for chemotherapy, it is essential to understand the genomic profile of unresectable advanced tumors and to clarify the response to approved/unapproved drugs.

Table 6.3 Clinical trials for RET fusion-positive NSCLC patients

Trial ID	Compound	Investigator	Enrollment and contents
NCT02540824	Apatinib	Tongji University	Ph II (2015~): For RET fusion-positive advanced NSCLC patients who failed to respond to previous treatment
NCT01639508	Cabozantinib	Memorial Sloan-Kettering Cancer Center	Ph II (2013~): For patients whose tumors have a gene called KIF5B-RET. Cabozantinib in patients with RET fusion-positive advanced non-small cell lung cancer
NCT01877083	Lenvatinib	Eisai Co., Ltd.	Ph II (2013~): For patients with KIF5B-RET-positive LADCs and other confirmed RET translocations
UMIN000010095	Vandetanib	National Cancer Center East	Ph II (2013~): For patients with RET fusion-positive unresectable locally advanced or metastatic NSCLC
NCT01823068	Vandetanib	Seoul National Univ. Hospital	Ph II (2013~): For patients with advanced NSCLC harboring RET gene fusion
NCT01813734	Ponatinib	Massachusetts General Hospital	Ph II (2013~): For patients with advanced NSCLC harboring RET translocations
NCT01829217	Sunitinib	Dana-Farber Cancer Institute	Ph II (2013~): For never-smoker LADCs w/o known cancer genes other than RET mutations
NCT02219711	MGCD516	Mirati Therapeutics, Inc.	Ph I/Ib (2014~): For non-small cell lung cancer and head and neck cancer positive for specific activating MET, NTRK2, NTRK3, or DDR2 mutations, MET or KIT/PDGFRA/KDR gene amplification, or selected gene rearrangements involving the MET, RET, AXL, NTRK1, or NTRK3 gene loci
NCT01922583	NVP-AUY922 (AUY922)	National Taiwan University Hospital	Ph II (2014~): For patients with stage IV EGFR T790 M, EGFR exon 20, and other uncommon, HER2, or BRAF mutated; ALK-, ROS1-, or RET-rearranged NSCLC

Acknowledgments We are grateful to Drs. S. Yoshiyama (Gunma University Initiative for Advanced Research), T. Kohno, M. Saito, M. Kato (National Cancer Center Research Institute), and A. Fujimoto (Kyoto University) for their insightful comments and suggestions.

References

1. Margulies M, Egholm M, Altman WE, Attiya S, Bader JS, Bemben LA, Berka J, Braverman MS, Chen YJ, Chen Z, Dewell SB, Du L, Fierro JM, Gomes XV, Godwin BC, He W, Helgesen S, Ho CH, Irzyk GP, Jando SC, Alenquer ML, Jarvie TP, Jirage KB, Kim JB, Knight JR, Lanza JR, Leamon JH, Lefkowitz SM, Lei M, Li J, Lohman KL, Lu H, Makhijani VB, McDade KE, McKenna MP, Myers EW, Nickerson E, Nobile JR, Plant R, Puc BP, Ronan MT, Roth GT, Sarkis GJ, Simons JF, Simpson JW, Srinivasan M, Tartaro KR, Tomasz A, Vogt KA, Volkmer GA, Wang SH, Wang Y, Weiner MP, Yu P, Begley RF, Rothberg JM (2005) Genome sequencing in microfabricated high-density picolitre reactors. Nature 437:376–380
2. Mardis ER (2008) Next-generation DNA sequencing methods. Annu Rev Genomics Hum Genet 9:387–402
3. Buermans HP, den Dunnen JT (2014) Next generation sequencing technology: advances and applications. Biochim Biophys Acta 842:1932–1941
4. Mardis ER (2011) A decade's perspective on DNA sequencing technology. Nature 470:198–203
5. Stephens PJ, Greenman CD, Fu B, Yang F, Bignell GR, Mudie LJ, Pleasance ED, Lau KW, Beare D, Stebbings LA et al (2011) Massive genomic rearrangement acquired in a single catastrophic event during cancer development. Cell 144:27–40
6. Nik-Zainal S, Alexandrov LB, Wedge DC, Van Loo P, Greenman CD, Raine K, Jones D, Hinton J, Marshall J, Stebbings LA, Menzies A, Martin S, Leung K, Chen L, Leroy C, Ramakrishna M, Rance R, Lau KW, Mudie LJ, Varela I, McBride DJ, Bignell GR, Cooke SL, Shlien A, Gamble J, Whitmore I, Maddison M, Tarpey PS, Davies HR, Papaemmanuil E, Stephens PJ, McLaren S, Butler AP, Teague JW, Jönsson G, Garber JE, Silver D, Miron P, Fatima A, Boyault S, Langerød A, Tutt A, Martens JW, Aparicio SA, Borg Å, Salomon AV, Thomas G, Børresen-Dale AL, Richardson AL, Neuberger MS, Futreal PA, Campbell PJ, Stratton MR, Breast Cancer Working Group of the International Cancer Genome Consortium (2012) Mutational processes molding the genomes of 21 breast cancers. Cell 149:979–993
7. Baca SC, Prandi D, Lawrence MS, Mosquera JM, Romanel A, Drier Y, Park K, Kitabayashi N, MacDonald TY, Ghandi M, Van Allen E, Kryukov GV, Sboner A, Theurillat JP, Soong TD, Nickerson E, Auclair D, Tewari A, Beltran H, Onofrio RC, Boysen G, Guiducci C, Barbieri CE, Cibulskis K, Sivachenko A, Carter SL, Saksena G, Voet D, Ramos AH, Winckler W, Cipicchio M, Ardlie K, Kantoff PW, Berger MF, Gabriel SB, Golub TR, Meyerson M, Lander ES, Elemento O, Getz G, Demichelis F, Rubin MA, Garraway LA (2013) Punctuated evolution of prostate cancer genomes. Cell 153:666–677
8. Roberts RJ, Carneiro MO, Schatz MC (2013) The advantages of SMRT sequencing. Genome Biol 14:401–404
9. Eid J, Fehr A, Gray J, Luong K, Lyle J, Otto G, Peluso P, Rank D, Baybayan P, Bettman B, Bibillo A, Bjornson K, Chaudhuri B, Christians F, Cicero R, Clark S, Dalal R, Dewinter A, Dixon J, Foquet M, Gaertner A, Hardenbol P, Heiner C, Hester K, Holden D, Kearns G, Kong X, Kuse R, Lacroix Y, Lin S, Lundquist P, Ma C, Marks P, Maxham M, Murphy D, Park I, Pham T, Phillips M, Roy J, Sebra R, Shen G, Sorenson J, Tomaney A, Travers K, Trulson M, Vieceli J, Wegener J, Wu D, Yang A, Zaccarin D, Zhao P, Zhong F, Korlach J, Turner S (2009) Real-time DNA sequencing from single polymerase molecules. Science 323:133–138

10. Kim KE, Peluso P, Babayan P, Yeadon PJ, Yu C, Fisher WW, Chin CS, Rapicavoli NA, Rank DR, Li J, Catcheside DE, Celniker SE, Phillippy AM, Bergman CM, Landolin JM (2014) Long-read, whole-genome shotgun sequence data for five model organisms. Sci Data 1:140045. doi:10.1038/sdata.2014.45 eCollection 2014
11. Rhoads A, Au KF. (2015) PacBio sequencing and its applications. Genomics Proteomics Bioinformatics: pii: S1672-0229(15)00134-5. doi:10.1016/j.gpb.2015.08.002
12. Chin CS, Alexander DH, Marks P, Klammer AA, Drake J, Heiner C, Clum A, Copeland A, Huddleston J, Eichler EE, Turner SW, Korlach J (2013) Nonhybrid, finished microbial genome assemblies from long-read SMRT sequencing data. Nat Methods 10:563–569
13. Chen L, Kostadima M, Martens JH, Canu G, Garcia SP, Turro E, Downes K, Macaulay IC, Bielczyk-Maczynska E, Coe S, Farrow S, Poudel P, Burden F, Jansen SB, Astle WJ, Attwood A, Bariana T, de Bono B, Breschi A, Chambers JC, BRIDGE Consortium, Choudry FA, Clarke L, Coupland P, van der Ent M, Erber WN, Jansen JH, Favier R, Fenech ME, Foad N, Freson K, van Geet C, Gomez K, Guigo R, Hampshire D, Kelly AM, Kerstens HH, Kooner JS, Laffan M, Lentaigne C, Labalette C, Martin T, Meacham S, Mumford A, Nürnberg S, Palumbo E, van der Reijden BA, Richardson D, Sammut SJ, Slodkowicz G, Tamuri AU, Vasquez L, Voss K, Watt S, Westbury S, Flicek P, Loos R, Goldman N, Bertone P, Read RJ, Richardson S, Cvejic A, Soranzo N, Ouwehand WH, Stunnenberg HG, Frontini M, Rendon A (2014) Transcriptional diversity during lineage commitment of human blood progenitors. Science 345:1251033
14. Koren S, Schatz MC, Walenz BP, Martin J, Howard JT, Ganapathy G, Wang Z, Rasko DA, McCombie WR, Jarvis ED, Phillippy AM (2012) Hybrid error correction and de novo assembly of single-molecule sequencing reads. Nat Biotechnol 30:693–700
15. Ritz A, Bashir A, Sindi S, Hsu D, Hajirasouliha I, Raphael BJ (2014) Characterization of structural variants with single molecule and hybrid sequencing approaches. Bioinformatics 30:3458–3466
16. Doi K, Monjo T, Hoang PH, Yoshimura J, Yurino H, Mitsui J, Ishiura H, Takahashi Y, Ichikawa Y, Goto J, Tsuji S, Morishita S (2014) Rapid detection of expanded short tandem repeats in personal genomics using hybrid sequencing. Bioinformatics 30:815–822
17. Tilgner H, Grubert F, Sharon D, Snyder MP (2014) Defining a personal, allele-specific, and single-molecule long-read transcriptome. Proc Natl Acad Sci U S A 111:9869–9874
18. Weirather JL, Afshar PT, Clark TA, Tseng E, Powers LS, Underwood JG, Zabner J, Korlach J, Wong WH, Au KF (2015) Characterization of fusion genes and the significantly expressed fusion isoforms in breast cancer by hybrid sequencing. Nucleic Acids Res 43:e116
19. Au KF, Sebastiano V, Afshar PT, Durruthy JD, Lee L, Williams BA, van Bakel H, Schadt EE, Reijo-Pera RA, Underwood JG, Wong WH (2013) Characterization of the human ESC transcriptome by hybrid sequencing. Proc Natl Acad Sci U S A 110:E4821–E4830
20. Pendleton M, Sebra R, Pang AW, Ummat A, Franzen O, Rausch T, Stütz AM, Stedman W, Anantharaman T, Hastie A, Dai H, Fritz MH, Cao H, Cohain A, Deikus G, Durrett RE, Blanchard SC, Altman R, Chin CS, Guo Y, Paxinos EE, Korbel JO, Darnell RB, McCombie WR, Kwok PY, Mason CE, Schadt EE, Bashir A (2015) Assembly and diploid architecture of an individual human genome via single-molecule technologies. Nat Methods 12:780–786
21. Fomenkov A, Lunnen KD, Zhu Z, Anton BP, Wilson GG, Vincze T, Roberts RJ (2015) Complete genome sequence and methylome analysis of bacillus strain x1. Genome Announc 3:e01593–e015914
22. Smith CC, Wang Q, Chin CS, Salerno S, Damon LE, Levis MJ, Perl AE, Travers KJ, Wang S, Hunt JP, Zarrinkar PP, Schadt EE, Kasarskis A, Kuriyan J, Shah NP (2012) Validation of ITD mutations in FLT3 as a therapeutic target in human acute myeloid leukaemia. Nature 485:260–263
23. Smith CC, Zhang C, Lin KC, Lasater EA, Zhang Y, Massi E, Damon LE, Pendleton M, Bashir A, Sebra R, Perl A, Kasarskis A, Shellooe R, Tsang G, Carias H, Powell B, Burton EA, Matusow B, Zhang J, Spevak W, Ibrahim PN, Le MH, Hsu HH, Habets G, West BL, Bollag G, Shah NP (2015) Characterizing and overriding the structural mechanism of the Quizartinib-Resistant FLT3 "Gatekeeper" F691 L mutation with PLX3397. Cancer Discov 5:668–679

24. Cavelier L, Ameur A, Haggqvist S, Hoijer I, Cahill N, Olsson-Stromberg U, Hermanson M (2015) Clonal distribution of BCR-ABL1 mutations and splice isoforms by single-molecule long-read RNA sequencing. BMC Cancer 15:41–45
25. Soverini S, De Benedittis C, Machova Polakova K, Brouckova A, Horner D, Iacono M, Castagnetti F, Gugliotta G, Palandri F, Papayannidis C, Iacobucci I, Venturi C, Bochicchio MT, Klamova H, Cattina F, Russo D, Bresciani P, Binotto G, Giannini B, Kohlmann A, Haferlach T, Roller A, Rosti G, Cavo M, Baccarani M, Martinelli G (2013) Unraveling the complexity of tyrosine kinase inhibitor-resistant populations by ultra-deep sequencing of the BCR-ABL kinase domain. Blood 122:1634–1648
26. Kastner R, Zopf A, Preuner S, Proll J, Niklas N, Foskett P, Valent P, Lion T, Gabriel C (2014) Rapid identification of compound mutations in patients with Philadelphia-positive leukaemias by long-range next generation sequencing. Eur J Cancer 50:793–800
27. Patel A, Schwab R, Liu YT, Bafna V (2014) Amplification and thrifty single-molecule sequencing of recurrent somatic structural variations. Genome Res 24:318–328
28. Mikheyev AS, Tin MM (2014) A first look at the Oxford Nanopore MinION sequencer. Mol Ecol Resour 14:1097–1102
29. Quick J, Quinlan AR, Loman NJ (2014) A reference bacterial genome dataset generated on the MinION portable single-molecule nanopore sequencer. Gigascience 3:22
30. Kilianski A, Haas JL, Corriveau EJ, Liem AT, Willis KL, Kadavy DR, Rosenzweig CN, Minot SS (2015) Bacterial and viral identification and differentiation by amplicon sequencing on the MinION nanopore sequencer. Gigascience 4:11–18
31. Rusk N (2015) MinION takes center stage. Nat Methods 12:12–13
32. H L, Handsaker B, Wysoker A, Fennell T, Ruan J, Homer N, Marth G, Abecasis G, Durbin R, 1000 Genome Project Data Processing Subgroup (2009) The sequence alignment/map format and SAMtools. Bioinformatics 25:2078–2079
33. Quinlan AR, Hall IM (2010) BEDTools: a flexible suite of utilities for comparing genomic features. Bioinformatics 26:841–842
34. Cibulskis K, Lawrence MS, Carter SL, Sivachenko A, Jaffe D, Sougnez C, Gabriel S, Meyerson M, Lander ES, Getz G (2013) Sensitive detection of somatic point mutations in impure and heterogeneous cancer samples. Nat Biotechnol 31:213–219
35. Koboldt DC, Zhang Q, Larson DE, Shen D, McLellan MD, Lin L, Miller CA, Mardis ER, Ding L, Wilson RK (2012) VarScan 2: somatic mutation and copy number alteration discovery in cancer by exome sequencing. Genome Res 22:568–576
36. Wang Q, Jia P, Li F, Chen H, Ji H, Hucks D, Dahlman KB, Pao W, Zhao Z (2013) Detecting somatic point mutations in cancer genome sequencing data: a comparison of mutation callers. Genitourin Med 5:91
37. Usuyama N, Shiraishi Y, Sato Y, Kume H, Homma Y, Ogawa S, Miyano S, Imoto S (2014) HapMuC: somatic mutation calling using heterozygous germ line variants near candidate mutations. Bioinformatics 30:3302–3309
38. Lawrence MS, Stojanov P, Polak P, Kryukov GV, Cibulskis K, Sivachenko A, Carter SL, Stewart C, Mermel CH, Roberts SA, Kiezun A, Hammerman PS, McKenna A, Drier Y, Zou L, Ramos AH, Pugh TJ, Stransky N, Helman E, Kim J, Sougnez C, Ambrogio L, Nickerson E, Shefler E, Cortés ML, Auclair D, Saksena G, Voet D, Noble M, DiCara D, Lin P, Lichtenstein L, Heiman DI, Fennell T, Imielinski M, Hernandez B, Hodis E, Baca S, Dulak AM, Lohr J, Landau DA, Wu CJ, Melendez-Zajgla J, Hidalgo-Miranda A, Koren A, McCarroll SA, Mora J, Lee RS, Crompton B, Onofrio R, Parkin M, Winckler W, Ardlie K, Gabriel SB, Roberts CW, Biegel JA, Stegmaier K, Bass AJ, Garraway LA, Meyerson M, Golub TR, Gordenin DA, Sunyaev S, Lander ES, Getz G (2013) Mutational heterogeneity in cancer and the search for new cancer-associated genes. Nature 499:214–218
39. Gonzalez-Perez A, Lopez-Bigas N (2012) Functional impact bias reveals cancer drivers. Nucleic Acids Res 40:e169

40. Tamborero D, Gonzalez-Perez A, Lopez-Bigas N (2013) OncodriveCLUST: exploiting the positional clustering of somatic mutations to identify cancer genes. Bioinformatics 29:2238–2244
41. Maher CA, Kumar-Sinha C, Cao X, Kalyana-Sundaram S, Han B, Jing X, Sam L, Barrette T, Palanisamy N, Chinnaiyan AM (2009) Transcriptome sequencing to detect gene fusions in cancer. Nature 458:97–101
42. Kim D, Pertea G, Trapnell C, Pimentel H, Kelley R, Salzberg SL (2013) TopHat2: accurate alignment of transcriptomes in the presence of insertions, deletions and gene fusions. Genome Biol 14:R36
43. Dobin A, Gingeras TR (2015) Mapping RNA-seq reads with STAR. Curr Protoc Bioinformatics 51:11.14.1–11.14.19
44. Sun J, Nishiyama T, Shimizu K, Kadota K (2013) TCC: an R package for comparing tag count data with robust normalization strategies. BMC Bioinformatics 14:219
45. Forbes SA, Beare D, Gunasekaran P, Leung K, Bindal N, Boutselakis H, Ding M, Bamford S, Cole C, Ward S, Kok CY, Jia M, De T, Teague JW, Stratton MR, McDermott U, Campbell PJ (2015) COSMIC: exploring the world's knowledge of somatic mutations in human cancer. Nucleic Acids Res 43:D805–D811
46. Gao J, Aksoy BA, Dogrusoz U, Dresdner G, Gross B, Sumer SO, Sun Y, Jacobsen A, Sinha R, Larsson E, Cerami E, Sander C, Schultz N (2013) Integrative analysis of complex cancer genomics and clinical profiles using the cBioPortal. Sci Signal 6:pl1
47. Cerami E, Gao J, Dogrusoz U, Gross BE, Sumer SO, Aksoy BA, Jacobsen A, Byrne CJ, Heuer ML, Larsson E, Antipin Y, Reva B, Goldberg AP, Sander C, Schultz N (2012) The cBio cancer genomics portal: an open platform for exploring multidimensional cancer genomics data. Cancer Discov 2:401–404
48. Huang da W, Sherman BT, Lempicki RA (2009) Systematic and integrative analysis of large gene lists using DAVID bioinformatics resources. Nat Protoc 4:44–57
49. Griffith M, Griffith OL, Coffman AC, Weible JV, McMichael JF, Spies NC, Koval J, Das I, Callaway MB, Eldred JM, Miller CA, Subramanian J, Govindan R, Kumar RD, Bose R, Ding L, Walker JR, Larson DE, Dooling DJ, Smith SM, Ley TJ, Mardis ER, Wilson RK (2013) DGIdb: mining the druggable genome. Nat Methods 10:1209–1210
50. Yang W, Soares J, Greninger P, Edelman EJ, Lightfoot H, Forbes S, Bindal N, Beare D, Smith JA, Thompson IR, Ramaswamy S, Futreal PA, Haber DA, Stratton MR, Benes C, McDermott U, Garnett MJ (2013) Genomics of Drug Sensitivity in Cancer (GDSC): a resource for therapeutic biomarker discovery in cancer cells. Nucleic Acids Res 41:D955–D961
51. Lizio M, Harshbarger J, Shimoji H, Severin J, Kasukawa T, Sahin S, Abugessaisa I, Fukuda S, Hori F, Ishikawa-Kato S, Mungall CJ, Arner E, Baillie JK, Bertin N, Bono H, de Hoon M, Diehl AD, Dimont E, Freeman TC, Fujieda K, Hide W, Kaliyaperumal R, Katayama T, Lassmann T, Meehan TF, Nishikata K, Ono H, Rehli M, Sandelin A, Schultes EA, 't Hoen PA, Tatum Z, Thompson M, Toyoda T, Wright DW, Daub CO, Itoh M, Carninci P, Hayashizaki Y, Forrest AR, Kawaji H; FANTOM consortium. (2015) Gateways to the FANTOM5 promoter level mammalian expression atlas. Genome Biol 16:22.
52. Mizuno H, Kitada K, Nakai K, Sarai A (2009) PrognoScan: a new database for meta-analysis of the prognostic value of genes. BMC Med Genet 2:18
53. Kohno T, Ichikawa H, Totoki Y, Yasuda K, Hiramoto M, Nammo T, Sakamoto H, Tsuta K, Furuta K, Shimada Y, Iwakawa R, Ogiwara H, Oike T, Enari M, Schetter AJ, Okayama H, Haugen A, Skaug V, Chiku S, Yamanaka I, Arai Y, Watanabe S, Sekine I, Ogawa S, Harris CC, Tsuda H, Yoshida T, Yokota J, Shibata T (2012) KIF5B-RET fusions in lung adenocarcinoma. Nat Med 18:375–377
54. Takeuchi K, Soda M, Togashi Y, Suzuki R, Sakata S, Hatano S, Asaka R, Hamanaka W, Ninomiya H, Uehara H, Lim Choi Y, Satoh Y, Okumura S, Nakagawa K, Mano H, Ishikawa Y (2012) RET, ROS1 and ALK fusions in lung cancer. Nat Med 18:378–381

55. Lipson D, Capelletti M, Yelensky R, Otto G, Parker A, Jarosz M, Curran JA, Balasubramanian S, Bloom T, Brennan KW, Donahue A, Downing SR, Frampton GM, Garcia L, Juhn F, Mitchell KC, White E, White J, Zwirko Z, Peretz T, Nechushtan H, Soussan-Gutman L, Kim J, Sasaki H, Kim HR, Park SI, Ercan D, Sheehan CE, Ross JS, Cronin MT, Jänne PA, Stephens PJ (2012) Identification of new ALK and RET gene fusions from colorectal and lung cancer biopsies. Nat Med 18:382–384
56. Ju YS, Lee WC, Shin JY, Lee S, Bleazard T, Won JK, Kim YT, Kim JI, Kang JH, Seo JS (2012) A transforming KIF5B and RET gene fusion in lung adenocarcinoma revealed from whole-genome and transcriptome sequencing. Genome Res 22:436–445
57. Tsuta K, Kohno T, Yoshida A, Shimada Y, Asamura H, Furuta K, Kushima R (2014) RET-rearranged non-small-cell lung carcinoma: a clinicopathological and molecular analysis. Br J Cancer 110:1571–1578
58. Imielinski M, Berger AH, Hammerman PS, Hernandez B, Pugh TJ, Hodis E, Cho J, Suh J, Capelletti M, Sivachenko A, Sougnez C, Auclair D, Lawrence MS, Stojanov P, Cibulskis K, Choi K, de Waal L, Sharifnia T, Brooks A, Greulich H, Banerji S, Zander T, Seidel D, Leenders F, Ansén S, Ludwig C, Engel-Riedel W, Stoelben E, Wolf J, Goparju C, Thompson K, Winckler W, Kwiatkowski D, Johnson BE, Jänne PA, Miller VA, Pao W, Travis WD, Pass HI, Gabriel SB, Lander ES, Thomas RK, Garraway LA, Getz G, Meyerson M (2012) Mapping the hallmarks of lung adenocarcinoma with massively parallel sequencing. Cell 150:1107–1120
59. Seo JS, Ju YS, Lee WC, Shin JY, Lee JK, Bleazard T, Lee J, Jung YJ, Kim JO, Shin JY, Yu SB, Kim J, Lee ER, Kang CH, Park IK, Rhee H, Lee SH, Kim JI, Kang JH, Kim YT (2012) The transcriptional landscape and mutational profile of lung adenocarcinoma. Genome Res 22:2109–2119
60. Cancer Genome Atlas Research Network (2014) Comprehensive molecular profiling of lung adenocarcinoma. Nature 511:543–550
61. Fernandez-Cuesta L, Plenker D, Osada H, Sun R, Menon R, Leenders F, Ortiz-Cuaran S, Peifer M, Bos M, Daßler J, Malchers F, Schöttle J, Vogel W, Dahmen I, Koker M, Ullrich RT, Wright GM, Russell PA, Wainer Z, Solomon B, Brambilla E, Nagy-Mignotte H, Moro-Sibilot D, Brambilla CG, Lantuejoul S, Altmüller J, Becker C, Nürnberg P, Heuckmann JM, Stoelben E, Petersen I, Clement JH, Sänger J, Muscarella LA, la Torre A, Fazio VM, Lahortiga I, Perera T, Ogata S, Parade M, Brehmer D, Vingron M, Heukamp LC, Buettner R, Zander T, Wolf J, Perner S, Ansén S, Haas SA, Yatabe Y, Thomas RK (2014) CD74-NRG1 fusions in lung adenocarcinoma. Cancer Discov 4:415–422
62. Jang JS, Lee A, Li J, Liyanage H, Yang Y, Guo L, Asmann YW, Li PW, Erickson-Johnson M, Sakai Y, Sun Z, Jeon HS, Hwang H, Bungum AO, Edell ES, Simon VA, Kopp KJ, Eckloff B, Oliveira AM, Wieben E, Aubry MC, Yi E, Wigle D, Diasio RB, Yang P, Jen J (2015) Common oncogene mutations and novel SND1-BRAF transcript fusion in lung adenocarcinoma from never smokers. Sci Rep 18(5):9755
63. Cancer Genome Atlas Research Network (2012) Comprehensive genomic characterization of squamous cell lung cancers. Nature 489:519–525
64. Kim Y, Hammerman PS, Kim J, Yoon JA, Lee Y, Sun JM, Wilkerson MD, Pedamallu CS, Cibulskis K, Yoo YK, Lawrence MS, Stojanov P, Carter SL, McKenna A, Stewart C, Sivachenko AY, Oh IJ, Kim HK, Choi YS, Kim K, Shim YM, Kim KS, Song SY, Na KJ, Choi YL, Hayes DN, Kim J, Cho S, Kim YC, Ahn JS, Ahn MJ, Getz G, Meyerson M, Park K (2014) Integrative and comparative genomic analysis of lung squamous cell carcinomas in East Asian patients. J Clin Oncol 32:121–128
65. Roth GJ, Binder R, Colbatzky F, Dallinger C, Schlenker-Herceg R, Hilberg F, Wollin SL, Kaiser R (2015) Nintedanib: from discovery to the clinic. J Med Chem 58:1053–1063
66. Durm G, Hanna N (2014) Targeting multiple angiogenic pathways simultaneously: experience with nintedanib in non-small-cell lung cancer. Future Oncol 10:1167–1173
67. Peifer M, Fernández-Cuesta L, Sos ML, George J, Seidel D, Kasper LH, Plenker D, Leenders F, Sun R, Zander T, Menon R, Koker M, Dahmen I, Müller C, Di Cerbo V, Schildhaus HU, Altmüller J, Baessmann I, Becker C, de Wilde B, Vandesompele J, Böhm D, Ansén S, Gabler

F, Wilkening I, Heynck S, Heuckmann JM, Lu X, Carter SL, Cibulskis K, Banerji S, Getz G, Park KS, Rauh D, Grütter C, Fischer M, Pasqualucci L, Wright G, Wainer Z, Russell P, Petersen I, Chen Y, Stoelben E, Ludwig C, Schnabel P, Hoffmann H, Muley T, Brockmann M, Engel-Riedel W, Muscarella LA, Fazio VM, Groen H, Timens W, Sietsma H, Thunnissen E, Smit E, Heideman DA, Snijders PJ, Cappuzzo F, Ligorio C, Damiani S, Field J, Solberg S, Brustugun OT, Lund-Iversen M, Sänger J, Clement JH, Soltermann A, Moch H, Weder W, Solomon B, Soria JC, Validire P, Besse B, Brambilla E, Brambilla C, Lantuejoul S, Lorimier P, Schneider PM, Hallek M, Pao W, Meyerson M, Sage J, Shendure J, Schneider R, Büttner R, Wolf J, Nürnberg P, Perner S, Heukamp LC, Brindle PK, Haas S, Thomas RK (2012) Integrative genome analyses identify key somatic driver mutations of small-cell lung cancer. Nat Genet 44:1104–1110

68. Rudin CM, Durinck S, Stawiski EW, Poirier JT, Modrusan Z, Shames DS, Bergbower EA, Guan Y, Shin J, Guillory J, Rivers CS, Foo CK, Bhatt D, Stinson J, Gnad F, Haverty PM, Gentleman R, Chaudhuri S, Janakiraman V, Jaiswal BS, Parikh C, Yuan W, Zhang Z, Koeppen H, Wu TD, Stern HM, Yauch RL, Huffman KE, Paskulin DD, Illei PB, Varella-Garcia M, Gazdar AF, de Sauvage FJ, Bourgon R, Minna JD, Brock MV, Seshagiri S (2012) Comprehensive genomic analysis identifies SOX2 as a frequently amplified gene in small-cell lung cancer. Nat Genet 44:1111–1116
69. George J, Lim JS, Jang SJ, Cun Y, Ozretić L, Kong G, Leenders F, Lu X, Fernández-Cuesta L, Bosco G, Müller C, Dahmen I, Jahchan NS, Park KS, Yang D, Karnezis AN, Vaka D, Torres A, Wang MS, Korbel JO, Menon R, Chun SM, Kim D, Wilkerson M, Hayes N, Engelmann D, Pützer B, Bos M, Michels S, Vlasic I, Seidel D, Pinther B, Schaub P, Becker C, Altmüller J, Yokota J, Kohno T, Iwakawa R, Tsuta K, Noguchi M, Muley T, Hoffmann H, Schnabel PA, Petersen I, Chen Y, Soltermann A, Tischler V, Choi CM, Kim YH, Massion PP, Zou Y, Jovanovic D, Kontic M, Wright GM, Russell PA, Solomon B, Koch I, Lindner M, Muscarella LA, la Torre A, Field JK, Jakopovic M, Knezevic J, Castaños-Vélez E, Roz L, Pastorino U, Brustugun OT, Lund-Iversen M, Thunnissen E, Köhler J, Schuler M, Botling J, Sandelin M, Sanchez-Cespedes M, Salvesen HB, Achter V, Lang U, Bogus M, Schneider PM, Zander T, Ansén S, Hallek M, Wolf J, Vingron M, Yatabe Y, Travis WD, Nürnberg P, Reinhardt C, Perner S, Heukamp L, Büttner R, Haas SA, Brambilla E, Peifer M, Sage J, Thomas RK (2015) Comprehensive genomic profiles of small cell lung cancer. Nature 524:47–53
70. Govindan R, Ding L, Griffith M, Subramanian J, Dees ND, Kanchi KL, Maher CA, Fulton R, Fulton L, Wallis J, Chen K, Walker J, McDonald S, Bose R, Ornitz D, Xiong D, You M, Dooling DJ, Watson M, Mardis ER, Wilson RK (2012) Genomic landscape of non-small cell lung cancer in smokers and never-smokers. Cell 150:1121–1134
71. Gou LY, Niu FY, Wu YL, Zhong WZ (2015) Differences in driver genes between smoking-related and non-smoking-related lung cancer in the Chinese population. Cancer 121:3069–3079
72. Rizvi NA, Hellmann MD, Snyder A, Kvistborg P, Makarov V, Havel JJ, Lee W, Yuan J, Wong P, Ho TS, Miller ML, Rekhtman N, Moreira AL, Ibrahim F, Bruggeman C, Gasmi B, Zappasodi R, Maeda Y, Sander C, Garon EB, Merghoub T, Wolchok JD, Schumacher TN, Chan TA (2015) Cancer immunology. Mutational landscape determines sensitivity to PD-1 blockade in non-small cell lung cancer. Science 348:124–128
73. Fox JL (2014) Master protocol for squamous cell lung cancer readies for launch. Nat Biotechnol 32:116–118
74. Song M, Kim SH, Yoon SK (2015) Cabozantinib for the treatment of non-small cell lung cancer with KIF5B-RET fusion. An example of swift repositioning. Arch Pharm Res. doi:10.1007/s12272-015-0660-1 [Epub ahead of print]

Chapter 7
Companion Diagnostics

Emi Noguchi

Abstract Recent advances in understanding molecular mechanisms of cancer have already brought about precision medicine that combines the individuals' characteristics or the genetic profiles of individual tumors and treatment selection using companion diagnostics.

This chapter will give a brief summary of regulation of companion diagnostics in the United States, the EU, and Japan and overview of clinical data of CoDx for lung cancer.

Keywords Companion diagnostics • In vitro diagnostics • Personalized medicine • Precision medicine

7.1 Introduction

Precision medicine garnered increasingly more attention worldwide after US President Obama announced the "Precision Medicine Initiative" on January 20, 2015 [1]. Precision medicine is a breakaway from the "one-size-fits-all-approach" designed for the average patient; it is a new approach to disease prevention and treatment that takes into account individual differences in people's genetic information, living environments, and lifestyles. In June 2013, the Cabinet in Japan also adopted "the Revitalization Strategy" that includes the promotion of individualized medicine [2]. Recent advance in understanding the molecular mechanisms of cancer has already brought about several therapies that combine the individuals' characteristics or the genetic profiles of individual tumors and treatment selection using companion diagnostics (CoDx).

CoDx are diagnostic devices used to pre-inspect whether a drug is appropriate for use by the target patient in order to further increase the efficacy and safety of a particular drug. As "companion" suggests, the drug and the diagnostic test are

E. Noguchi (✉)
National Cancer Center Hospital, Breast and Medical Oncology,
5-1-1 Tsukiji, Chuo-ku, Tokyo 104-0045, Japan
e-mail: enoguchi@ncc.go.jp

© Springer Science+Business Media Singapore 2017

Y. Takiguchi (ed.), *Molecular Targeted Therapy of Lung Cancer*,
DOI 10.1007/978-981-10-2002-5_7

developed in parallel. In a narrow sense, diagnostics that were developed independently as in vitro diagnostics (IVD) (diagnostics intended to be used for detecting diseases and not to used directly on the human body) and used for treatment selection in later years are not included in CoDx; yet in a broad sense, all diagnostics that take into account efficacy and safety and are used for treatment selection are referred to as CoDx. CoDx are being used as a routine practice for patients with lung cancer, breast cancer, colorectal cancer, malignant melanoma, and other diseases.

7.2 Overview of Regulation of CoDx

The regulations of IVD, including CoDx, and clinical testing laboratories in each country and region are shown below.

7.2.1 United States

In the United States, the Department of Health and Human Services (HHS) is in charge of government administration related to protecting the health of citizens. Among the federal agencies under HHS jurisdiction, the Food and Drug Administration (FDA) is in charge of regulations related to approval of pharmaceutical products and medical devices, and the Centers for Medicare and Medicaid Services (CMS) is in charge of Medicare and Medicaid and regulations for clinical testing laboratories. Among the FDA departments, the Center for Drug Evaluation and Research (CDER) is in charge of submission and approval of drugs, and the Center for Devices and Radiological Health (CDRH) is in charge of submission and approval of IVD and medical devices.

The Food, Drug, and Cosmetic Act of 1938 (FD&C Act of 1938), an amendment of the Food and Drug Act of 1906, has become the basis of laws on pharmaceutical affairs in the United States. According to the Medical Device Amendments of 1976, which were established after amendments to the FD&C Act of 1938, IVD became a medical device under the charge of the FDA, and the submission, review, and approval/clearance process for Premarket Approval (PMA) or Premarket Notification 510(k) came to be required. In the Medical Device Amendments of 1976, medical devices are classified by risk into classes I, II, and III. The device classification defines the regulatory requirements for a general device type as below. Most Class I devices are exempt from Premarket Notification 510(k). Most Class II devices require Premarket Notification 510(k). High-risk medical devices are classified into class III, and PMA submission is required.

It should be noted that when the Medical Device Amendments of 1976 was enacted, the reagents that had been manufactured internally at hospitals' test laboratories for use within their facilities were found to be excluded from the scope of regulations as laboratory-developed tests (LDTs). Additionally, the Clinical Laboratory Improvement Amendments (CLIA) of 1988 was enacted as a law on regulations of American clinical test laboratories, and LDTs were regulated and managed under the same law. However, only analytical validity is within the range of management, and clinical validity/utility is beyond the scope of management.

Recently, the FDA has tried to strengthen regulations on LDTs and published draft guidances in October 2014 [3, 4]. In the guidance plan, it was sought to impose a PMA submission for an LDT that has the same intended use as CoDx that the FDA gave regulatory approval to and cleared. In December 2014, olaparib was approved in the United States for the treatment of advanced ovarian cancer with *BRCA1/2* mutation [5], and Myriad's genetic test called BRACAnalysis CDx was also given regulatory approval [6]. It was the first time that the FDA approved an LDT under the PMA process and was the first-ever approval of an LDT as CoDx. Similar trends deserve attention in the future.

Table 7.1 shows the guidance list for CoDx in the United States. The FDA encouraged voluntary genomic data submission with the submission of pharmaceutical products since the mid-2000s and expressed the opinion that when reviewing submissions for approval of new drugs, it is useful to simultaneously submit diagnostic agents to screen for patients who will use it effectively prior to administering the new drug to patients. The FDA finalized the guidance on CoDx in July 2014 [7]. In the FDA's guidance, CoDx is defined as follows.

An IVD companion diagnostic device could be essential for the safe and effective use of a corresponding therapeutic product to:

- *Identify patients who are most likely to benefit from a particular therapeutic result of product.*
- *Identify patients likely to be at increased risk for serious adverse reactions as a result of treatment with a particular therapeutic product.*
- *Monitor response to treatment for the purpose of adjusting treatment* (e.g., *schedule, dose, discontinuation) to achieve improved safety or effectiveness.*
- *Identify patients in the population for whom the therapeutic product has been adequately studied, and found safe and effective,* i.e., *there is insufficient*

Table 7.1 List of the guidance for CoDx in the United States

Release date	Title
April 2005	Draft preliminary concept paper: Drug-Diagnostic Co-development Concept Paper
July 2011	Draft Guidance for Industry and Food and Drug Administration Staff: In Vitro Companion Diagnostic Devices
December 2012	Draft Guidance for Industry: Enrichment Strategies for Clinical Trials to Support Approval of Human Drugs and Biological Products
January 2013	Guidance for Industry: Clinical Pharmacogenomics: Premarket Evaluation in Early-Phase Clinical Studies and Recommendations for Labeling
July 2014	Guidance for Industry and Food and Drug Administration Staff: In Vitro Companion Diagnostic Devices [7]

information about the safety and effectiveness of the therapeutic product in any other population.

The guidance describes that as a rule, pharmaceutical products and CoDx need to be developed and approved or cleared at the same time; for CoDx, PMA or 510(k) need to be submitted; and that for both pharmaceutical products and CoDx, regulations on usage of each are stated on their labels (package inserts).

As stated in the guidance, CoDx is not unequivocally defined as class III; however, the CoDx that have been approved by the FDA thus far are all approved as class III. A list of FDA-approved pharmaceutical products and the corresponding CoDx can be found on the website at http://www.fda.gov/CompanionDiagnostics.

Figure 7.1 shows one example of provided CoDx information for medical facilities – the FDA-approved label for osimertinib which was approved in the United States in November 2015 and its CoDx, the cobas® EGFR Mutation Test v2 [8, 9]. In the label for osimertinib, in the Indications and Usage section, the administration targets patients who were found to have the EGFR T790M mutation based on an FDA-approved testing method. The Dosage and Administration section describes FDA-approved testing methods for detecting EGFR T790M mutation, and the Clinical Studies section describes how patients targeted in the clinical trials were patients found to have EGFR T790M mutation through the cobas® EGFR Mutation Test. In the label for cobas® EGFR Mutation Test v2, the Intended Use section describes the definition of EGFR mutation, which established the drugs' (erlotinib or osimertinib) safety and efficacy, and the Clinical Performance Evaluation section describes the clinical trial results that confirmed the clinical performance of the cobas® EGFR Mutation Test.

7.2.2 EU

In the EU, the approval of pharmaceutical products is carried out by the European Medicine Agency (EMA) in a unified manner. The approval of IVD, however, is beyond the charge of the EMA, and based on the In Vitro Diagnostic Medical Device Directive 98/79/CE, they are certified by a notified body certification system. By acquiring the CE mark (standard conformity mark), they are able to be manufactured and marketed within EU member countries, additional countries in the European Economic Area, as well as in Switzerland. In other words, in the EU, there is no established system for the evaluation of pharmaceutical products and CoDx at the same time. Therefore, guidance on CoDx in the EU does not include content on the review and approval of CoDx. However, recently, we have seen movements toward revising regulations, such as recommendations for guidance from the EMA regarding IVD. Table 7.2 shows the guidance list for CoDx in the EU.

Excerpt from the FDA-approved label for osimertinib (Tagrisso®)

1 INDICATIONS AND USAGE

TAGRISSO is indicated for the treatment of patients with metastatic epidermal growth factor receptor (EGFR) T790M mutation-positive non-small cell lung cancer (NSCLC), as detected by an FDA-approved test, who have progressed on or after EGFR tyrosine kinase inhibitor (TKI) therapy.

2 DOSAGE AND ADMINISTRATION

2.1 Patient Selection

Confirm the presence of a T790M EGFR mutation in tumor specimens prior to initiation of treatment with TAGRISSO *[see Indications and Usage (1) and Clinical Studies (14)]*. Information on FDA-approved tests for the detection of T790M mutations is available at http://www.fda.gov/companiondiagnostics.

14 CLINICAL STUDIES

… All patients were required to have EGFR T790M mutation-positive NSCLC as detected by the cobas® EGFR mutation test and received TAGRISSO 80 mg once daily. …

Excerpt from the FDA-approved label for cobas® EGFR Mutation Test v2

Intended Use

The cobas® EGFR Mutation Test v2 is a real -time PCR test for the qualitative detection of defined mutations of the epidermal growth factor receptor (EGFR) gene in DNA derived from formalin-fixed paraffin-embedded tumor tissue (FFPET) from non-small cell lung cancer (NSCLC) patients. The test is intended to aid in identifying patients with NSCLC whose tumors have defined EGFR mutations and for whom safety and efficacy of a drug have been established as follows:

Tarceva® (erlotinib)	Exon 19 deletions and L858R
Tagrisso® (osimertinib)	T790M

Drug safety and efficacy have not been established for the following EGFR mutations also detected by the cobas® EGFR Mutation Test v2:

Tarceva® (erlotinib)	G719X, exon 20 insertions, T790M, S768I and L861Q
Tagrisso® (osimertinib)	G719X, exon 19 deletions, L858R, exon 20 insertions, S768I, and L861Q

Fig. 7.1 Example of provided CoDx information – osimertinib and its CoDx (Adapted from Refs. [8, 9])

Table 7.2 List of the guidance for CoDx in the EU

Release date	Title
June 2010	Draft; Reflection paper on co-development of pharmacogenomics biomarkers and assays in the context of drug development
June 2011	Draft; Reflection paper on methodological issues associated with pharmacogenomic biomarkers in relation to clinical development and patient selection
August 2012	Guideline on the use of pharmacogenetic methodologies in the pharmacokinetic evaluation of medicinal products

7.2.3 *Japan*

In Japan, the Ministry of Health, Labour, and Welfare (MHLW) has jurisdiction over administration related to medicine, and the Pharmaceuticals and Medical Devices Agency (PMDA) carries out approval of pharmaceutical products and medical devices. Japanese laws on pharmaceutical affairs are based on the "Law for Ensuring the Quality, Efficacy, and Safety of Drugs and Medical Devices," which was a 2014 amendment to the "Pharmaceutical Affairs Law" enacted in 1961. In July 2013, the MHLW published guidance on CoDx [10], and now simultaneous development and approval of pharmaceuticals and CoDx are required. The definition of CoDx in the Japanese guidance is nearly the same as in the FDA's guidance.

7.3 CoDx for Lung Cancer

This section will summarize the FDA-approved pharmaceutical products and its corresponding CoDx for lung cancer. Table 7.3 shows the list of the FDA-approved CoDx for lung cancer.

Table 7.3 List of the FDA-approved CoDx for lung cancer

Drug trade name (generic name)	Drug approval date	Device trade name	Device manufacturer	Device approval date
Xalkori (crizotinib)	08/26/2011	VYSIS ALK Break-Apart FISH Probe Kit	Abbott Molecular Inc.	08/26/2011
Tarceva (erlotinib)	11/18/2004	cobas EGFR Mutation Test	Roche Molecular Systems, Inc.	05/14/2013
Gilotrif (afatinib)	07/12/2013	therascreen EGFR RGQ PCR Kit	Qiagen Manchester, Ltd.	07/12/2013
Xalkori (crizotinib)	08/26/2011	VENTANA ALK (D5F3) CDx Assay	Ventana Medical Systems, Inc.	06/12/2015
Iressa (gefitinib)	05/05/2003[a] 07/13/2015	therascreen EGFR RGQ PCR Kit	Qiagen Manchester, Ltd.	07/10/2015
Keytruda (pembrolizumab)	10/02/2015	PD-L1 IHC 22C3 pharmDx	Dako, North America, Inc.	10/02/2015
Tagrisso (osimertinib)	11/13/2015	cobas EGFR Mutation Test v2	Dako, North America, Inc.	11/13/2015

[a]Quote modified from Ref. [77]

7.3.1 *CoDx for Epidermal Growth Factor Receptor-Tyrosine Kinase Inhibitors (EGFR-TKIs)*

The development of first-generation EGFR-TKIs – gefitinib and erlotinib, which have been approved as first-line treatment of advanced NSCLC harboring activating *EGFR* mutations today – was full of twists and turns. When these drugs were introduced into the clinical setting, the mechanism of drug action and the drug targets was not well known. The subsequent discovery of the activating mutations of the *EGFR* gene as a biomarker to select patients who are most likely to have benefit from the therapy have had a great impact on the treatment of non-small cell lung cancer (NSCLC). It has taught us that the identification and validation of biomarkers was important factor in targeted therapy [11].

Second-generation EGFR-TKIs – afatinib and dacomitinib – irreversibly bind to tyrosine kinase of EGFR and other ErbB-family members. Afatinib also has been approved as first-line treatment of advanced NSCLC harboring activating *EGFR* mutations. Dacomitinib is under development. Third-generation EGFR-TKIs – osimertinib, rociletinib (CO-1686), BI 1482694 (previously named HM61713), and other agents – inhibit both *EGFR*-activating and resistance mutations, such as *EGFR* T790M. Osimertinib has been granted accelerated approval for the treatment of advancer NSCLC harboring *EGFR* T790M mutation.

7.3.1.1 Gefitinib (Iressa)/Therascreen® EGFR RGQ PCR Kit

Gefitinib was first-ever approved in Japan on July 2002 for "the treatment of the patients with inoperable or recurrent NSCLC" based on response rates from two phase II IDEAL 1 and 2 studies [12, 13], which were conducted with unselected NSCLC populations. In the United States, gefitinib had originally been granted accelerated approval in May 2003 for "patients with locally advanced or metastatic NSCLC after failure of both platinum-based and docetaxel" [14] based on phase II IDEAL 2 study. Other data included results of the phase III INTACT 1 and 2 studies [15, 16]; however, results from these studies showed no benefit from adding gefitinib to chemotherapy (cisplatin plus gemcitabine or carboplatin plus paclitaxel, respectively) in unselected NSCLC patients. Consequently, gefitinib was only approved for use as monotherapy. Confirmatory trial was conducted but failed to verify clinical benefit from gefitinib; the phase III ISEL study which investigated the efficacy of gefitinib as second-line or third-line treatment in unselected patients did not demonstrate improved overall survival (OS) for gefitinib compared with best supportive care [17], although OS benefit was observed in the subgroup of Asian patients. On June 2005, following that erlotinib had been approved by the FDA based on phase III NCIC CTG BR.21 study [18], the FDA limited the indication to "patients who were currently receiving and benefiting, or had previously benefited, from gefitinib" [19]. Finally, on April 2012, the FDA withdrew approval of a new drug application for gefitinib [20].

In parallel with these clinical trials, many biomarkers were tested such as EGFR expression detected by immunohistochemistry (IHC), *EGFR* amplification or polysomy detected by fluorescence in situ hybridization (FISH) or quantitative polymerase chain reaction (PCR), or activating *EGFR* mutations. It was not until 2004 when it was discovered that NSCLC patients with activating *EGFR* mutations in the EGFR kinase domain yielded better response to EGFR-TKI [21–23]. The discovery of *EGFR* mutations has great implications in NSCLC treatment. *EGFR* mutations are more common with adenocarcinoma histology, Asian, women, and nonsmokers [23–25], which brought the Phase III IPASS study [26]. The IPASS study was conducted in East Asia in NSCLC patients with clinical characteristics known to be associated with higher prevalence of *EGFR* mutations (adenocarcinoma histology, never of light smoker) and met the primary endpoint of demonstrating non-inferiority of gefitinib compared with carboplatin plus paclitaxel as first-line treatment in these patients in terms of progression-free survival (PFS), but Kaplan-Meier curves for PFS crossed, indicating nonproportional hazards. What was important was the result of subgroup analyses. In the subgroup of patients who were positive for *EGFR* mutations detected by using the scorpion ARMS method, PFS was significantly longer among those who received gefitinib than among those who received chemotherapy, whereas in the subgroup of patients who were negative for the mutation, PFS was significantly longer among those who received chemotherapy. In Korea, the First-SIGNAL study was conducted in NSCLC patients with similar clinical characteristics to the IPASS study but failed to demonstrate superiority of gefitinib compared with gemcitabine plus cisplatin as first-line treatment in terms of OS [27]. In the subgroup of patients with negative for EGFR mutations detected by using direct sequencing method, overall response rate was relatively high, which suggests higher false-negative rates.

After that, the phase III WJTOG3405 and NEJ002 studies comparing gefitinib to chemotherapy as first-line treatment in patients selected for the presence of tumor harboring EGFR mutations were conducted [28, 29]. In Japan, approval application of gefitinib changed to "the treatment of the patients with inoperable or recurrent NSCLC with EGFR mutations" based on the subgroup analysis from IPASS study, WJTOG3405, and NEJ002 study on November 2011 [30]. In USA, the FDA approved gefitinib in July 2015 [31] for "the first-line treatment of patients with metastatic non-small cell lung cancer (NSCLC) whose tumors harbor specific types of epidermal growth factor receptor (EGFR) gene mutations, as detected by an FDA-approved test" based on the result from the single-arm phase IV IFUM study [32] and the subgroup analysis from IPASS study. The IFUM study, tumor samples were tested retrospectively using the therascreen® EGFR RGQ PCR Kit, which was a real-time PCR test for qualitative detection of exon 19 deletions and exon 21 (L858R) substitution mutations of the *EGFR* gene in DNA derived from formalin-fixed paraffin-embedded (FFPE) NSCLC tumor tissue. The FDA also approved the therascreen® EGFR RGQ PCR as CoDx for gefitinib in July 2015 [33].

7.3.1.2 Erlotinib (Tarceva)/Cobas EGFR Mutation Test: Cobas® EGFR Mutation Test v2

Erlotinib was originally approved by the FDA in November 2004 [34] for "the treatment of patients with locally advanced or metastatic non-small cell lung cancer after failure of at least one prior chemotherapy regimen" based on data from the phase III BR.21 study comparing erlotinib to platinum-based doublet chemotherapy in term of OS in unselected previously treated patients with advanced NSCLC [35]. In the BR.21 study, the expression of EGFR protein but *EGFR* mutations had been reported that was associated with response [36]. There was possibly selection bias since the EGFR expression was evaluated in less than half of the study population. As well as gefitinib, other data included results of the phase III TALENT and TRIBUTE studies was also submitted; however, results from these studies showed no benefit from adding erlotinib to chemotherapy (cisplatin plus gemcitabine or carboplatin plus paclitaxel, respectively) in unselected NSCLC patients [37, 38]. Consequently, erlotinib was also only approved for use as monotherapy.

In case of erlotinib, probably because the initial phase III BR.21 study met the primary endpoint in unselected NSCLC patients, there was not so much confirmatory trial as gefitinib. In May 2013, the FDA approved erlotinib for "the first-line treatment of patients with metastatic non-small cell lung cancer (NSCLC) whose tumors have epidermal growth factor receptor (EGFR) exon 19 deletions or exon 21 (L858R) substitution mutations as detected by an FDA-approved test [39]." This approval was based on data from the phase III EURTAC study comparing erlotinib to platinum-based standard chemotherapy in untreated patients with advanced *EGFR* mutation-positive NSCLC whose tumors had EGFR exon 19 deletions or exon 21 (L858R) substitution mutations determined by a clinical trial assay [40]. Tumor samples were retrospectively tested by the cobas EGFR Mutation Test, which was a real-time PCR test for the qualitative detection of defined mutations of *EGFR* gene in DNA derived from FFPE tissue. This indication for erlotinib was approved concurrently with the cobas EGFR Mutation Test, the CoDx for patient selection [9].

7.3.1.3 Afatinib (Gilotrif)/Therascreen EGFR RGQ PCR Kit

Afatinib is a second-generation EGFR-TKI, which was approved by the FDA for "the first-line treatment of patients with metastatic non-small cell lung cancer (NSCLC) whose tumors have epidermal growth factor receptor (EGFR) exon 19 deletions or exon 21 (L858R) substitution mutations as detected by an FDA-approved test" in July 2013 [41]. This approval was based on data from the phase III LUX-Lung 3 study comparing afatinib to cisplatin plus pemetrexed in patients with metastatic NSCLC with *EGFR* mutations determined by a clinical trial assay [42]. Tumor samples were retrospectively tested by the therascreen EGFR RGQ PCR Kit. Concurrent with this application, the FDA approved the therascreen EGFR RGQ PCR Kit for detection of EGFR exon 19 deletions or exon 21 (L858R) substitution mutations as the CoDx [43].

7.3.1.4 Osimertinib (Tagrisso)/Cobas® EGFR Mutation Test v2

Osimertinib is a third-generation EGFR-TKI. The FDA granted accelerated approval to osimertinib for "the treatment of patients with metastatic epidermal growth factor receptor (EGFR) T790M mutation-positive non-small cell lung cancer (NSCLC), as detected by an FDA-approved test, who have progressed on or after EGFR tyrosine kinase inhibitor (TKI) therapy" in November 2015 [8]. This accelerated approval was based on objective response rate from two single-arm studies (AURA Extension and AURA 2) in patients with metastatic *EGFR* T790M mutation-positive NSCLC who had progressed on prior systemic therapy, including an EGFR-TKI. All patients were required to have *EGFR* T790M mutation-positive NSCLC as detected by the cobas® EGFR Mutation Test. The FDA simultaneously approved the cobas® EGFR Mutation Test v2.

7.3.1.5 Liquid Biopsy-Based CoDx for *EGFR* Mutations Detection

On November 2014, the EMA extended the label of gefitinib to include the detection of *EGFR* mutations in circulating tumor DNA (ctDNA) from a blood (plasma) sample if a tumor sample is not evaluable [44]. This update was based on data from the IFUM study, which assessed *EGFR* mutation status in tumor and ctDNA samples derived from plasma, using the therascreen EGFR RGQ Plasma PCR kit. Following that, on January 2015, the therascreen EGFR RGQ Plasma PCR kit received CE mark as the first-ever liquid biopsy-based CoDx for *EGFR* mutations detection.

The utility of liquid biopsy methods is still being developed, and it is to be noted that the current gold standard in detecting *EGFR* mutation is tissue biopsy. The liquid biopsy methods have not yet been approved as CoDx for lung cancer in the USA and Japan.

7.3.2 *CoDx for Anaplastic Lymphoma Kinase (ALK) Inhibitors*

Crizotinib is first-generation ALK inhibitor approved for *ALK*-positive NSCLC. Second-generation ALK inhibitors – ceritinib, alectinib, and brigatinib – are designed to have more potency against ALK. Ceritinib has received accelerated approval, alectinib has been approved, and brigatinib is under development. Third-generation ALK inhibitors – lorlatinib and other agents – are designed to have efficacy against *ALK* mutants. Lorlatinib and other agents are also under development.

7.3.2.1 Crizotinib (Xalkori)/The Vysis ALK Break-Apart FISH Probe Kit

In August 2011, the FDA granted accelerated approval to crizotinib for "the treatment of patients with locally advanced or metastatic non-small cell lung cancer (NSCLC) that is anaplastic lymphoma kinase (ALK)-positive as detected by an FDA-approved test [45]." This accelerated approval was based on response rates from two single-arm PROFILE 1005 and 1001 [46] studies.

In PROFILE1005 study, *ALK*-positive NSCLC was identified using the Vysis ALK Break-Apart FISH Probe Kit, which was a qualitative test to detect rearrangements involving the *ALK* gene via FISH in FFPE NSCLC tissue specimens. In PROFILE 1001 study, *ALK*-positive NSCLC was identified using a number of local clinical trial assays. The FDA approved the Vysis ALK Break-Apart FISH Probe Kit to detect the presence of an *ALK* gene rearrangement concurrently with the approval of crizotinib.

Subsequently, on November 2013, the FDA has granted regular approval for crizotinib for "the treatment of patients with metastatic non-small cell lung cancer (NSCLC) whose tumors are anaplastic lymphoma kinase (ALK)-positive as detected by an FDA-approved test" based on data from the phase III PROFILE 1007 study comparing crizotinib to pemetrexed or docetaxel in patients with *ALK*-positive metastatic NSCLC who were previously treated with one platinum-based chemotherapy regimen [47]. On September 2015, the FDA approved the label update including data from the phase III PROFILE 1014 study comparing crizotinib to pemetrexed with cisplatin or carboplatin in previously untreated ALK-positive metastatic NSCLC patients [48]. In both PROFILE 1007 and 1014 studies, patients were required to have *ALK*-positive NSCLC as identified by the Vysis ALK Break-Apart FISH Probe Kit.

7.3.2.2 Ceritinib (Zykadia)

Ceritinib is a second-generation ALK inhibitor. The FDA approved ceritinib under accelerated approval for "anaplastic lymphoma kinase (ALK)-positive metastatic non-small cell lung cancer (NSCLC) who have progressed on or are intolerant to crizotinib" on April 2014 [49]. This approval was based on data from the phase I ASCEND-1 study [50]. The X2101 study enrolled patients who had previously received crizotinib; therefore, all patients had been tested for *ALK* rearrangement by the Vysis ALK Break-Apart FISH Probe Kit. In X2101 study, ALK positivity was verified retrospectively by local test. This indication for ceritinib was approved without CoDx.

7.3.2.3 Alectinib (Alecensa)

Alectinib was first approved in Japan on July 2014 for "*ALK* fusion gene-positive, unresectable, recurrent or advanced non-small cell lung cancer" [51] based on response rates from the Japanese phase I/II AF-001JP study [52]. In AF-001JP study, ALK inhibitor-naïve patients were enrolled, and *ALK* fusion gene was detected using IHC for FFPE specimens, reverse transcription PCR (RT-PCR) for unfixed specimens, or both [53]. Alectinib has been approved as first-line treatment for *ALK* rearrangement NSCLC, and both IHC and RT-PCR methods were approved as CoDx for alectinib in Japan.

The FDA granted accelerated approval to alectinib for "the treatment of patients with anaplastic lymphoma kinase (ALK)-positive, metastatic non-small cell lung cancer (NSCLC) who have progressed on or are intolerant to crizotinib" on December 2015 [54]. This approval was based on data from the two phase II studies (NP28761 [55] and NP28673 [56]), both of which enrolled patients who had crizotinib-resistant *ALK* rearrangement NSCLC; therefore, all patients had been tested for *ALK* rearrangement by the Vysis ALK Break-Apart FISH Probe Kit. This indication for alectinib is USA was approved without CoDx.

7.3.2.4 Optimal *ALK* Rearrangement Detection Methods

ALK rearrangement can be detected by several methods, including FISH, IHC, and RT-PCR [57]. The College of American Pathologists, International Association for the Study of Lung Cancer, and Association for Molecular Pathology published molecular testing guidelines for EGFR- and ALK-TKIs [58]. It is recommended that laboratories should use an ALK FISH assay for selecting patients for ALK-TKI therapy; ALK immunohistochemistry may be used as a screening method to select specimens for ALK FISH testing. RT-PCR is no longer recommended as an alternative to FISH for selecting patients for treatment with ALK-TKI.

7.3.3 CoDx for Programmed Death-1/Programmed Death-Ligand 1 (PD-1/PD-L1) Inhibitors

Immune checkpoint inhibitors, especially antibodies to PD-1/PD-L1, have provided promising clinical activity in NSCLC. Expression of PD-L1 in tumor tissue, explored in the majority of cases by IHC, has emerged as a putative biomarker in several studies. Table 7.4 shows PD-L1 expression levels and efficacy of anti-PD-1/PD-L1 inhibitors for NSCLC.

Table 7.4 PD-L1 expression and the efficacy of anti-PD-1/PD-L1 therapies in NSCLC

Drug (target)		Pembrolizumab (PD-1)				Nivolumab (PD-1)				Atezolizumab [MPDL3280A] (PD-L1)				Durvalumab [MEDI4736] (PD-L1)
Device Manufacturer		Dako				Dako				Ventana				Ventana
mAb clone		22C3				28–8				SP142				SP263
Staining cells scored		Tumor cells (and stroma)				Tumor cell membrane				Tumor or tumor-infiltrating immune cells				Tumor cell membrane
Tissue		Recent				Archival				Archival/recent				Archival/recent
Setting		1st [60][a]		2nd ++ [60][a]		1st [79]	2nd ++ [67, 68]			1st [80]		2nd ++ [80]		2nd ++ [73]
Cut-point		50 %	1 %	50 %	1 %	5 %	1 %	5 %	10 %	TC3 or IC3	TC2/3 or IC2/3	TC3 or IC3	TC2/3 or IC2/3	25 %
ORR in PD-L1 + % (*n*)	non-sq	47 (8/17)	29 (14/48)	41 (25/61)	26 (37/142)	50 (5/10)	31 (38/123)	36 (34/95)	37 (32/86)	26 (17/65)	19 (26/139)	38 (9/24)	22 (11/50)	27 (23)
	sq						17 (11/63)	21 (9/42)	19 (7/36)					
ORR in PD-L1– % (*n*)	non-sq	18 (7/38)	14 (1/7)	13 (14/104)	9 (2/23)	0 (0/7)	9 (10/108)	10 (14/136)	11 (16/145)	–	–	10 (12/120)	11 (10/94)	5 (5)
	sq						17 (9/54)	15 (11/75)	16 (13/81)					
Regulation		CoDx				Complementary Dx				Possible CoDx				Possible CoDx

Adapted from Ref. [78]

mAb monoclonal antibody, *non-sq* non-squamous cell non-small cell lung cancer, *sq* squamous-cell non-small cell lung cancer, *IC* immune cells

[a]All treated patients from the validation sets, including patients without baseline measurable disease

7.3.3.1 Pembrolizumab (Keytruda)/PD-L1 IHC 22C3 PharmDx

Pembrolizumab is a monoclonal antibody that binds to PD-1 receptor and blocks the interaction between PD-1 and its ligands, PD-L1 and PD-L2, releasing PD-1 immune-checkpoint pathway-mediated inhibition of the tumor immune response. The FDA granted accelerated approval to pembrolizumab for "the treatment of patients with metastatic non-small cell lung cancer (NSCLC) whose tumors express PD-L1 as determined by an FDA-approved test with disease progression on or after platinum-containing chemotherapy" on October 2015 [59]. This approval was based on data from the phase I KEYNOTE-001 study [60].

The KEYNOTE-001 trial validated the PD-L1 IHC 22C3 pharmDx assay as well as assessed safety and efficacy of pembrolizumab and demonstrated a correlation between PD-L1 expression and clinical efficacy in NSCLC. The FDA also approved the PD-L1 IHC 22C3 pharmDx as the first CoDx designed to assess the PD-L1 expression in NSCLC [61].

Several phase III confirmatory clinical trials of pembrolizumab are ongoing, and all studies enroll patients with PD-L1-positive tumor [62–64].

7.3.3.2 Nivolumab (Opdivo)

Nivolumab is a monoclonal antibody that binds to PD-1 receptor. The FDA approved nivolumab for "the treatment of patients with metastatic squamous non-small cell lung cancer (NSCLC) with progression on or after platinum-based chemotherapy" on April 2015 [65]. Subsequently, a new indication for "the treatment of patients with metastatic non-squamous non-small cell lung cancer (NSCLC) with progression on or after platinum-based chemotherapy" were provided in October 2015 [66]. These approvals were based on data from the phase III CheckMate 017 and 057 studies [67, 68].

In the CheckMate 017 study, PD-L1 expression levels seemed not to correlate with increased efficacy of nivolumab in patients with squamous-cell NSCLC. However, the CheckMate 057 study demonstrated improved efficacy for nivolumab in patients with non-squamous cell NSCLC whose tumor expressed PD-L1, across all the PD-L1 expression levels (at least 1, 5, or 10 % of tumor cells with PD-L1 staining). Based on these findings, the FDA approved nivolumab for non-squamous cell NSCLC regardless of tumor PD-L1 expression and also approved the PD-L1 IHC 28–8 pharmDx which was used to assess PD-L1 expression in the CheckMate 057 study as a complementary diagnostic not a companion diagnostic at the same time [69].

The complementary diagnostic is recognized that it is not typically linked to a specific drug but rather to a class of drugs, in order to improve development in the field of personalized medicine and is distinct from companion diagnostic, which is essential for safe and effective use of a drug, although the FDA have not issued the guidance for complementary diagnostics. In the label for PD-L1 IHC 28–8 pharmDx [70], the Intended Use section describes that "PD-L1 expression as detected by PD-L1 IHC 28-8 pharmDx in non-squamous NSCLC may be associated with enhanced survival from OPDIVO® (nivolumab)."

7.3.3.3 Atezolizumab (MPDL3280A)

Atezolizumab is an agent under development, a monoclonal antibody that binds to PD-L1 and prevents it from binding to PD-1 and B7.1. Preliminary results from the phase II POPLAR study evaluating the efficacy and safety of atezolizumab compared with docetaxel in patients with previously treated NSCLC have been presented at the European Cancer Congress 2015 [71]. The PD-L1 expression on tumor cells (TCs) or tumor-infiltrating immune cells (ICs) was associated with improvement in survival with atezolizumab. Preliminary results from another phase II BIRCH trial evaluating the safety and efficacy of atezolizumab in patients with PD-L1-positive NSCLC have also been presented at the European Cancer Congress 2015 [72]. In the BIRCH trial, PD-L1 positivity was defined as at least 5 percent of TC or IC expression for PD-L1.

7.3.3.4 Durvalumab (MEDI4736)

Durvalumab is an agent under development, a monoclonal antibody that binds to the PD-L1. Preliminary results from the NSCLC cohort of the phase I study evaluating the safety and tolerability of durvalumab have been presented at the 2015 American Society of Clinical Oncology (ASCO) Annual Meeting [73]. PD-L1 positivity which was defined as at least 25 % staining on tumor cell membrane was associated with response.

7.3.3.5 Limitations of the PD-L1 Biomarker

Several potential limitations with assessment of PD-L1 expression are acknowledged [74]; intratumoral heterogeneity in PD-L1 expression in NSCLC is well known, especially when tumor samples are obtained by a needle biopsy and cannot be representative for the entire tumor; timing of tumor biopsy could also influence PD-L1 expression, since PD-L1 expression may vary over time and treatment.

And, as well as the CoDx for EGFR-TKIs, a unique antibody utilized for IHC is developing for each PD-1 or PD-L1 checkpoint inhibitor. Furthermore, for each of the agents, the definition and threshold of PD-L1 positivity is different (see Table 7.4). It is not practical and may be impossible to use each of the diagnostic tests to select one of the several available targeted agents in the same class. To overcome these circumstances, the US FDA, the American Association for Cancer Research (AACR), and ASCO convened a workshop "Complexities in Personalized Medicine: Harmonizing Companion Diagnostics Across a Class of Targeted Therapies," and a blueprint proposal was developed by pharmaceutical and diagnostic companies

[75]. Their blueprint aims to compare the analytical performance of each diagnostic test and standardize. The International Association for the Study of Lung Cancer (IASLC) is coordinating the PD-L1 Blueprint Project and is planning a new PD-L1 Characterization Project ("PCP-Study") [76].

References

1. The White House (2015) FACT SHEET: President Obama's precision medicine initiative. Available at https://www.whitehouse.gov/the-press-office/2015/01/30/fact-sheet-president-obama-s-precision-medicine-initiative. Accessed 31 Mar 2016
2. The Prime Minister of Japan and His Cabinet (2013) Japan revitalization strategy. Available at https://www.kantei.go.jp/jp/singi/keizaisaisei/pdf/en_saikou_jpn_hon.pdf. Accessed 31 Mar 2016
3. The Food and Drug Administration (2015) Draft Guidance for Industry, Food and Drug Administration Staff, and Clinical Laboratories: Framework for Regulatory Oversight of Laboratory Developed Tests (LDTs). Available at http://www.fda.gov/downloads/medicaldevices/deviceregulationandguidance/guidancedocuments/ucm416685.pdf. Accessed 31 Mar 2016
4. The Food and Drug Administration (2015) Draft Guidance for Industry, Food and Drug Administration Staff, and Clinical Laboratories: FDA Notification and Medical Device Reporting for Laboratory Developed Tests (LDTs). Available at http://www.fda.gov/downloads/medicaldevices/deviceregulationandguidance/guidancedocuments/ucm416684.pdf. Accessed 31 Mar 2016
5. Food and Drug Administration (2014) LYNPARZA. In: Drugs@FDA: FDA approved drug products. Available at http://www.accessdata.fda.gov/drugsatfda_docs/nda/2014/206162Orig1s000TOC.cfm. Accessed 31 Mar 2016
6. Food and Drug Administration (2014) BRACANALYSIS CDX. In: Premarket Approval (PMA). Available at http://www.accessdata.fda.gov/scripts/cdrh/cfdocs/cfPMA/pma.cfm?start_search=1&PMANumber=P140020. Accessed 31 Mar 2016
7. The Food and Drug Administration (2014) Guidance for Industry and Food and Drug Administration Staff: in vitro companion diagnostic devices. Available at http://www.fda.gov/downloads/medicaldevices/deviceregulationandguidance/guidancedocuments/ucm262327.pdf. Accessed 31 Mar 2016
8. Food and Drug Administration (2015) TAGRISSO. In: Drugs@FDA: FDA approved drug products. Available at http://www.accessdata.fda.gov/drugsatfda_docs/label/2015/208065s000lbl.pdf. Accessed 31 Mar 2016
9. Food and Drug Administration (2015) COBAS EGFR MUTATION TEST V2. In: Premarket Approval (PMA). Available at http://www.accessdata.fda.gov/scripts/cdrh/cfdocs/cfPMA/pma.cfm?start_search=1&pmanumber=p120019. Accessed 31 Mar 2016
10. The Ministry of Health, Labour and Welfare (2013) Notification on approval application for in vitro companion diagnostics and corresponding therapeutic products. In: In vitro companion diagnostic devices WG. Available via the Pharmaceuticals and Medical Devices Agency. https://www.pmda.go.jp/files/000153914.pdf. Accessed 31 Mar 2016
11. Pao W, Miller VA (2005) Epidermal growth factor receptor mutations, small-molecule kinase inhibitors, and non-small-cell lung cancer: current knowledge and future directions. J Clin Oncol 23(11):2556–2568
12. Fukuoka M, Yano S, Giaccone G et al (2003) Multi-institutional randomized phase II trial of gefitinib for previously treated patients with advanced non-small-cell lung cancer (The IDEAL 1 Trial) [corrected]. J Clin Oncol 21(12):2237–2246

13. Kris MG, Natale RB, Herbst RS et al (2003) Efficacy of gefitinib, an inhibitor of the epidermal growth factor receptor tyrosine kinase, in symptomatic patients with non-small cell lung cancer: a randomized trial. JAMA 290(16):2149–2158
14. Food and Drug Administration (2003) IRESSA. In: Drugs@FDA: FDA approved drug products. Available at http://www.accessdata.fda.gov/drugsatfda_docs/nda/2003/021399_iressa.cfm. Accessed 31 Mar 2016
15. Giaccone G, Herbst RS, Manegold C et al (2004) Gefitinib in combination with gemcitabine and cisplatin in advanced non-small-cell lung cancer: a phase III trial—INTACT 1. J Clin Oncol 22(5):777–784
16. Herbst RS, Giaccone G, Schiller JH et al (2004) Gefitinib in combination with paclitaxel and carboplatin in advanced non-small-cell lung cancer: a phase III trial—INTACT 2. J Clin Oncol 22(5):785–794
17. Thatcher N, Chang A, Parikh P et al (2005) Gefitinib plus best supportive care in previously treated patients with refractory advanced non-small-cell lung cancer: results from a randomised, placebo-controlled, multicentre study (Iressa Survival Evaluation in Lung Cancer). Lancet 366(9496):1527–1537
18. Food and Drug Administration (2004) TARCEVA. In: Drugs@FDA: FDA approved drug products. Available at http://www.accessdata.fda.gov/drugsatfda_docs/nda/2004/21-743_Tarceva.cfm. Accessed 31 Mar 2016
19. Food and Drug Administration (2005) IRESSA. In: Drugs@FDA: FDA approved drug products. Available at http://www.accessdata.fda.gov/drugsatfda_docs/appletter/2005/021399s008ltr.pdf. Accessed 31 Mar 2016
20. Federal Register (2012) AstraZeneca Pharmaceuticals LP; withdrawal of approval of a new drug application for IRESSA. Available at https://www.federalregister.gov/articles/2012/04/25/2012-9944/astrazeneca-pharmaceuticals-lp-withdrawal-of-approval-of-a-new-drug-application-for-iressa#h-6. Accessed 31 Mar 2016
21. Lynch TJ, Bell DW, Sordella R et al (2004) Activating mutations in the epidermal growth factor receptor underlying responsiveness of non-small-cell lung cancer to gefitinib. N Engl J Med 350(21):2129–2139
22. Paez JG, Jänne PA, Lee JC et al (2004) EGFR mutations in lung cancer: correlation with clinical response to gefitinib therapy. Science 304(5676):1497–1500
23. Pao W, Miller V, Zakowski M et al (2004) EGF receptor gene mutations are common in lung cancers from "never smokers" and are associated with sensitivity of tumors to gefitinib and erlotinib. Proc Natl Acad Sci U S A 101(36):13306–13311
24. Kosaka T, Yatabe Y, Endoh H et al (2004) Mutations of the epidermal growth factor receptor gene in lung cancer: biological and clinical implications. Cancer Res 64(24):8919–8923
25. Shigematsu H, Lin L, Takahashi T et al (2005) Clinical and biological features associated with epidermal growth factor receptor gene mutations in lung cancers. J Natl Cancer Inst 97(5):339–346
26. Mok TS, Wu YL, Thongprasert S et al (2009) Gefitinib or carboplatin-paclitaxel in pulmonary adenocarcinoma. N Engl J Med 361(10):947–957
27. Han JY, Park K, Kim SW et al (2012) First-SIGNAL: first-line single-agent iressa versus gemcitabine and cisplatin trial in never-smokers with adenocarcinoma of the lung. J Clin Oncol 30(10):1122–1128
28. Mitsudomi T, Morita S, Yatabe Y et al (2010) Gefitinib versus cisplatin plus docetaxel in patients with non-small-cell lung cancer harbouring mutations of the epidermal growth factor receptor (WJTOG3405): an open label, randomised phase 3 trial. Lancet Oncol 11(2):121–128
29. Maemondo M, Inoue A, Kobayashi K et al (2010) Gefitinib or chemotherapy for non-small-cell lung cancer with mutated EGFR. N Engl J Med 362(25):2380–2388
30. Pharmaceuticals and Medical Devices Agency (2011) IRESSA: review report. [Japanese] Available at http://www.pmda.go.jp/drugs/2011/P201100179/670227000_21400AMY00188_A100_1.pdf. Accessed 31 Mar 2016

31. Food and Drug Administration (2015) IRESSA. In: Drugs@FDA: FDA approved drug products. Available at http://www.accessdata.fda.gov/drugsatfda_docs/label/2015/206995s000lbl.pdf. Accessed 31 Mar 2016
32. Douillard JY, Ostoros G, Cobo M et al (2014) First-line gefitinib in Caucasian EGFR mutation-positive NSCLC patients: a phase-IV, open-label, single-arm study. Br J Cancer 110(1):55–62
33. Food and Drug Administration (2015) therascreen® EGFR RGQ PCR kit. In: Premarket Approval (PMA). Available at http://www.accessdata.fda.gov/scripts/cdrh/cfdocs/cfPMA/pma.cfm?id=351444. Accessed 31 Mar 2016
34. Food and Drug Administration (2004) TARCEVA. In: Drugs@FDA: FDA approved drug products. Available at http://www.accessdata.fda.gov/drugsatfda_docs/nda/2004/21-743_Tarceva.cfm. Accessed 31 Mar 2016
35. Shepherd FA, Rodrigues Pereira J et al (2005) Erlotinib in previously treated non-small-cell lung cancer. N Engl J Med. 353(2):123–132
36. Tsao MS, Sakurada A, Cutz JC et al (2005) Erlotinib in lung cancer – molecular and clinical predictors of outcome. N Engl J Med 353(2):133–144
37. Gatzemeier U, Pluzanska A, Szczesna A et al (2007) Phase III study of erlotinib in combination with cisplatin and gemcitabine in advanced non-small-cell lung cancer: the Tarceva Lung Cancer Investigation Trial. J Clin Oncol 25(12):1545–1552
38. Herbst RS, Prager D, Hermann R et al (2005) TRIBUTE: a phase III trial of erlotinib hydrochloride (OSI-774) combined with carboplatin and paclitaxel chemotherapy in advanced non-small-cell lung cancer. J Clin Oncol 23(25):5892–5899
39. Food and Drug Administration (2013) TARCEVA. In: Drugs@FDA: FDA approved drug products. Available at http://www.accessdata.fda.gov/drugsatfda_docs/appletter/2013/021743Orig1s018ltr.pdf. Accessed 31 Mar 2016
40. Rosell R, Carcereny E, Gervais R et al (2012) Erlotinib versus standard chemotherapy as first-line treatment for European patients with advanced EGFR mutation-positive non-small-cell lung cancer (EURTAC): a multicentre, open-label, randomised phase 3 trial. Lancet Oncol 13(3):239–246
41. Food and Drug Administration (2013) GILOTRIF. In: Drugs@FDA: FDA approved drug products. Available at http://www.accessdata.fda.gov/drugsatfda_docs/nda/2013/201292Orig1s000TOC.cfm. Accessed 31 Mar 2016
42. Sequist LV, Yang JC, Yamamoto N et al (2013) Phase III study of afatinib or cisplatin plus pemetrexed in patients with metastatic lung adenocarcinoma with EGFR mutations. J Clin Oncol 31(27):3327–3334
43. Food and Drug Administration (2015) therascreen® EGFR RGQ PCR kit. In: Premarket Approval (PMA). Available at http://www.accessdata.fda.gov/scripts/cdrh/cfdocs/cfTopic/pma/pma.cfm?num=P120022. Accessed 31 Mar 2016
44. European Medicines Agency (2014) IRESSA: summary of product characteristics. Available at http://www.ema.europa.eu/docs/en_GB/document_library/EPAR_-_Product_Information/human/001016/WC500036358.pdf. Accessed 31 Mar 2016
45. Food and Drug Administration (2011) XALKORI. In: Drugs@FDA: FDA approved drug products. Available at http://www.accessdata.fda.gov/drugsatfda_docs/nda/2011/202570Orig1s000TOC.cfm. Accessed 31 Mar 2016
46. Kwak EL, Bang YJ, Camidge DR et al (2010) Anaplastic lymphoma kinase inhibition in non-small-cell lung cancer. N Engl J Med 363(18):1693–1703
47. Food and Drug Administration (2013) XALKORI. In: Drugs@FDA: FDA approved drug products. Available at http://www.accessdata.fda.gov/drugsatfda_docs/appletter/2013/202570Orig1s006ltr.pdf. Accessed 31 Mar 2016
48. Food and Drug Administration (2015) XALKORI. In: Drugs@FDA: FDA approved drug products. Available at http://www.accessdata.fda.gov/drugsatfda_docs/appletter/2015/202570Orig1s014ltr.pdf. Accessed 31 Mar 2016

49. Food and Drug Administration (2014) ZYKADIA. In: Drugs@FDA: FDA approved drug products. Available at http://www.accessdata.fda.gov/drugsatfda_docs/nda/2014/205755Orig1s000TOC.cfm. Accessed 31 Mar 2016
50. Shaw AT, Kim DW, Mehra R et al (2014) Ceritinib in ALK-rearranged non-small-cell lung cancer. N Engl J Med 370(13):1189–1197
51. Pharmaceuticals and Medical Devices Agency (2014) ALECENSA: review report. Available at http://www.pmda.go.jp/files/000208811.pdf. Accessed 31 Mar 2016
52. Seto T, Kiura K, Nishio M et al (2013) CH5424802 (RO5424802) for patients with ALK-rearranged advanced non-small-cell lung cancer (AF-001JP study): a single-arm, open-label, phase 1-2 study. Lancet Oncol 14(7):590–598
53. Takeuchi K, Togashi Y, Kamihara Y et al (2016) Prospective and clinical validation of ALK immunohistochemistry: results from the phase I/II study of alectinib for ALK-positive lung cancer (AF-001JP study). Ann Oncol 27(1):185–192
54. Food and Drug Administration (2015) ALECENSA. In: Drugs@FDA: FDA approved drug products. Available at http://www.accessdata.fda.gov/drugsatfda_docs/nda/2015/208434Orig1s000TOC.cfm. Accessed 31 Mar 2016
55. Gadgeel SM, Gandhi L, Riely GJ et al (2014) Safety and activity of alectinib against systemic disease and brain metastases in patients with crizotinib-resistant ALK-rearranged non-small-cell lung cancer (AF-002JG): results from the dose-finding portion of a phase 1/2 study. Lancet Oncol 15(10):1119–1128
56. Ou SH, Ahn JS, De Petris L et al (2016) Alectinib in crizotinib-refractory ALK-rearranged non-small-cell lung cancer: a phase II global study. J Clin Oncol 34(7):661–668
57. Weickhardt AJ, Aisner DL, Franklin WA et al (2013) Diagnostic assays for identification of anaplastic lymphoma kinase-positive non-small cell lung cancer. Cancer. 119(8):1467–1477
58. Lindeman NI, Cagle PT, Beasley MB et al (2013) Molecular testing guideline for selection of lung cancer patients for EGFR and ALK tyrosine kinase inhibitors: guideline from the College of American Pathologists, International Association for the Study of Lung Cancer, and Association for Molecular Pathology. Arch Pathol Lab Med. 137(6):828–860
59. Food and Drug Administration (2015) KEYTRUDA. In: Drugs@FDA: FDA approved drug products. Available at http://www.accessdata.fda.gov/drugsatfda_docs/nda/2014/125514Orig1s000TOC.cfm. Accessed 31 Mar 2016
60. Garon EB, Rizvi NA, Hui R et al (2015) Pembrolizumab for the treatment of non-small-cell lung cancer. N Engl J Med. 372(21):2018–2028
61. Food and Drug Administration (2015) PD-L1 IHC 22C3 pharmDx. In: Premarket Approval (PMA). Available at http://www.accessdata.fda.gov/scripts/cdrh/cfdocs/cfTopic/pma/pma.cfm?num=P150013. Accessed 31 Mar 2016
62. Study of MK-3475 (Pembrolizumab) versus platinum-based chemotherapy for participants with PD-L1-positive advanced or metastatic non-small cell lung cancer (MK-3475-042/KEYNOTE-042). Available at https://clinicaltrials.gov/ct2/show/NCT02220894. Accessed 31 Mar 2016
63. Study of Pembrolizumab (MK-3475) compared to platinum-based chemotherapies in participants with metastatic non-small cell lung cancer (MK-3475-024/KEYNOTE-024). Available at https://clinicaltrials.gov/ct2/show/NCT02142738. Accessed 31 Mar 2016
64. Study of two doses of MK-3475 (Pembrolizumab) versus docetaxel in previously-treated participants with non-small cell lung cancer (MK-3475-010/KEYNOTE-010). Available at https://clinicaltrials.gov/ct2/show/NCT01905657. Accessed 31 Mar 2016
65. Food and Drug Administration (2015) OPDIVO. In: Drugs@FDA: FDA approved drug products. Available at http://www.accessdata.fda.gov/drugsatfda_docs/label/2015/125527s000lbl.pdf. Accessed 31 Mar 2016
66. Food and Drug Administration (2015) OPDIVO. In: Drugs@FDA: FDA approved drug products. Available at http://www.accessdata.fda.gov/drugsatfda_docs/label/2015/125554s005lbl.pdf. Accessed 31 Mar 2016

67. Brahmer J, Reckamp KL, Baas P et al (2015) Nivolumab versus docetaxel in advanced squamous-cell non-small-cell lung cancer. N Engl J Med. 373(2):123–135
68. Borghaei H, Paz-Ares L, Horn L et al (2015) Nivolumab versus docetaxel in advanced non-squamous non-small-cell lung cancer. N Engl J Med. 373(17):1627–1639
69. Bristol-Myers Squibb (2015) Bristol-Myers Squibb's Opdivo (nivolumab) receives expanded FDA approval in previously-treated metastatic non-small cell lung cancer (NSCLC), offering improved survival to more patients. Available at http://news.bms.com/press-release/bristol-myers-squibbs-opdivo-nivolumab-receives-expanded-fda-approval-previously-treat. Accessed 31 Mar 2016
70. Food and Drug Administration (2015) PD-L1 IHC 22C3 pharmDx. In: Premarket Approval (PMA). Available at http://www.accessdata.fda.gov/scripts/cdrh/cfdocs/cfTopic/pma/pma.cfm?num=P150025. Accessed 31 Mar 2016
71. Vansteenkiste J, Fehrenbacher L, Spira AI, et al Atezolizumab monotherapy versus docetaxelin 2 L/3 L non-small cell lung cancer: Primary analysis for efficacy, safety, and predictive biomarkers from a randomized Phase II study (POPLAR). European Cancer Congress annual meeting 2015. Abstract number 14LBA
72. Besse B, Johnson ML, Janne PA, et al Phase II, single-arm trial (BIRCH) of atezolizumab as first-line or subsequent therapy for locally advanced or metastatic PD-L1–selected non-small cell lung cancer (NSCLC). European Cancer Congress 2015 annual meeting. Abstract number 16LBA
73. Rizvi NA, Brahmer JR, Ou S-HI et al (2015) Safety and clinical activity of MEDI4736, an anti-programmed cell death-ligand 1 (PD-L1) antibody, in patients with non-small cell lung cancer (NSCLC). J Clin Oncol 33(S1):8032
74. Ramalingam S (2015) Immune checkpoint inhibitors: the dawn of a new era for lung cancer therapy. Available at http://www.ascopost.com/issues/august-25-2015/immune-checkpoint-inhibitors-the-dawn-of-a-new-era-for-lung-cancer-therapy/. Accessed 31 Mar 2016
75. Food and Drug Administration (2015) A blueprint proposal for companion diagnostic comparability. Available at http://www.fda.gov/downloads/MedicalDevices/NewsEvents/WorkshopsConferences/UCM439440.pdf. Accessed 31 Mar 2016
76. Hirsch FR (2015) IASLC: PDL-1 CHARACTERIZATION PROJECT ("PCP-Study"). Available at http://www.fda.gov/downloads/MedicalDevices/NewsEvents/WorkshopsConferences/UCM439878.pdf. Accessed 31 Mar 2016
77. Food and Drug Administration (2016) A list of cleared or approved companion diagnostic devices (in vitro and imaging tools). Available at http://www.fda.gov/medicaldevices/productsandmedicalprocedures/invitrodiagnostics/ucm301431.htm. Accessed 31 Mar 2016
78. Blumenthal G: The looming PD1/PDL1 storm: CDER oncology perspective: complexities in personalized medicine: harmonizing companion diagnostics across a class of targeted therapies, March 24, 2015. Available at http://www.fda.gov/downloads/MedicalDevices/NewsEvents/WorkshopsConferences/UCM439878.pdf. Accessed 31 Mar 2016
79. Gettinger SN, Shepherd FA, Antonia SJ, et al (2014) First-line nivolumab (anti-PD-1; BMS-936558, ONO-4538) monotherapy in advanced NSCLC: safety, efficacy, and correlation of outcomes with PD-L1 status. J Clin Oncol 32:5 s, (suppl; abstr 8024)
80. F. Hoffmann-La Roche, Ltd. (2015) Media release: two positive studies of Roche's investigational cancer immunotherapy atezolizumab in specific type of lung cancer presented at 2015 European Cancer Congress, September 27, 2015. Available at http://www.roche.com/med-cor-2015-09-27-e.pdf. Accessed 31 Mar 2016

Part II
Treatment

Chapter 8
Small Cell Lung Cancer and Molecular Targeted Therapy

Shunichiro Iwasawa

Abstract Small cell lung cancer (SCLC) is one of the major pathological types of lung cancer. Most of the patients with SCLC remain uncured with standard treatment consisting of platinum-based chemotherapy and have poor prognosis. The genomic characterization of SCLC has not been fully elucidated, resulting that SCLC has significantly lagged behind NSCLC for the development of molecular targeted therapies. However, the development of sequencing technologies makes it possible to identify potent targets for the treatment in SCLC.

A number of studies with various types of molecular targeted therapies have been conducted in patients with SCLC. Several molecular targeted agents showed promising efficacy with good tolerability but have not proven survival benefit in large clinical studies. Unfortunately, no molecular targeted agents have been approved for SCLC to date. Recently, immune checkpoint inhibitors have emerged as new promising therapeutic agents also in SCLC.

These novel approaches are expected to improve clinical outcomes. In this chapter, currently available data of the molecular targeted therapies for SCLC and the future directions will be discussed.

Keywords Small cell lung cancer (SCLC) • Chemotherapy • Molecular targeted therapy • Immune checkpoint inhibitor • Mutation

8.1 Introduction

Small cell lung cancer (SCLC) is one of the major pathological types of lung cancer and accounts for approximately 15 % of all cases of the disease worldwide [1, 2]. The aggressive behavior of SCLC leads to high rate of late stage at diagnosis, recurrence, and mortality. The 5-year survival rate of SCLC is only 3.5–6.8 % [3, 4]. Since chemotherapy with cyclophosphamide showed a survival benefit over best

S. Iwasawa (✉)
Department of Medical Oncology, Graduate School of Medicine, Chiba University,
1-8-1 Inohana, Chuo-ku, Chiba 260-8677, JAPAN
e-mail: iwasawas@chiba-u.jp

© Springer Science+Business Media Singapore 2017

Y. Takiguchi (ed.), *Molecular Targeted Therapy of Lung Cancer*,
DOI 10.1007/978-981-10-2002-5_8

supportive care in the 1960s [5], chemotherapy has played an important role in the treatment of SCLC. For patients with limited disease (LD), concurrent chemoradiotherapy is the standard treatment which demonstrated a median survival time of –24 months [1, 6, 7]. In addition, prophylactic cranial irradiation (PCI) after complete response to chemoradiotherapy improved overall survival [8, 9]. For patients with extensive disease (ED), chemotherapy is the only standard modality of treatment. Platinum-based chemotherapy, including etoposide or irinotecan, showed a median survival time of 8–13 months [10, 11]. Although these treatment approaches have been established, most of the patients with SCLC remain uncured. A number of studies have been undertaken to overcome this situation. However, only slight progress has been made in the past decade. It seems that therapeutic ceiling has been reached in SCLC with chemotherapy.

Like other types of cancer, it is thought that molecular targeted therapy for SCLC has a potential to solve the challenging problem. Based on identification of some promising targets for treatment of SCLC, various types of molecular targeted therapy have been evaluated in clinical trials. While there are no molecular targeted drugs approved for SCLC at the present time, novel drugs are expected to improve clinical outcomes.

8.2 Molecular Targets of Treatment for SCLC

SCLC has one of the highest rates of mutations. Chronic cigarette smoking, a major risk factor for SCLC, results in the accumulation of mutations and carcinogenesis [12]. The complex genomic landscape makes it difficult to identify key driver mutations. Therefore, SCLC has significantly lagged behind non-small cell lung cancer (NSCLC) for the development of molecular targeted therapies.

Comprehensive genome-wide study of SCLC has been scarcely performed mainly due to the low frequency of surgical resection and insufficient amount of tumor tissue [13]. A sequencing study of 110 patients with SCLC revealed that TP53 and RB1, both are tumor suppressor genes, were inactivated in almost all of cases, and BRAF, KIT19, and PIK3CA mutations were found in several cases [14]. Moreover, inactivating mutations in NOTCH family genes were observed in 25 % of the cases. These genetic alterations may be new potential targets for treatment of SCLC.

Immunotherapy is an active area of cancer research and is showing great promise. In recent years, the survival benefit of the immune checkpoint blockade has been demonstrated in patients with advanced NSCLC [15, 16]. Some of the immune checkpoint inhibitors, which target PD-1 and PD-L1, are now in clinical use and have been evaluated in SCLC. A number of studies with various types of molecular targeted therapies have been conducted in patients with SCLC and remain under active clinical evaluation, as described below.

8.2.1 *Angiogenesis*

8.2.1.1 VEGF-Targeted Agents

Blood supply is essential for the development and growth of cancer cells, and angiogenesis is a necessary process in the progression of cancer [17]. SCLC has high vascularization and has shown that increased serum levels of vascular endothelial growth factor (VEGF) correlate with poor outcome [18, 19]. VEGF-targeted therapy is considered to be a potent candidate for treatment of SCLC.

The studied agents targeting angiogenesis are listed in Table 8.1.

Bevacizumab

Bevacizumab, which is a humanized IgG1 monoclonal antibody targeting VEGF-A, showed the most promising results among VEGF-targeted agents. Although this agent, in combination with platinum-based chemotherapy, has been investigated in three phase II studies, the results were controversial.

A single-arm phase II study, Eastern Cooperative Oncology Group (ECOG) 3501, reported that the addition of bevacizumab to standard cisplatin and etoposide (EP) regimen followed by maintenance bevacizumab resulted in improved PFS and OS relative to historical controls who received EP regimen without bevacizumab: ORR of 63.5 %, 6 month PFS rate of 30.2 %, median PFS 4.7 months, and median

Table 8.1 Angiogenesis-targeted agents in SCLC

Target	Agent	Study phase	Result	References
VEGF-A	Bevacizumab	II	Promising	[20]
		II	Promising	[21]
		II	Positive for PFS, but no benefit in OS	[22]
VEGFR-1, VEGFR-2, VEGFR-3	Cediranib	II	Negative	[23]
VEGFR, EGFR, RET	Vandetanib	II	Negative	[24]
VEGFR-2, VEGFR-3, PDGFR, BRAF, KIT, FLT3, RET	Sorafenib	II	Negative	[25]
VEGFR-1, VEGFR-2, VEGFR-3, PDGFR, KIT, FLT3, RET, CSF-1R	Sunitinib	II	Negative	[26]
		II	Promising	[27]
		II	Positive for PFS	[28]
VEGF-A, B	Ziv-aflibercept	II	Positive for PFS only in platinum-refractory patients	[30]
VEGFR, EGFR, RET	Thalidomide	III	Negative	[32]
		III	Negative	[33]
MMPs	Marimastat	II	Negative	[36]

VEGF vascular endothelial growth factor, *VEGFR* VEGF receptor, *EGFR* epidermal growth factor receptor, *PDGFR* platelet-derived growth factor receptor, *CSF-1R* colony-stimulating factor 1 receptor, *MMPs* matrix metalloproteinases

OS of 10.9 months [20]. Another single-arm phase II study combining bevacizumab with carboplatin and irinotecan regimen reported more promising results, ORR of 84 % and median OS of 12.1 months [21]. A randomized phase II study, SALUTE, showed that the addition of bevacizumab to cisplatin or carboplatin plus etoposide regimen improved PFS (5.5 vs 4.4 months; HR: 0.53; 95 % CI, 0.32–0.86) with acceptable toxicity but failed to improve OS (9.4 vs 10.9 months, HR: 1.16; 95 % CI, 0.66–2.04) [22].

Cediranib
Cediranib is a potent and selective inhibitor of the VEGF receptors (VEGFR) 1, 2, and 3. A single-arm phase II study of cediranib for second-line therapy of SCLC reported that nine cases out of 25 had stable disease, but none had objective responses [23].

Vandetanib
Vandetanib is a multi-targeted kinase inhibitor of mainly VEGFR, epidermal growth factor receptor (EGFR), and RET-tyrosine kinases. A randomized phase II study (NCIC CTG BR.20) examined the improvement in PFS with vandetanib as maintenance therapy after objective responses to platinum-based chemotherapy with or without radiation therapy [24]. Vandetanib failed to prolong PFS, but limited-stage (LD) patients treated with vandetanib tended to have longer OS in planned subgroup analyses (HR: 0.45; $p = .07$).

Sorafenib
Sorafenib is a multi-targeted kinase inhibitor of VEGFR-2 and VEGFR-3, platelet-derived growth factor receptor (PDGFR), BRAF, KIT, FLT3, and RET. A single-arm phase II study (SWOG 0435) was conducted to evaluate the efficacy and safety of sorafenib in patients with SCLC who previously received platinum-based chemotherapy [25]. As the results were disappointing, the single-agent sorafenib was determined to be not recommended for SCLC.

Sunitinib
Sunitinib is a multi-targeted kinase inhibitor of VEGFR-1, VEGFR-2, and VEGFR-3, PDGFR, KIT, FLT3, RET, and colony-stimulating factor 1 receptor (CSF-1R). A single-arm phase II study was conducted to evaluate the efficacy and safety of sorafenib in previously treated patients with SCLC [26]. The efficacy was disappointing: ORR of 9 % (95 % CI, 1–28 %), median PFS 1.4 months (95 % CI, 1.1–1.7), and median OS 5.6 months (95 % CI, 3.5–7.7). Moreover, most of the patients were unable to tolerate sunitinib. Another single-arm phase II study evaluated the efficacy and safety of sunitinib as maintenance therapy following carboplatin and irinotecan regimen in patients with ED SCLC [27]. Sunitinib was well tolerated and showed encouraging results with 1-year OS of 54 % and median time to progression of 7.6 months.

A randomized phase II study (CALGB 30504, ALLIANCE) also evaluated the efficacy of maintenance sunitinib after cisplatin and etoposide regimen. Compared with placebo, sunitinib significantly improved PFS (2.1 vs 3.7 months; HR: 1.62; 95 % CI, 1.02–2.60, $P = .02$) [28]. The result supports the strategy of investigating the use of sunitinib in the maintenance setting and the future study.

Ziv-Aflibercept

Ziv-aflbercept is a recombinant fusion protein consisting of VEGF-binding portions from the extracellular domains of human VEGFR-1 and VEGFR-2 that are fused to the Fc portion of the human IgG1 immunoglobulin. The agent acts as a soluble decoy receptor that inhibits angiogenesis by targeting VEGF-A and VEGF-B [29]. A randomized phase II study (S0802) evaluated the efficacy of ziv-aflibercept with topotecan in patients with previously treated SCLC [30]. Ziv-aflibercept improved the 3-month PFS only in patients who had platinum-refractory SCLC (27 % v 10 %; *P* = .02) but increased toxicity. The addition of ziv-aflibercept to topotecan did not improve OS.

Thalidomide

Thalidomide, a glutamic acid derivative, inhibits angiogenesis by repression of key angiogenic genes and downregulation of VEGF and basic fibroblast growth factor [31, 32], while the mechanism is not fully understood. Two phase III studies have been conducted to evaluate the efficacy of thalidomide in patients with SCLC. An intergroup phase III study (FNCLCC cleo04 IFCT 00–01) failed to show survival benefit of adding thalidomide to platinum-based chemotherapy following response to induction chemotherapy (median OS, 11.7 vs 8.7 months; HR, 0.74; 95 % CI, 0.49–1.12, *P* = .16), but some benefit was observed among patients with a performance (PS) of 1 or 2 in exploratory analyses (HR, 0.59; 95 % CI, 0.37–0.92, *P* = .02) [32]. Moreover, sensory neuropathy occurred more frequently in the thalidomide group compared with the placebo group (33 % v 12 %, respectively). The other phase III trials evaluated thalidomide in combination with carboplatin and etoposide regimen [33]. The study also did not meet its primary endpoint of OS (10.5 vs 10.1 months; HR, 1.09; 95 % CI, 0.93–1.27, *P* = .28) and showed that thalidomide increased risk of thrombotic events.

8.2.1.2 MMP Inhibitors

Matrix metalloproteinases (MMPs) are a family of secreted proteins remodeling the extracellular matrix. The remodeling process is necessary for physiological events, such as wound repair, organismal growth and development, and mediation of immune responses. MMPs have been shown to contribute to angiogenesis and cell migration and found to be dysregulated in human cancers [34]. Overexpression of MMPs has been shown to correlate with poor prognosis in SCLC [35]. Therefore, MMPs has been proposed to be a potential target for treatment of SCLC.

Marimastat

Marimastat is a broad-spectrum MMP inhibitor. A phase III trial was conducted to evaluate efficacy of this agent as maintenance therapy after response to platinum-based chemotherapy in patients with SCLC [36]. Marimastat failed to improve survival (median PFS, 4.3 vs 4.4 months; HR, 0.977; 95 % CI, 0.807–1.184, *P* = .81; median OS, 9.3 vs 9.7 months; HR, 1.013; 95 % CI, 0.831–1.235, *P* = .90) and had a negative impact on quality of life.

Table 8.2 Proliferative signaling pathway-targeted agents in SCLC

Target	Agent	Study phase	Result	References
KIT, PDGFR, BCR-ABL	Imatinib	II	Negative	[38]
		II	Negative	[39]
		II	Negative	[40]
		II	Negative	[41]
KIT, SRC, BCR-ABL	Dasatinib	II	Negative	[42]
EGFR	Gefitinib	II	Negative	[43]
MET	Rilotumumab	Ib/II	Ongoing	NCT00791154
IGF-1R	Ganitumab	Ib/II	Ongoing	NCT00791154
mTOR	Temsirolimus	II	Negative	[47]
	Everolimus	II	Negative	[48]
SHH	Vismodegib	II	Negative	[50]
NOTCH	Tarextumab	Ib/II	Ongoing	[52] NCT01859741

PDGFR platelet-derived growth factor receptor, *EGFR* epidermal growth factor receptor, *IGF-1R* insulin-like growth factor 1 receptor, *SHH* sonic hedgehog

8.2.2 *Proliferative Signaling Pathways*

The studied drugs targeting angiogenesis are listed in Table 8.2.

8.2.2.1 c-KIT

A transmembrane tyrosine kinase receptor c-KIT (CD117), which plays an important role in hematopoiesis, spermatogenesis, melanogenesis, and carcinogenesis, has shown to be overexpressed in SCLC [37]. It was considered that c-KIT may be a potential target for treatment of SCLC.

Imatinib
Imatinib is a multi-targeted kinase inhibitor of KIT, PDGFR, and BCR-ABL. Four phase II studies with imatinib in SCLC have been conducted, but none of them showed promising results. A single-arm phase II study with imatinib at a dose of 600 mg once daily showed no objective response in 19 patients, while the KIT receptor was detected in four of them (21 %) [38]. A single-arm phase II study with high-dose imatinib in patients with KIT receptor expression also failed to show antitumor activity [39]. In the other two phase II studies, imatinib failed to demonstrate any clinical activity in spite of patient selection for c-KIT receptor expression [40, 41].

Dasatinib
Dasatinib is a multi-targeted kinase inhibitor of KIT, SRC, and BCR-ABL. SRC also plays an important role in neuropeptide-induced proliferation of cancer cells and is overexpressed in SCLC. A single-arm phase II study with dasatinib was conducted in previously treated patients with SCLC [42]. Dasatinib did not meet the efficacy criteria, and the study was terminated early.

8.2.2.2 EGFR

Gefitinib is the only EGFR-targeted agent investigated in phase II setting, while EGFR is considered to be not overexpressed in SCLC. In a single-arm phase II study, there was no objective response in previously treated 19 patients [43].

8.2.2.3 MET

MET is involved in mediating tumorigenesis, cell motility, scattering, and invasion. A preclinical study demonstrated that gain-of-function mutations in MET were explored in SCLC [44]. The fact suggests that MET signaling in SCLC plays an essential role in cytoskeletal functions and metastasis and may be a therapeutic target against SCLC.

Rilotumumab, a human monoclonal antibody against human hepatocyte growth factor (HGF) inhibiting MET pathway, is currently evaluated in a phase I/II study in combination with platinum-based chemotherapy (NCT00791154).

8.2.2.4 IGF-R

The insulin-like growth factor 1 receptor (IGF-1R) signaling promotes cell differentiation and proliferation through the Ras/mitogen-activated protein kinase (MAPK) pathway and the phosphoinositide 3-kinase/protein kinase B (PI3K/AKT) pathway [45].

Ganitumab, a human monoclonal antibody against IGF-1R, is currently evaluated in a phase I/II study in combination with platinum-based chemotherapy (NCT00791154).

8.2.2.5 mTOR

The mammalian target of rapamycin (mTOR) is a ubiquitous protein kinase and a key regulator of the PI3K/AKT pathway that is implicated in cell cycle control. The pathway regulates the protein synthesis necessary for cell growth, proliferation, angiogenesis, and other cellular endpoints and is implicated in an increasing number of pathological conditions, including cancer [46]. Therefore, mTOR is proposed to be an attractive target for anticancer therapy.

Temsirolimus

Temsirolimus is a specific inhibitor of mTOR and inhibits the synthesis of proteins regulating proliferation, growth, and survival of tumor cells. A phase II study evaluated the efficacy of temsirolimus as a maintenance therapy after response to platinum-based chemotherapy in patients with SCLC [47]. The study showed that temsirolimus seemed not to improve PFS.

Everolimus

Everolimus is a novel macrolide and has potent antiproliferative effects with applications as immunosuppressant and anticancer agent. A single-arm phase II study evaluated the efficacy and safety of everolimus in patients with previously treated SCLC [48]. The study showed that everolimus was well tolerated but had limited antitumor activity with ORR of 3 % and disease control rate (DCR) of 23 %. The median PFS and OS were 1.4 and 5.5 months, respectively.

8.2.2.6 Sonic Hedgehog

The sonic hedgehog (Shh) pathway is critical in embryogenesis and homeostasis of airway epithelial cells. The activation of SHH pathway is thought to be the process of carcinogenesis and progression in SCLC [49].

Vismodegib binds to the transmembrane protein Smoothened (SMO) and inhibits the Shh pathway. A randomized phase II study (E1508) evaluated the efficacy of vismodegib in patients with SCLC and showed that there were no significant improvements in PFS and OS with the addition of vismodegib to platinum-based chemotherapy [50].

8.2.2.7 NOTCH

The NOTCH pathway mediates self-replication, proliferation, and differentiation of cancer stem cells [51]. A sequencing study of 110 patients with SCLC revealed that inactivating mutations in NOTCH family genes were observed in 25 % of the cases [14]. The genetic alterations may be new potential targets for treatment of SCLC. Tarextumab is a human monoclonal IgG2 antibody against NOTCH2 and NOTCH3. A randomized phase Ib/II study with tarextumab (PINNACLE) has been conducted in combination with platinum-based chemotherapy as first-line therapy. In the phase Ib part of the study, tarextumab was well tolerated and showed promising antitumor activity [52] (NCT01859741).

Other types of molecular targeted agents are listed in Table 8.3.

8.2.3 DNA Damage Repair

Histone deacetylases (HDACs) regulate the expression and activity of numerous proteins implicated in carcinogenesis and proliferation of cancer cells, while they are tightly controlled in normal cells [53]. Several studies demonstrated that HDAC inhibitors had cytotoxic effects in SCLC cell lines [54, 55]. Some HDAC inhibitors, including vorinostat and belinostat, are currently evaluated in clinical studies (NCT00702962, NCT00926640).

Table 8.3 Other types of molecular targeted agents in SCLC

Target	Agent	Study phase	Result	References
HDACs	Vorinostat	I/II	Ongoing	NCT00702962
	Belinostat	I	Ongoing	NCT00926640
Bcl-2	Oblimersen	II	Negative	[58]
Bcl-2	Obatoclax	II	Negative	[59, 60]
Bcl-2	Navitoclax	II	Negative	[61]
Bcl-2	Gossypol	I/II	Negative	[62, 63]
Proteasome	Bortezomib	II	Negative	[64]
P-gp, MRP-1	Biricodar	II	Negative	[66]
CTLA-4	Ipilimumab	II	Promising	[67]
		II	Ongoing	NCT01331525
		III	Ongoing	NCT01450761
PD-1	Nivolumab	I/II	Promising	[68]

HDACs histone deacetylases, *P-gp* P-glycoprotein, *MRP-1* multidrug resistance protein 1, *CTLA-4* cytotoxic T lymphocyte antigen 4, *PD-1* programmed cell death-1

8.2.4 *Regulators of Apoptosis*

8.2.4.1 Bcl-2

Bcl-2 family proteins regulate all major types of cell death, including apoptosis, necrosis, and autophagy. The overexpression of Bcl-2 has been demonstrated to inhibit cell death induced by many stimuli, including growth factor deprivation, hypoxia, and oxidative stress, resulting in resistance to chemotherapy [56]. Moreover, it has been reported that Bcl-2 is overexpressed in SCLC [57]. Several agents targeting Bcl-2-family have been evaluated in clinical studies.

Oblimersen is an antisense oligonucleotide to a section of the Bcl-2 mRNA and the first anti-Bcl-2 agent evaluated in SCLC. A randomized phase II study with oblimersen (CALGB 30103) was conducted in combination with carboplatin and etoposide regimen as first-line therapy [58]. However, the results were disappointing. The percentage of patients alive at 1 year was 24 % (95 % CI, 12–40 %) with oblimersen and 47 % (95 % CI, 21–73 %) without oblimersen. Hazard ratios for failure-free survival (1.79; $P = .07$) and overall survival (2.13; $P = .02$) suggested worse outcome for patients receiving oblimersen.

In addition, small-molecule BH3 mimetics, which antagonize Bcl-2 family pro-survival proteins, and Bcl-2 inhibitor have been investigated in clinical studies. Obatoclax mesylate has been evaluated in two phase II studies. The agent was well tolerated, but the efficacy was not promising and failed to significantly improve ORR, PFS, or OS [59, 60]. Navitoclax was evaluated in a single-arm phase II study in patients with previously treated SCLC [61]. The study showed navitoclax had limited antitumor activity with ORR of 2.6 %. Gossypol, a pan Bcl-2 family protein inhibitor, has been evaluated in two early phase studies [62, 63]. The agent was well tolerated but showed no antitumor activity in patients with SCLC.

8.2.4.2 Ubiquitin-Proteasome System

The ubiquitin-proteasome pathway controls the amounts of proteins, such as nuclear factor-kappa B (NF-κB), and regulates cell death and survival. NF-κB promotes tumor cell survival and resistance to therapy in many cell types, regulating the expression of Bcl-2. Bortezomib is a proteasome inhibitor and decreases activity of survival factors, such as NF-κB, Akt, and Bcl-2. A phase II trial of bortezomib as a monotherapy in patients with previously treated SCLC failed to demonstrate efficacy [64].

8.2.5 Multidrug Resistance

Although SCLC is highly sensitive to initial chemotherapy, the disease often relapses and acquires a chemoresistant phenotype. Multidrug resistance (MDR) factors include genes encoding for P-glycoprotein (P-gp) and the MDR-associated protein (MRP-1). Both of them prevent accumulation of some chemotherapeutic agents through adenosine triphosphate-dependent transport out of cells. MRP-1 gene expression has been reported to be present in SCLC both before and after chemotherapy [65]. The presence of MRP-1 gene expression correlated with poor survival and no response to chemotherapy. Therefore, an agent targeting MDR through P-gp or MRP-1 could provide clinical benefit for patients with SCLC who are resistant to chemotherapy.

Biricodar is a multidrug resistance inhibitor that acts on P-gp and MRP-1. A single-arm phase II study evaluated the efficacy and safety of biricodar combined with chemotherapy in patients with recurrent SCLC after response to first-line chemotherapy [66]. The addition of biricodar to chemotherapy did not significantly improve antitumor activity or survival and increased hematologic toxicities. The study was terminated early.

8.2.6 Immune Checkpoint Inhibitors

Immunotherapy is one of the most active areas of cancer research and is showing great promise in various types of cancer. T-cell activation including cytotoxic T lymphocyte antigen 4 (CTLA-4) and programmed cell death-1 (PD-1) have been well investigated as a target for cancer treatment. Some of immune checkpoint inhibitors, which target CTLA-4 and PD-1, are now in clinical use and have been evaluated also in SCLC.

A randomized phase II study was conducted in patients with chemotherapy-naïve ED SCLC [67]. Ipilimumab was evaluated in two alternative regimens, concurrent ipilimumab (ipilimumab + paclitaxel/carboplatin followed by placebo + paclitaxel/carboplatin) or phased ipilimumab (placebo + paclitaxel/carboplatin followed by ipilimumab + paclitaxel/carboplatin). The primary endpoint was immune-related progression-free survival (irPFS, time from randomization to immune-related progressive disease or death). Although no improvements in PFS and OS were observed, phased ipilimumab, but not concurrent ipilimumab, improved irPFS versus control (6.4 vs 5.3 months; HR, 0.64; 95 % CI, 0.40–1.02, *P* = .03). The limitation of the study is that paclitaxel/carboplatin regimen is not standard chemotherapy for SCLC.

A phase I/II study with nivolumab monotherapy and nivolumab plus ipilimumab (CheckMate 032) showed promising results with durable responses and manageable safety profiles in previously treated patients with SCLC [68]. The objective response ranged from 10 to 33 %.

Other clinical studies are under evaluation:

Phase II: the addition of ipilimumab to carboplatin and etoposide chemotherapy for extensive stage small cell lung cancer (NCT01331525).

Phase III: trial in extensive-disease small cell lung cancer (ED SCLC) subjects comparing ipilimumab plus etoposide and platinum therapy to etoposide and platinum therapy alone (NCT01450761).

8.3 Summary

Most of the patients with SCLC remain uncured with standard treatment consisting of platinum-based chemotherapy and have poor prognosis. To overcome the situation, a number of studies with various types of molecular targeted therapies have been conducted in patients with SCLC. Several molecular targeted agents showed promising efficacy with good tolerability but have not proven survival benefit in large clinical studies. Unfortunately, no molecular targeted agents have been approved for SCLC to date.

SCLC has complex mutational signatures mostly caused by smoking. The genomic characterization has not been fully elucidated, resulting that SCLC has significantly lagged behind NSCLC for the development of molecular targeted therapies. Although SCLC is rarely resected by surgery and has limited availability of tumor tissue samples, the development of sequencing technologies makes it possible to identify potent targets for the treatment using nonsurgical samples of SCLC. In addition, emerging immunotherapies will hopefully provide a paradigm shift in the treatment of SCLC. These novel approaches are expected to improve clinical outcomes.

References

1. Jackman DM, Johnson BE (2005) Small-cell lung cancer. Lancet 366(9494):1385–1396. doi:10.1016/S0140-6736(05)67569-1
2. Govindan R, Page N, Morgensztern D, Read W, Tierney R, Vlahiotis A, Spitznagel EL, Piccirillo J (2006) Changing epidemiology of small-cell lung cancer in the United States over

the last 30 years: analysis of the surveillance, epidemiologic, and end results database. J Clin Oncol 24(28):4539–4544. doi:10.1200/JCO.2005.04.4859

3. Lassen U, Osterlind K, Hansen M, Dombernowsky P, Bergman B, Hansen HH (1995) Long-term survival in small-cell lung cancer: posttreatment characteristics in patients surviving 5 to 18+ years--an analysis of 1,714 consecutive patients. J Clin Oncol 13(5):1215–1220
4. Shepherd FA, Crowley J, Van Houtte P, Postmus PE, Carney D, Chansky K, Shaikh Z, Goldstraw P (2007) The International Association for the Study of Lung Cancer lung cancer staging project: proposals regarding the clinical staging of small cell lung cancer in the forthcoming (seventh) edition of the tumor, node, metastasis classification for lung cancer. J Thorac Oncol 2(12):1067–1077. doi:10.1097/JTO.0b013e31815bdc0d
5. Green RA, Humphrey E, Close H, Patno ME (1969) Alkylating agents in bronchogenic carcinoma. Am J Med 46(4):516–525
6. Turrisi AT 3rd, Kim K, Blum R, Sause WT, Livingston RB, Komaki R, Wagner H, Aisner S, Johnson DH (1999) Twice-daily compared with once-daily thoracic radiotherapy in limited small-cell lung cancer treated concurrently with cisplatin and etoposide. N Engl J Med 340(4):265–271. doi:10.1056/NEJM199901283400403
7. Janne PA, Freidlin B, Saxman S, Johnson DH, Livingston RB, Shepherd FA, Johnson BE (2002) Twenty-five years of clinical research for patients with limited-stage small cell lung carcinoma in North America. Cancer 95(7):1528–1538. doi:10.1002/cncr.10841
8. Auperin A, Arriagada R, Pignon JP, Le Pechoux C, Gregor A, Stephens RJ, Kristjansen PE, Johnson BE, Ueoka H, Wagner H, Aisner J (1999) Prophylactic cranial irradiation for patients with small-cell lung cancer in complete remission. Prophylactic Cranial Irradiation Overview Collaborative Group. N Engl J Med 341(7):476–484. doi:10.1056/NEJM199908123410703
9. Slotman B, Faivre-Finn C, Kramer G, Rankin E, Snee M, Hatton M, Postmus P, Collette L, Musat E, Senan S (2007) Prophylactic cranial irradiation in extensive small-cell lung cancer. N Engl J Med 357(7):664–672. doi:10.1056/NEJMoa071780
10. Roth BJ, Johnson DH, Einhorn LH, Schacter LP, Cherng NC, Cohen HJ, Crawford J, Randolph JA, Goodlow JL, Broun GO et al (1992) Randomized study of cyclophosphamide, doxorubicin, and vincristine versus etoposide and cisplatin versus alternation of these two regimens in extensive small-cell lung cancer: a phase III trial of the Southeastern Cancer Study Group. J Clin Oncol 10(2):282–291
11. Noda K, Nishiwaki Y, Kawahara M, Negoro S, Sugiura T, Yokoyama A, Fukuoka M, Mori K, Watanabe K, Tamura T, Yamamoto S, Saijo N (2002) Irinotecan plus cisplatin compared with etoposide plus cisplatin for extensive small-cell lung cancer. N Engl J Med 346(2):85–91. doi:10.1056/NEJMoa003034
12. Hecht SS (2002) Cigarette smoking and lung cancer: chemical mechanisms and approaches to prevention. Lancet Oncol 3(8):461–469
13. Semenova EA, Nagel R, Berns A (2015) Origins, genetic landscape, and emerging therapies of small cell lung cancer. Genes Dev 29(14):1447–1462. doi:10.1101/gad.263145.115
14. George J, Lim JS, Jang SJ, Cun Y, Ozretic L, Kong G, Leenders F, Lu X, Fernandez-Cuesta L, Bosco G, Muller C, Dahmen I, Jahchan NS, Park KS, Yang D, Karnezis AN, Vaka D, Torres A, Wang MS, Korbel JO, Menon R, Chun SM, Kim D, Wilkerson M, Hayes N, Engelmann D, Putzer B, Bos M, Michels S, Vlasic I, Seidel D, Pinther B, Schaub P, Becker C, Altmuller J, Yokota J, Kohno T, Iwakawa R, Tsuta K, Noguchi M, Muley T, Hoffmann H, Schnabel PA, Petersen I, Chen Y, Soltermann A, Tischler V, Choi CM, Kim YH, Massion PP, Zou Y, Jovanovic D, Kontic M, Wright GM, Russell PA, Solomon B, Koch I, Lindner M, Muscarella LA, la Torre A, Field JK, Jakopovic M, Knezevic J, Castanos-Velez E, Roz L, Pastorino U, Brustugun OT, Lund-Iversen M, Thunnissen E, Kohler J, Schuler M, Botling J, Sandelin M, Sanchez-Cespedes M, Salvesen HB, Achter V, Lang U, Bogus M, Schneider PM, Zander T, Ansen S, Hallek M, Wolf J, Vingron M, Yatabe Y, Travis WD, Nurnberg P, Reinhardt C, Perner S, Heukamp L, Buttner R, Haas SA, Brambilla E, Peifer M, Sage J, Thomas RK (2015) Comprehensive genomic profiles of small cell lung cancer. Nature 524(7563):47–53. doi:10.1038/nature14664

15. Brahmer J, Reckamp KL, Baas P, Crino L, Eberhardt WE, Poddubskaya E, Antonia S, Pluzanski A, Vokes EE, Holgado E, Waterhouse D, Ready N, Gainor J, Aren Frontera O, Havel L, Steins M, Garassino MC, Aerts JG, Domine M, Paz-Ares L, Reck M, Baudelet C, Harbison CT, Lestini B, Spigel DR (2015) Nivolumab versus Docetaxel in advanced squamous-cell non-small-cell lung cancer. N Engl J Med 373(2):123–135. doi:10.1056/NEJMoa1504627
16. Borghaei H, Paz-Ares L, Horn L, Spigel DR, Steins M, Ready NE, Chow LQ, Vokes EE, Felip E, Holgado E, Barlesi F, Kohlhaufl M, Arrieta O, Burgio MA, Fayette J, Lena H, Poddubskaya E, Gerber DE, Gettinger SN, Rudin CM, Rizvi N, Crino L, Blumenschein GR Jr, Antonia SJ, Dorange C, Harbison CT, Graf Finckenstein F, Brahmer JR (2015) Nivolumab versus docetaxel in advanced nonsquamous non-small-cell lung cancer. N Engl J Med 373(17):1627–1639. doi:10.1056/NEJMoa1507643
17. Carmeliet P, Jain RK (2000) Angiogenesis in cancer and other diseases. Nature 407(6801):249–257. doi:10.1038/35025220
18. Lucchi M, Mussi A, Fontanini G, Faviana P, Ribechini A, Angeletti CA (2002) Small cell lung carcinoma (SCLC): the angiogenic phenomenon. Eur J Cardiothorac Surg 21(6):1105–1110
19. Salven P, Ruotsalainen T, Mattson K, Joensuu H (1998) High pre-treatment serum level of vascular endothelial growth factor (VEGF) is associated with poor outcome in small-cell lung cancer. Int J Cancer 79(2):144–146
20. Horn L, Dahlberg SE, Sandler AB, Dowlati A, Moore DF, Murren JR, Schiller JH (2009) Phase II study of cisplatin plus etoposide and bevacizumab for previously untreated, extensive-stage small-cell lung cancer: Eastern Cooperative Oncology Group Study E3501. J Clin Oncol 27(35):6006–6011. doi:10.1200/JCO.2009.23.7545
21. Spigel DR, Greco FA, Zubkus JD, Murphy PB, Saez RA, Farley C, Yardley DA, Burris HA 3rd, Hainsworth JD (2009) Phase II trial of irinotecan, carboplatin, and bevacizumab in the treatment of patients with extensive-stage small-cell lung cancer. J Thorac Oncol 4(12):1555–1560. doi:10.1097/JTO.0b013e3181bbc540
22. Spigel DR, Townley PM, Waterhouse DM, Fang L, Adiguzel I, Huang JE, Karlin DA, Faoro L, Scappaticci FA, Socinski MA (2011) Randomized phase II study of bevacizumab in combination with chemotherapy in previously untreated extensive-stage small-cell lung cancer: results from the SALUTE trial. J Clin Oncol 29(16):2215–2222. doi:10.1200/JCO.2010.29.3423
23. Ramalingam SS, Belani CP, Mack PC, Vokes EE, Longmate J, Govindan R, Koczywas M, Ivy SP, Gandara DR (2010) Phase II study of Cediranib (AZD 2171), an inhibitor of the vascular endothelial growth factor receptor, for second-line therapy of small cell lung cancer (National Cancer Institute #7097). J Thorac Oncol 5(8):1279–1284. doi:10.1097/JTO.0b013e3181e2fcb0
24. Arnold AM, Seymour L, Smylie M, Ding K, Ung Y, Findlay B, Lee CW, Djurfeldt M, Whitehead M, Ellis P, Goss G, Chan A, Meharchand J, Alam Y, Gregg R, Butts C, Langmuir P, Shepherd F, National Cancer Institute of Canada Clinical Trials Group Study BR (2007) Phase II study of vandetanib or placebo in small-cell lung cancer patients after complete or partial response to induction chemotherapy with or without radiation therapy: National Cancer Institute of Canada Clinical Trials Group Study BR.20. J Clin Oncol 25(27):4278–4284. doi:10.1200/JCO.2007.12.3083
25. Gitlitz BJ, Moon J, Glisson BS, Reimers HJ, Bury MJ, Floyd JD, Schulz TK, Sundaram PK, Ho C, Gandara DR (2010) Sorafenib in platinum-treated patients with extensive stage small cell lung cancer: a Southwest Oncology Group (SWOG 0435) phase II trial. J Thorac Oncol 5(11):1835–1840. doi:10.1097/JTO.0b013e3181f0bd78
26. Han JY, Kim HY, Lim KY, Han JH, Lee YJ, Kwak MH, Kim HJ, Yun T, Kim HT, Lee JS (2013) A phase II study of sunitinib in patients with relapsed or refractory small cell lung cancer. Lung Cancer 79(2):137–142. doi:10.1016/j.lungcan.2012.09.019
27. Spigel DR, Greco FA, Rubin MS, Shipley D, Thompson DS, Lubiner ET, Eakle JF, Quinn R, Burris HA, Hainsworth JD (2012) Phase II study of maintenance sunitinib following irinotecan and carboplatin as first-line treatment for patients with extensive-stage small-cell lung cancer. Lung Cancer 77(2):359–364. doi:10.1016/j.lungcan.2012.03.009

28. Ready NE, Pang HH, Gu L, Otterson GA, Thomas SP, Miller AA, Baggstrom M, Masters GA, Graziano SL, Crawford J, Bogart J, Vokes EE (2015) Chemotherapy with or without maintenance sunitinib for untreated extensive-stage small-cell lung cancer: a randomized, double-blind, placebo-controlled phase II study-cALGB 30504 (Alliance). J Clin Oncol 33(15):1660–1665. doi:10.1200/JCO.2014.57.3105
29. Holash J, Davis S, Papadopoulos N, Croll SD, Ho L, Russell M, Boland P, Leidich R, Hylton D, Burova E, Ioffe E, Huang T, Radziejewski C, Bailey K, Fandl JP, Daly T, Wiegand SJ, Yancopoulos GD, Rudge JS (2002) VEGF-Trap: a VEGF blocker with potent antitumor effects. Proc Natl Acad Sci U S A 99(17):11393–11398. doi:10.1073/pnas.172398299
30. Allen JW, Moon J, Redman M, Gadgeel SM, Kelly K, Mack PC, Saba HM, Mohamed MK, Jahanzeb M, Gandara DR (2014) Southwest Oncology Group S0802: a randomized, phase II trial of weekly topotecan with and without ziv-aflibercept in patients with platinum-treated small-cell lung cancer. J Clin Oncol 32(23):2463–2470. doi:10.1200/JCO.2013.51.4109
31. D'Amato RJ, Loughnan MS, Flynn E, Folkman J (1994) Thalidomide is an inhibitor of angiogenesis. Proc Natl Acad Sci U S A 91(9):4082–4085
32. Pujol JL, Breton JL, Gervais R, Tanguy ML, Quoix E, David P, Janicot H, Westeel V, Gameroff S, Geneve J, Maraninchi D (2007) Phase III double-blind, placebo-controlled study of thalidomide in extensive-disease small-cell lung cancer after response to chemotherapy: an intergroup study FNCLCC cleo04 IFCT 00-01. J Clin Oncol 25(25):3945–3951. doi:10.1200/JCO.2007.11.8109
33. Lee SM, Woll PJ, Rudd R, Ferry D, O'Brien M, Middleton G, Spiro S, James L, Ali K, Jitlal M, Hackshaw A (2009) Anti-angiogenic therapy using thalidomide combined with chemotherapy in small cell lung cancer: a randomized, double-blind, placebo-controlled trial. J Natl Cancer Inst 101(15):1049–1057. doi:10.1093/jnci/djp200
34. Cathcart J, Pulkoski-Gross A, Cao J (2015) Targeting matrix metalloproteinases in cancer: bringing new life to old ideas. Genes Dis 2(1):26–34. doi:10.1016/j.gendis.2014.12.002
35. Michael M, Babic B, Khokha R, Tsao M, Ho J, Pintilie M, Leco K, Chamberlain D, Shepherd FA (1999) Expression and prognostic significance of metalloproteinases and their tissue inhibitors in patients with small-cell lung cancer. J Clin Oncol 17(6):1802–1808
36. Shepherd FA, Giaccone G, Seymour L, Debruyne C, Bezjak A, Hirsh V, Smylie M, Rubin S, Martins H, Lamont A, Krzakowski M, Sadura A, Zee B (2002) Prospective, randomized, double-blind, placebo-controlled trial of marimastat after response to first-line chemotherapy in patients with small-cell lung cancer: a trial of the National Cancer Institute of Canada-Clinical Trials Group and the European Organization for Research and Treatment of Cancer. J Clin Oncol 20(22):4434–4439
37. Potti A, Moazzam N, Ramar K, Hanekom DS, Kargas S, Koch M (2003) CD117 (c-KIT) overexpression in patients with extensive-stage small-cell lung carcinoma. Ann Oncol 14(6):894–897
38. Johnson BE, Fischer T, Fischer B, Dunlop D, Rischin D, Silberman S, Kowalski MO, Sayles D, Dimitrijevic S, Fletcher C, Hornick J, Salgia R, Le Chevalier T (2003) Phase II study of imatinib in patients with small cell lung cancer. Clin Cancer Res 9(16 Pt 1):5880–5887
39. Krug LM, Crapanzano JP, Azzoli CG, Miller VA, Rizvi N, Gomez J, Kris MG, Pizzo B, Tyson L, Dunne M, Heelan RT (2005) Imatinib mesylate lacks activity in small cell lung carcinoma expressing c-kit protein: a phase II clinical trial. Cancer 103(10):2128–2131. doi:10.1002/cncr.21000
40. Dy GK, Miller AA, Mandrekar SJ, Aubry MC, Langdon RM Jr, Morton RF, Schild SE, Jett JR, Adjei AA (2005) A phase II trial of imatinib (ST1571) in patients with c-kit expressing relapsed small-cell lung cancer: a CALGB and NCCTG study. Ann Oncol 16(11):1811–1816. doi:10.1093/annonc/mdi365
41. Schneider BJ, Kalemkerian GP, Ramnath N, Kraut MJ, Wozniak AJ, Worden FP, Ruckdeschel JC, Zhang X, Chen W, Gadgeel SM (2010) Phase II trial of imatinib maintenance therapy after irinotecan and cisplatin in patients with c-Kit-positive, extensive-stage small-cell lung cancer. Clin Lung Cancer 11(4):223–227. doi:10.3816/CLC.2010.n.028

42. Miller AA, Pang H, Hodgson L, Ramnath N, Otterson GA, Kelley MJ, Kratzke RA, Vokes EE, Cancer, Leukemia Group B (2010) A phase II study of dasatinib in patients with chemosensitive relapsed small cell lung cancer (Cancer and Leukemia Group B 30602). J Thorac Oncol 5(3):380–384. doi:10.1097/JTO.0b013e3181cee36e
43. Moore AM, Einhorn LH, Estes D, Govindan R, Axelson J, Vinson J, Breen TE, Yu M, Hanna NH (2006) Gefitinib in patients with chemo-sensitive and chemo-refractory relapsed small cell cancers: a Hoosier Oncology Group phase II trial. Lung Cancer 52(1):93–97. doi:10.1016/j.lungcan.2005.12.002
44. Ma PC, Kijima T, Maulik G, Fox EA, Sattler M, Griffin JD, Johnson BE, Salgia R (2003) c-MET mutational analysis in small cell lung cancer: novel juxtamembrane domain mutations regulating cytoskeletal functions. Cancer Res 63(19):6272–6281
45. Denduluri SK, Idowu O, Wang Z, Liao Z, Yan Z, Mohammed MK, Ye J, Wei Q, Wang J, Zhao L, Luu HH (2015) Insulin-like growth factor (IGF) signaling in tumorigenesis and the development of cancer drug resistance. Genes Dis 2(1):13–25. doi:10.1016/j.gendis.2014.10.004
46. Laplante M, Sabatini DM (2012) mTOR signaling in growth control and disease. Cell 149(2):274–293. doi:10.1016/j.cell.2012.03.017
47. Pandya KJ, Dahlberg S, Hidalgo M, Cohen RB, Lee MW, Schiller JH, Johnson DH, Eastern Cooperative Oncology G (2007) A randomized, phase II trial of two dose levels of temsirolimus (CCI-779) in patients with extensive-stage small-cell lung cancer who have responding or stable disease after induction chemotherapy: a trial of the Eastern Cooperative Oncology Group (E1500). J Thorac Oncol 2(11):1036–1041. doi:10.1097/JTO.0b013e318155a439
48. Tarhini A, Kotsakis A, Gooding W, Shuai Y, Petro D, Friedland D, Belani CP, Dacic S, Argiris A (2010) Phase II study of everolimus (RAD001) in previously treated small cell lung cancer. Clin Cancer Res 16(23):5900–5907. doi:10.1158/1078-0432.CCR-10-0802
49. Velcheti V, Govindan R (2007) Hedgehog signaling pathway and lung cancer. J Thorac Oncol 2(1):7–10. doi:10.1097/JTO.0b013e31802c0276
50. Belani CP, Dahlberg SE, Rudin CM, Fleisher M, Chen HX, Takebe N, Velasco MR Jr, Tester WJ, Sturtz K, Hann CL, Shanks JC, Monga M, Ramalingam SS, Schiller JH (2016) Vismodegib or cixutumumab in combination with standard chemotherapy for patients with extensive-stage small cell lung cancer: a trial of the ECOG-ACRIN Cancer Research Group (E1508). Cancer 122(15):2371–2378. doi:10.1002/cncr.30062
51. Pannuti A, Foreman K, Rizzo P, Osipo C, Golde T, Osborne B, Miele L (2010) Targeting Notch to target cancer stem cells. Clin Cancer Res 16(12):3141–3152. doi:10.1158/1078-0432.CCR-09-2823
52. Pietanza MC, Spira AI, Jotte RM, Gadgeel SM, Mita AC, Hart LL, Gluck WL, Chiang AC, Liu SV, Kapoun AM (2015) Final results of phase Ib of tarextumab (TRXT, OMP-59R5, anti-Notch2/3) in combination with etoposide and platinum (EP) in patients (pts) with untreated extensive-stage small-cell lung cancer (ED-SCLC). In: ASCO annual meeting proceedings, vol 15_suppl, p 7508
53. Glozak MA, Seto E (2007) Histone deacetylases and cancer. Oncogene 26(37):5420–5432. doi:10.1038/sj.onc.1210610
54. Luchenko VL, Salcido CD, Zhang Y, Agama K, Komlodi-Pasztor E, Murphy RF, Giaccone G, Pommier Y, Bates SE, Varticovski L (2011) Schedule-dependent synergy of histone deacetylase inhibitors with DNA damaging agents in small cell lung cancer. Cell Cycle 10(18):3119–3128. doi:10.4161/cc.10.18.17190
55. Gray J, Cubitt CL, Zhang S, Chiappori A (2012) Combination of HDAC and topoisomerase inhibitors in small cell lung cancer. Cancer Biol Ther 13(8):614–622. doi:10.4161/cbt.19848
56. Yip KW, Reed JC (2008) Bcl-2 family proteins and cancer. Oncogene 27(50):6398–6406. doi:10.1038/onc.2008.307
57. Ikegaki N, Katsumata M, Minna J, Tsujimoto Y (1994) Expression of bcl-2 in small cell lung carcinoma cells. Cancer Res 54(1):6–8
58. Rudin CM, Salgia R, Wang X, Hodgson LD, Masters GA, Green M, Vokes EE (2008) Randomized phase II Study of carboplatin and etoposide with or without the bcl-2 antisense

oligonucleotide oblimersen for extensive-stage small-cell lung cancer: CALGB 30103. J Clin Oncol 26(6):870–876. doi:10.1200/JCO.2007.14.3461

59. Paik PK, Rudin CM, Pietanza MC, Brown A, Rizvi NA, Takebe N, Travis W, James L, Ginsberg MS, Juergens R, Markus S, Tyson L, Subzwari S, Kris MG, Krug LM (2011) A phase II study of obatoclax mesylate, a Bcl-2 antagonist, plus topotecan in relapsed small cell lung cancer. Lung Cancer 74(3):481–485. doi:10.1016/j.lungcan.2011.05.005
60. Langer CJ, Albert I, Ross HJ, Kovacs P, Blakely LJ, Pajkos G, Somfay A, Zatloukal P, Kazarnowicz A, Moezi MM, Schreeder MT, Schnyder J, Ao-Baslock A, Pathak AK, Berger MS (2014) Randomized phase II study of carboplatin and etoposide with or without obatoclax mesylate in extensive-stage small cell lung cancer. Lung Cancer 85(3):420–428. doi:10.1016/j.lungcan.2014.05.003
61. Rudin CM, Hann CL, Garon EB, Ribeiro de Oliveira M, Bonomi PD, Camidge DR, Chu Q, Giaccone G, Khaira D, Ramalingam SS, Ranson MR, Dive C, McKeegan EM, Chyla BJ, Dowell BL, Chakravartty A, Nolan CE, Rudersdorf N, Busman TA, Mabry MH, Krivoshik AP, Humerickhouse RA, Shapiro GI, Gandhi L (2012) Phase II study of single-agent navitoclax (ABT-263) and biomarker correlates in patients with relapsed small cell lung cancer. Clin Cancer Res 18(11):3163–3169. doi:10.1158/1078-0432.CCR-11-3090
62. Heist RS, Fain J, Chinnasami B, Khan W, Molina JR, Sequist LV, Temel JS, Fidias P, Brainerd V, Leopold L, Lynch TJ (2010) Phase I/II study of AT-101 with topotecan in relapsed and refractory small cell lung cancer. J Thorac Oncol 5(10):1637–1643. doi:10.1097/JTO.0b013e3181e8f4dc
63. Baggstrom MQ, Qi Y, Koczywas M, Argiris A, Johnson EA, Millward MJ, Murphy SC, Erlichman C, Rudin CM, Govindan R (2011) A phase II study of AT-101 (Gossypol) in chemotherapy-sensitive recurrent extensive-stage small cell lung cancer. J Thorac Oncol 6(10):1757–1760. doi:10.1097/JTO.0b013e31822e2941
64. Lara PN Jr, Chansky K, Davies AM, Franklin WA, Gumerlock PH, Guaglianone PP, Atkins JN, Farneth N, Mack PC, Crowley JJ, Gandara DR (2006) Bortezomib (PS-341) in relapsed or refractory extensive stage small cell lung cancer: a Southwest Oncology Group phase II trial (S0327). J Thorac Oncol 1(9):996–1001
65. Savaraj N, Wu CJ, Xu R, Lampidis T, Lai S, Donnelly E, Solomon J, Feun LG (1997) Multidrug-resistant gene expression in small-cell lung cancer. Am J Clin Oncol 20(4):398–403
66. Gandhi L, Harding MW, Neubauer M, Langer CJ, Moore M, Ross HJ, Johnson BE, Lynch TJ (2007) A phase II study of the safety and efficacy of the multidrug resistance inhibitor VX-710 combined with doxorubicin and vincristine in patients with recurrent small cell lung cancer. Cancer 109(5):924–932. doi:10.1002/cncr.22492
67. Reck M, Bondarenko I, Luft A, Serwatowski P, Barlesi F, Chacko R, Sebastian M, Lu H, Cuillerot JM, Lynch TJ (2013) Ipilimumab in combination with paclitaxel and carboplatin as first-line therapy in extensive-disease-small-cell lung cancer: results from a randomized, double-blind, multicenter phase 2 trial. Ann Oncol 24(1):75–83. doi:10.1093/annonc/mds213
68. Antonia SJ, Lopez-Martin JA, Bendell J, Ott PA, Taylor M, Eder JP, Jager D, Pietanza MC, Le DT, de Braud F, Morse MA, Ascierto PA, Horn L, Amin A, Pillai RN, Evans J, Chau I, Bono P, Atmaca A, Sharma P, Harbison CT, Lin CS, Christensen O, Calvo E (2016) Nivolumab alone and nivolumab plus ipilimumab in recurrent small-cell lung cancer (CheckMate 032): a multi-centre, open-label, phase 1/2 trial. Lancet Oncol 17(7):883–895. doi:10.1016/S1470-2045(16)30098-5

Chapter 9
Locally Advanced Non-small Cell Lung Cancer and Targeted Therapy

Ikuo Sekine

Abstract Locally advanced unresectable non-small cell lung cancer (NSCLC), stage IIIA with bulky N2 and stage IIIB diseases, has been treated with concurrent chemoradiotherapy using a platinum doublet, but the effect of this conventional therapy has reached a plateau. Current research focuses on molecular targeted agents, especially epidermal growth factor receptor (EGFR)-directed agents and angiogenesis inhibitors. Although many preclinical experiments showed promising synergistic effects of EGFR-directed agents and radiation, no clinical trials have yet demonstrated the reproducibility of the preclinical results. Numerous preclinical models also showed synergistic effects of angiogenesis inhibitors and radiation without excessive toxicity. However, early clinical investigations of bevacizumab and chemoradiotherapy were closed early due to serious and unacceptable toxicities such as tracheoesophageal fistula and potentially fatal pneumonitis. The current review disclosed and discussed many issues on incorporation of molecular targeted agents into the treatment of unresectable stage III NSCLC.

Keywords Chemoradiotherapy • Epidermal growth factor receptor • Angiogenesis inhibitors • Tracheoesophageal fistula

9.1 Standard Treatment of Locally Advanced Unresectable Non-small Cell Lung Cancer

Locally advanced unresectable non-small cell lung cancer (NSCLC), stage IIIA with bulky N2 and stage IIIB diseases, accounts for approximately one fourth of all patients with NSCLC [1, 2]. The disease of this stage is characterized by a large primary lesion and/or involvement of the mediastinal or supraclavicular lymph nodes as well as occult systemic micrometastases in the majority of patients. The standard treatment for patients with a good performance status has been concurrent

---

I. Sekine (✉)
Department of Medical Oncology, Faculty of Medicine, University of Tsukuba, 1-1-1 Tennodai, Tsukuba, Ibaraki 305-8575, Japan
e-mail: isekine@md.tsukuba.ac.jp

© Springer Science+Business Media Singapore 2017

Y. Takiguchi (ed.), *Molecular Targeted Therapy of Lung Cancer*,
DOI 10.1007/978-981-10-2002-5_9

chemoradiotherapy [3, 4]. A platinum doublet using a third-generation anticancer agent combined with thoracic radiotherapy has yielded a median overall survival time of 22–30 months and long-term survivors of about 20 % [5], but the effect of platinum-based chemotherapy has reached a plateau [6–8]. Since about one third of patients had a relapse within a radiation field, enhanced local treatment may improve survival of these patients. However, a phase III trial of high-dose versus standard-dose thoracic radiotherapy using three-dimensional conformal radiotherapy or IMRT concurrently combined with chemotherapy (RTOG0617) showed poorer survival in the high-dose arm probably due to excessive toxicity to normal tissues [9]. Thus, new types of agents are needed for patients with locally advanced NSCLC to lead a longer and fuller life. Current research focuses on molecular targeted agents, especially epidermal growth factor receptor (EGFR)-directed agents and angiogenesis inhibitors.

9.2 Monoclonal Antibodies Against EGFR

In tumor cells, EGFR has an important role in cellular proliferation, inhibition of apoptosis, migration and invasion, and angiogenesis through activation of many signaling pathways including the RAS-mitogen-activated protein kinase pathway and phosphatidylinositol-3-kinase-AKT pathway [10]. Activation of EGFR signaling can be mediated by ionizing radiation as well as oncogenic EGFR. The activated EGFR signaling leads to radioresistance by inducing cell proliferation and apoptosis inhibition and enhancing DNA repair [11, 12]. The relationship between EGFR expression and radioresistance was shown among several murine carcinoma cell lines [13, 14], and a transfection study confirmed this relationship [15]. Several in vitro and in vivo studies showed synergistic effect of anti-EGFR antibody cetuximab and radiation [13, 16, 17]. This synergistic activity was obtained only in cetuximab-sensitive cell lines [17].

A benefit of the cetuximab and radiation combination was also demonstrated in a clinical setting; a combination of cetuximab and radiotherapy resulted in a significant improvement in 5-year overall over radiotherapy alone (45.6 % vs. 36.4 %) in a randomized phase III trial of locally advanced head and neck cancer [18]. However, the addition of cetuximab to chemoradiotherapy failed to show any survival benefits in either head and neck cancer (RTOG 0522) [19] or esophageal cancer (RTOG 0436) [20].

Safety of cetuximab combined with thoracic radiotherapy was firstly evaluated in SCRATCH study ($n = 12$), showing that the early and late toxicities of concurrent cetuximab and thoracic radiotherapy were acceptable, although one patient died of bronchopneumonia [21]. The following phase II trials of concurrent cetuximab and thoracic radiation with induction or consolidation platinum-doublet chemotherapy showed a median OS of 17.0 months or 19.4 months, respectively, with one death of pneumonitis in each study [22, 23]. A phase I trial of cetuximab in addition to chemoradiotherapy with cisplatin and vinorelbine showed that cetuximab could be safely added to chemoradiotherapy with grade 3–4 toxicity in 12 of 25 patients and one treatment-related death of hemoptysis 4 months after radiotherapy [24]. Phase II trials confirmed the toxicity profile of cetuximab combined with chemoradiotherapy, but median overall survival times seemed no improvement when compared with chemoradiotherapy without cetuximab (Table 9.1) [22, 23, 25–28]. A randomized phase II trial of carboplatin, pemetrexed, and thoracic radiotherapy with or without cetuximab (CALGB30407) showed no difference in failure-free survival, overall survival, or grade 3 or severe toxicity except for skin rash between the arms [27]. A landmark phase III trial of paclitaxel and carboplatin chemotherapy combined with thoracic radiotherapy of 60 Gy ($n = 151$), 74 Gy ($n = 107$), 60 Gy with cetuximab ($n = 137$), or 74 Gy with cetuximab ($n = 100$) (RTOG0617) showed an identical median overall survival of 25.0 months in patients receiving cetuximab and 24.0 months in patients who did not (HR 1.07, 95 % CI 0.84–1.35; $p = 0.29$). The overall incidence of grade 3 or worse toxicity for patients receiving chemoradiotherapy with cetuximab and chemoradiotherapy alone was 86 % and 70 %, respectively ($P < 0.0001$).

9.3 EGFR Tyrosine Kinase Inhibitors

EGFR tyrosine kinase inhibitors (TKIs) are another class of agents that inhibit EGFR, especially EGFR with somatic mutations in its tyrosine kinase domain [29, 30]. The EGFR mutations sensitize tumor cells to ionizing radiation by 500- to 1000-fold through the delayed repair of DNA double-strand breaks when compared with EGFR wild-type tumor cells [31, 32]. Several lines of studies showed that gefitinib and erlotinib potentiated radiation effects in NSCLC with EGFR wild type in vitro and in vivo by suppressing cellular DNA repair capacity and G2/M phase cell cycle arrest [33–38]. For EGFR-mutated cells, there are no experimental studies that tried to investigate the interaction between EGFR-TKIs and radiation.

There are several feasibility and phase II trials of EGFR-TKIs combined with thoracic radiotherapy or chemoradiotherapy in patients with unresectable stage III NSCLC (Table 9.2) [39–45]. These studies showed that these attempts were all feasible with acceptable toxicity, but the efficacy was disappointing in all but one trial. Komaki R et al. reported a phase II trial of erlotinib concurrently with weekly carboplatin and paclitaxel chemotherapy and thoracic intensity-modulated radiation therapy at a total dose of 63 Gy in 35 fractions followed by carboplatin and paclitaxel chemotherapy in 46 patients with previously untreated unresectable stage III NSCLC [45]. In the amended protocol, the primary endpoint of this study was time to progression, and the authors hypothesized that combining erlotinib and chemoradiation would increase the median time to progression from 15 to 25 months. The survival results looked promising at a glance; the median OS in this study was 36.5 months, and 2- and 5-year OS rates were 67.4 % and 35.9 %, respectively. None of these survival parameters differed by EGFR mutation status. The time to progression, however, was only 14 months with a distant failure noted in 59 % of

Table 9.1 Phase II trials of cetuximab and thoracic radiotherapy or chemoradiotherapy in patients with non-small cell lung cancer

Author (study name)	Year	Radiation (Gy/fractions)	Chemotherapy			N of patients	Overall survival		Grade 3–4 pneumonitis (%)	Treatment-related death (%)
			Induction	Concurrent	Consolidation		Median (month)	2y rate (%)		
Hallqvist A (SATELLITE)	2011	68/34	CDDP+DTX	Cetuximab	None	75	17.0	37.0	4.2	1.4
Ramalingam SS	2013	73.5/35	None	Cetuximab	CBDCA + PTX	38	19.4	25.0	0	2.6
van den Heuvel MM	2014	66/24	None	Daily cisplatin	None	51	NR	58.0	0	0
		66/24	None	Daily cisplatin + cetuximab	None	51	NR	62.0	11.8	3.9
Blumenschein GR (RTOG 0324)	2011	63/35	None	CBDCA + PTX + cetuximab	CBDCA + PTX + cetuximab	93	22.7	49.3	16.1	6.5
Govindan R (CALGB 30407)	2011	70/35	None	CBDCA + PEM	PEM	48	21.2	48.0	12.0	4.0
		70/35	None	CBDCA + PEM + cetuximab	PEM	53	25.2	50.0	11.3	5.7
Liu D	2015	60–66/30–33	CDDP + DTX + cetuximab	CDDP + DTX + cetuximab	None	27	26.7	51.9	0	0

CBDCA carboplatin, *CDDP* cisplatin, *DTX* docetaxel, *NR* not reported, *PEM* pemetrexed, *PTX* paclitaxel

Table 9.2 Epidermal growth factor receptor-tyrosine kinase inhibitors: feasibility and phase II trials

Author (study name)	Year	Radiation (Gy/fractions)	Chemotherapy			N of patients	Mutated EGFR (%)	Overall survival		Grade 3–4 pneumonitis (%)	Treatment-related death (%)
			Induction	Concurrent	Consolidation			Median (month)	2y rate (%)		
Feasibility trials											
Stinchcombe TE	2008	74/37	CBDCA + PTX + CPT	CBDCA + PTX + gefitinib	Gefitinib	23	NR	16.0	20.0	4.8	0
Okamoto I	2011	60/30	Gefitinib	Gefitinib	Gefitinib	9	25	6.3	33.0	11.1	0
Rothschild S	2011	63/34	CDDP + GEM or CBDCA +PTX	Gefitinib	Gefitinib	5	NR	12.6	NR	0.0	0
				CDDP + gefitinib	Gefitinib	9	NR			11.1	0
Choong NW	2008	66/33	None	CDDP + ETP + erlotinib	DTX	17	NR	10.2	25.0	5.9	0
			CBDCA + PTX	CBDCA + PTX + erlotinib	none	17	NR	13.7	20.0	0.0	0
Phase II trials											
Ready N (CALGB30106)	2010	66/33	CBDCA + PTX	Gefitinib	Gefitinib	(Poor risk) 21	28.9	19.0	32.0	9.5	4.8
		66/33	CBDCA + PTX	CBDCA + PTX + gefitinib	Gefitinib	(Good risk) 39		13.0	26.0	10.3	5.1
Niho S (JCOG0402)	2012	60/30	CDDP + VNR	Gefitinib	Gefitinib	38	NR	28.1	65.4	3.0	0
Komaki R	2015	63/32	None	CBDCA + PTX + erlotinib	CBDCA + PTX	48	9.8	36.6	67.4	8.3	0

CBDCA carboplatin, *CDDP* cisplatin, *CPT* irinotecan, *DTX* docetaxel, *EGFR* epidermal growth factor receptor, *ETP* etoposide, *GEM* gemcitabine, *NR* not reported, *PEM* pemetrexed, *PTX* paclitaxel, *VNR* vinorelbine patients. They concluded that the time to progression did not meet the assumption and more effective systemic therapy was needed.

Another approach to unresectable stage III NSCLC is to add an EGFR-TKI as a maintenance therapy after completion of a standard chemoradiotherapy. A phase III trial of maintenance gefitinib or placebo after concurrent chemoradiotherapy and docetaxel consolidation in unresectable stage III NSCLC showed that the gefitinib arm was inferior in overall survival to the placebo arm (median survival: 23 months versus 35 months, $p = 0.013$) [46]. This unexpected outcome could not be explained by excessive toxicity in the gefitinib arm, because grade 4 toxicity was noted only 2 % of patients and no toxic death in the gefitinib arm. One possible explanation was an imbalance in prognostic factors including smoking history, tumor EGFR mutation status, and K-ras mutation status, which might have contribution in poorer outcome in the gefitinib arm. Finally, the possibility that gefitinib somehow stimulated tumor growth either directly or indirectly cannot be excluded [47]. Erlotinib as maintenance treatment after concurrent chemoradiotherapy seemed also not promising in patients with stage III NSCLC not selected by EGFR mutations [48].

9.4 Angiogenesis Inhibitors

Angiogenesis is the essential process for further growth after tumors reach a diameter of 1–2 mm to maintain blood supply to the tumors. Angiogenesis is also critical for the efficacy of radiotherapy through several mechanisms. Tumor vascular bed is abnormal and irregular in its structure and function with the incomplete and heterogeneous oxygen supply. This leads to hypoxic radioresistance of tumor cells through lack of oxygen to facilitate DNA damage by radiation-induced free radicals and upregulation of hypoxia-inducible factor-1α (HIF-1α) [49]. In addition, radiation directly induces HIF-1α expression in tumor cells. The HIF-1α renders tumor cell phenotype suitable for proliferation by transcriptionally activating several genes, as well as induces tumor cells to secrete vascular endothelial growth factor (VEGF) [50], which serves to enhance endothelial cell radioresistance and angiogenesis after radiation [51, 52]. It was shown that tumor response to radiotherapy was closely related to endothelial cell apoptosis [53].

The rationale of combining angiogenesis inhibitors and radiation is vascular normalization, the remodeling of a dysfunctional tumor vasculature to a normal phenotype to restore tumor perfusion and oxygenation, and inhibition of radiation-induced angiogenesis signaling for repopulation of tumor cells after radiation [54]. Numerous preclinical models showed synergistic effects of the two modalities in a dose- and schedule-dependent manner, probably because disruption of tumor vessels hampers proper perfusion and aggravates tumor tissue hypoxia [55–58]. Thus, the vascular normalization window, the transient period of vessel normalization during anti-angiogenesis therapy, is important for the clinical application of angiogenesis inhibitors during radiotherapy, but it is difficult to determine when the normalization window occurs in patients, because the tumor growth kinetics in patients differ from those in animal models [54].

Early clinical investigations of bevacizumab and chemoradiotherapy were closed early due to severe toxicity. A phase II trial of carboplatin, pemetrexed, and bevacizumab induction for two cycles followed by thoracic radiotherapy at a dose of 61.2 Gy in 34 fractions concurrently combined with the same chemotherapy in patients with stage III NSCLC showed that of five patients enrolled, two developed tracheoesophageal fistula, and one died of bilateral pulmonary hemorrhage, left ventricular dysfunction, and subsequent pneumonia [59]. Socinski et al. reported a phase I/II trial of carboplatin, paclitaxel, and bevacizumab for two cycles followed by chemoradiotherapy with weekly carboplatin, paclitaxel and biweekly bevacizumab, and thoracic radiotherapy at a dose of 74 Gy in 37 fractions. Patients in cohort 1 received no erlotinib, whereas patients in cohorts 2 and 3 also received erlotinib at 100 and 150 mg, respectively. Of 45 patients enrolled, one developed grade 3 pulmonary hemorrhage and another developed tracheoesophageal fistula [60]. A phase I trial of induction cisplatin-based doublet chemotherapy and subsequent thoracic radiotherapy to a total dose of 66 Gy in 33 fractions concurrently combined with escalating doses of bevacizumab showed that four of six patients developed pneumonitis [61]. These trials clearly showed that concurrent bevacizumab and thoracic radiotherapy was too toxic. Another feasibility trial of chemoradiotherapy followed by consolidation docetaxel and bevacizumab resulted in two grade 3 pneumonitis and two fatal pulmonary hemoptysis among 21 patients assessable for safety [62]. Thus, even after completion of chemoradiotherapy, bevacizumab develops serious toxicity.

9.5 The Current Issues and Future Directions

The current review disclosed many issues on incorporation of molecular-targeted agents into the treatment of unresectable stage III NSCLC. One strategy for the treatment of stage III disease has been selecting a drug with a survival benefit demonstrated in patients with stage IV NSCLC. In addition, special importance has been placed on synergistic effects shown by preclinical studies. However, these strategies used for conventional cytotoxic chemotherapy require amendment because little is known about combined effects and optimal schedule of molecular-targeted agents and radiation. Identification of patient populations most likely to benefit is also an important subject for both clinical and basic researchers. The EGFR mutation status is crucial when selecting treatment for patients with stage IV NSCLC, but its significance in the treatment of stage III NSCLC remains unknown, because no preclinical or clinical studies showed combined effects of EGFR-TKIs and radiation on EGFR-mutated tumors. Toxicity enhancement by the combination of molecular-targeted agents and radiation also requires further investigation. Observation of tumor-bearing mice treated with a molecular-targeted agent and radiation to the tumor is not enough to evaluate toxicity. Precise experiments focused on toxicity may be more helpful to predict toxicity of the combination in humans.

References

1. Goldstraw P, Crowley J, Chansky K et al (2007) The IASLC Lung Cancer Staging Project: proposals for the revision of the TNM stage groupings in the forthcoming (seventh) edition of the TNM classification of malignant tumours. J Thorac Oncol 2:706–714
2. Groome PA, Bolejack V, Crowley JJ et al (2007) The IASLC Lung Cancer Staging Project: validation of the proposals for revision of the T, N, and M descriptors and consequent stage groupings in the forthcoming (seventh) edition of the TNM classification of malignant tumours. J Thorac Oncol 2:694–705
3. Furuse K, Fukuoka M, Kawahara M et al (1999) Phase III study of concurrent versus sequential thoracic radiotherapy in combination with mitomycin, vindesine, and cisplatin in unresectable stage III non-small-cell lung cancer. J Clin Oncol 17:2692–2699
4. Curran WJ Jr, Paulus R, Langer CJ et al (2011) Sequential vs. concurrent chemoradiation for stage III non-small cell lung cancer: randomized phase III trial RTOG 9410. J Natl Cancer Inst 103:1452–1460
5. Horinouchi H, Sekine I, Sumi M et al (2013) Long-term results of concurrent chemoradiotherapy using cisplatin and vinorelbine for stage III non-small-cell lung cancer. Cancer Sci 104:93–97
6. Segawa Y, Kiura K, Takigawa N et al (2010) Phase III trial comparing docetaxel and cisplatin combination chemotherapy with mitomycin, vindesine, and cisplatin combination chemotherapy with concurrent thoracic radiotherapy in locally advanced non-small-cell lung cancer: OLCSG 0007. J Clin Oncol 28:3299–3306
7. Yamamoto N, Nakagawa K, Nishimura Y et al (2010) Phase III study comparing second- and third-generation regimens with concurrent thoracic radiotherapy in patients with unresectable stage III non-small-cell lung cancer: West Japan Thoracic Oncology Group WJTOG0105. J Clin Oncol 28:3739–3745
8. Senan S, Brade A, Wang LH et al (2016) PROCLAIM: randomized Phase III trial of Pemetrexed-Cisplatin or Etoposide-Cisplatin Plus thoracic radiation therapy followed by consolidation chemotherapy in locally advanced nonsquamous non-small-cell lung cancer. J Clin Oncol 34:953–962
9. Bradley JD, Paulus R, Komaki R et al (2015) Standard-dose versus high-dose conformal radiotherapy with concurrent and consolidation carboplatin plus paclitaxel with or without cetuximab for patients with stage IIIA or IIIB non-small-cell lung cancer (RTOG 0617): a randomised, two-by-two factorial phase 3 study. Lancet Oncol 16:187–199
10. Roskoski R Jr (2014) The ErbB/HER family of protein-tyrosine kinases and cancer. Pharmacol Res 79:34–74
11. Nyati MK, Morgan MA, Feng FY, Lawrence TS (2006) Integration of EGFR inhibitors with radiochemotherapy. Nat Rev Cancer 6:876–885
12. Baumann M, Krause M, Dikomey E et al (2007) EGFR-targeted anti-cancer drugs in radiotherapy: preclinical evaluation of mechanisms. Radiother Oncol 83:238–248
13. Milas L, Fan Z, Andratschke NH, Ang KK (2004) Epidermal growth factor receptor and tumor response to radiation: in vivo preclinical studies. Int J Radiat Oncol Biol Phys 58:966–971
14. Akimoto T, Hunter NR, Buchmiller L et al (1999) Inverse relationship between epidermal growth factor receptor expression and radiocurability of murine carcinomas. Clin Cancer Res 5:2884–2890
15. Liang K, Ang KK, Milas L et al (2003) The epidermal growth factor receptor mediates radioresistance. Int J Radiat Oncol Biol Phys 57:246–254
16. Huang SM, Harari PM (2000) Modulation of radiation response after epidermal growth factor receptor blockade in squamous cell carcinomas: inhibition of damage repair, cell cycle kinetics, and tumor angiogenesis. Clin Cancer Res 6:2166–2174
17. Raben D, Helfrich B, Chan DC et al (2005) The effects of cetuximab alone and in combination with radiation and/or chemotherapy in lung cancer. Clin Cancer Res 11:795–805

18. Bonner JA, Harari PM, Giralt J et al (2006) Radiotherapy plus cetuximab for squamous-cell carcinoma of the head and neck. N Engl J Med 354:567–578
19. Ang KK, Zhang Q, Rosenthal DI et al (2014) Randomized phase III trial of concurrent accelerated radiation plus cisplatin with or without cetuximab for stage III to IV head and neck carcinoma: RTOG 0522. J Clin Oncol 32:2940–2950
20. Suntharalingam M, Winter K, Ilson D et al (2014) The initial report of RTOG 0436: a phase III trial evaluating the addition of cetuximab to paclitaxel, cisplatin, and radiation for patients with esophageal cancer treated without surgery. J Clin Oncol 32:(suppl 3; abstr LBA6)
21. Hughes S, Liong J, Miah A et al (2008) A brief report on the safety study of induction chemotherapy followed by synchronous radiotherapy and cetuximab in stage III non-small cell lung cancer (NSCLC): SCRATCH study. J Thorac Oncol 3:648–651
22. Hallqvist A, Wagenius G, Rylander H et al (2011) Concurrent cetuximab and radiotherapy after docetaxel-cisplatin induction chemotherapy in stage III NSCLC: satellite—a phase II study from the Swedish Lung Cancer Study Group. Lung Cancer 71:166–172
23. Ramalingam SS, Kotsakis A, Tarhini AA et al (2013) A multicenter phase II study of cetuximab in combination with chest radiotherapy and consolidation chemotherapy in patients with stage III non-small cell lung cancer. Lung Cancer 81:416–421
24. Dingemans AM, Bootsma G, van Baardwijk A et al (2014) A phase I study of concurrent individualized, isotoxic accelerated radiotherapy and cisplatin-vinorelbine-cetuximab in patients with stage III non-small-cell lung cancer. J Thorac Oncol 9:710–716
25. van den Heuvel MM, Uyterlinde W, Vincent AD et al (2014) Additional weekly Cetuximab to concurrent chemoradiotherapy in locally advanced non-small cell lung carcinoma: efficacy and safety outcomes of a randomized, multi-center phase II study investigating. Radiother Oncol 110:126–131
26. Blumenschein GR Jr, Paulus R, Curran WJ et al (2011) Phase II study of cetuximab in combination with chemoradiation in patients with stage IIIA/B non-small-cell lung cancer: RTOG 0324. J Clin Oncol 29:2312–2318
27. Govindan R, Bogart J, Stinchcombe T et al (2011) Randomized phase II study of pemetrexed, carboplatin, and thoracic radiation with or without cetuximab in patients with locally advanced unresectable non-small-cell lung cancer: Cancer and Leukemia Group B trial 30407. J Clin Oncol 29:3120–3125
28. Liu D, Zheng X, Chen J et al (2015) Induction chemotherapy with cetuximab, vinorelbine-cisplatin followed by thoracic radiotherapy and concurrent cetuximab, vinorelbine-cisplatin in patients with unresectable stage III non-small cell lung cancer. Lung Cancer 89:249–254
29. Lynch TJ, Bell DW, Sordella R et al (2004) Activating mutations in the epidermal growth factor receptor underlying responsiveness of non-small-cell lung cancer to gefitinib. N Engl J Med 350:2129–2139
30. Paez JG, Janne PA, Lee JC et al (2004) EGFR mutations in lung cancer: correlation with clinical response to gefitinib therapy. Science 304:1497–1500
31. Das AK, Sato M, Story MD et al (2006) Non-small-cell lung cancers with kinase domain mutations in the epidermal growth factor receptor are sensitive to ionizing radiation. Cancer Res 66:9601–9608
32. Das AK, Chen BP, Story MD et al (2007) Somatic mutations in the tyrosine kinase domain of epidermal growth factor receptor (EGFR) abrogate EGFR-mediated radioprotection in non-small cell lung carcinoma. Cancer Res 67:5267–5274
33. Bianco C, Tortora G, Bianco R et al (2002) Enhancement of antitumor activity of ionizing radiation by combined treatment with the selective epidermal growth factor receptor-tyrosine kinase inhibitor ZD1839 (Iressa). Clin Cancer Res 8:3250–3258
34. She Y, Lee F, Chen J et al (2003) The epidermal growth factor receptor tyrosine kinase inhibitor ZD1839 selectively potentiates radiation response of human tumors in nude mice, with a marked improvement in therapeutic index. Clin Cancer Res 9:3773–3778
35. Tanaka T, Munshi A, Brooks C et al (2008) Gefitinib radiosensitizes non-small cell lung cancer cells by suppressing cellular DNA repair capacity. Clin Cancer Res 14:1266–1273

36. Zhuang HQ, Sun J, Yuan ZY et al (2009) Radiosensitizing effects of gefitinib at different administration times in vitro. Cancer Sci 100:1520–1525
37. Park SY, Kim YM, Pyo H (2010) Gefitinib radiosensitizes non-small cell lung cancer cells through inhibition of ataxia telangiectasia mutated. Mol Cancer 9:222
38. Chinnaiyan P, Huang S, Vallabhaneni G et al (2005) Mechanisms of enhanced radiation response following epidermal growth factor receptor signaling inhibition by erlotinib (Tarceva). Cancer Res 65:3328–3335
39. Stinchcombe TE, Morris DE, Lee CB et al (2008) Induction chemotherapy with carboplatin, irinotecan, and paclitaxel followed by high dose three-dimension conformal thoracic radiotherapy (74 Gy) with concurrent carboplatin, paclitaxel, and gefitinib in unresectable stage IIIA and stage IIIB non-small cell lung cancer. J Thorac Oncol 3:250–257
40. Okamoto I, Takahashi T, Okamoto H et al (2011) Single-agent gefitinib with concurrent radiotherapy for locally advanced non-small cell lung cancer harboring mutations of the epidermal growth factor receptor. Lung Cancer 72:199–204
41. Rothschild S, Bucher SE, Bernier J et al (2011) Gefitinib in combination with irradiation with or without cisplatin in patients with inoperable stage III non-small cell lung cancer: a phase I trial. Int J Radiat Oncol Biol Phys 80:126–132
42. Choong NW, Mauer AM, Haraf DJ et al (2008) Phase I trial of erlotinib-based multimodality therapy for inoperable stage III non-small cell lung cancer. J Thorac Oncol 3:1003–1011
43. Ready N, Janne PA, Bogart J et al (2010) Chemoradiotherapy and gefitinib in stage III non-small cell lung cancer with epidermal growth factor receptor and KRAS mutation analysis: cancer and leukemia group B (CALEB) 30106, a CALGB-stratified phase II trial. J Thorac Oncol 5:1382–1390
44. Niho S, Ohe Y, Ishikura S et al (2012) Induction chemotherapy followed by gefitinib and concurrent thoracic radiotherapy for unresectable locally advanced adenocarcinoma of the lung: a multicenter feasibility study (JCOG 0402). Ann Oncol 23:2253–2258
45. Komaki R, Allen PK, Wei X et al (2015) Adding erlotinib to chemoradiation improves overall survival but not progression-free survival in Stage III non-small cell lung cancer. Int J Radiat Oncol Biol Phys 92:317–324
46. Kelly K, Chansky K, Gaspar LE et al (2008) Phase III trial of maintenance gefitinib or placebo after concurrent chemoradiotherapy and docetaxel consolidation in inoperable stage III non-small-cell lung cancer: SWOG S0023. J Clin Oncol 26:2450–2456
47. Keedy VL, Arteaga CL, Johnson DH (2008) Does gefitinib shorten lung cancer survival? Chaos redux. J Clin Oncol 26:2428–2430
48. Casal Rubio J, Firvida-Perez JL, Lazaro-Quintela M et al (2014) A phase II trial of erlotinib as maintenance treatment after concurrent chemoradiotherapy in stage III non-small-cell lung cancer (NSCLC): a Galician Lung Cancer Group (GGCP) study. Cancer Chemother Pharmacol 73:451–457
49. Wachsberger P, Burd R, Dicker AP (2003) Tumor response to ionizing radiation combined with antiangiogenesis or vascular targeting agents: exploring mechanisms of interaction. Clin Cancer Res 9:1957–1971
50. Semenza GL (2003) Targeting HIF-1 for cancer therapy. Nat Rev Cancer 3:721–732
51. Moeller BJ, Cao Y, Li CY, Dewhirst MW (2004) Radiation activates HIF-1 to regulate vascular radiosensitivity in tumors: role of reoxygenation, free radicals, and stress granules. Cancer Cell 5:429–441
52. Moeller BJ, Dreher MR, Rabbani ZN et al (2005) Pleiotropic effects of HIF-1 blockade on tumor radiosensitivity. Cancer Cell 8:99–110
53. Garcia-Barros M, Paris F, Cordon-Cardo C et al (2003) Tumor response to radiotherapy regulated by endothelial cell apoptosis. Science 300:1155–1159
54. Kleibeuker EA, Griffioen AW, Verheul HM et al (2012) Combining angiogenesis inhibition and radiotherapy: a double-edged sword. Drug Resist Updat 15:173–182

55. Winkler F, Kozin SV, Tong RT et al (2004) Kinetics of vascular normalization by VEGFR2 blockade governs brain tumor response to radiation: role of oxygenation, angiopoietin-1, and matrix metalloproteinases. Cancer Cell 6:553–563
56. Dings RP, Loren M, Heun H et al (2007) Scheduling of radiation with angiogenesis inhibitors anginex and Avastin improves therapeutic outcome via vessel normalization. Clin Cancer Res 13:3395–3402
57. Cao C, Albert JM, Geng L et al (2006) Vascular endothelial growth factor tyrosine kinase inhibitor AZD2171 and fractionated radiotherapy in mouse models of lung cancer. Cancer Res 66:11409–11415
58. Geng L, Donnelly E, McMahon G et al (2001) Inhibition of vascular endothelial growth factor receptor signaling leads to reversal of tumor resistance to radiotherapy. Cancer Res 61:2413–2419
59. Spigel DR, Hainsworth JD, Yardley DA et al (2010) Tracheoesophageal fistula formation in patients with lung cancer treated with chemoradiation and bevacizumab. J Clin Oncol 28:43–48
60. Socinski MA, Stinchcombe TE, Moore DT et al (2012) Incorporating bevacizumab and erlotinib in the combined-modality treatment of stage III non-small-cell lung cancer: results of a phase I/II trial. J Clin Oncol 30:3953–3959
61. Lind JS, Senan S, Smit EF (2012) Pulmonary toxicity after bevacizumab and concurrent thoracic radiotherapy observed in a phase I study for inoperable stage III non-small-cell lung cancer. J Clin Oncol 30:e104–e108
62. Wozniak AJ, Moon J, Thomas CR Jr et al (2015) A pilot trial of cisplatin/etoposide/radiotherapy followed by consolidation docetaxel and the combination of bevacizumab (NSC-704865) in patients with inoperable locally advanced stage III non-small-cell lung cancer: SWOG S0533. Clin Lung Cancer 16:340–347

Chapter 10
EGFR Mutant

Kunihiko Kobayashi and Hiroshi Kagamu

Abstract Before 2009, advanced non-small cell lung cancer (NSCLC) was regarded as a single disease entity, which was treated by cytotoxic chemotherapy and provided a response rate of 20–35 % and a median survival time (MST) of 10–12 months. In 2004, it was found that activating mutations of the epidermal growth factor receptor gene (*EGFR*) were present in a subset of NSCLCs and that tumors with *EGFR* mutations were highly sensitive to EGFR tyrosine kinase inhibitors (TKIs). Four phase III studies (NEJ 002, WJTOG3405, OPTIMAL, and EUROTAC) have opened the possibility of precision medicine for advanced NSCLC. These studies prospectively compared a TKI of gefitinib or erlotinib with cytotoxic chemotherapies as the first-line therapy in NSCLC harboring *EGFR* mutations. They reported that progression-free survival (PFS) as the primary endpoint was significantly longer with TKIs than with the standard chemotherapies (hazard ratios = 0.16–0.49). Although these studies indicated identical overall survival between the treatments, quality of life (QoL) was maintained for longer by patients treated with gefitinib in NEJ 002. Therefore, TKIs should be considered as the standard first-line therapy for advanced *EGFR*-mutated NSCLCs. Since 2009, a new step has been introduced to the treatment *algorithm* for advanced NSCLC. In this chapter, both the road to precision medicine for advanced NSCLC and the present knowledge of new-generation TKIs (afatinib or third-generation TKIs) and of TKIs in combination treatments are reviewed.

Keywords Non-small cell lung cancer (NSCLC) • *EGFR* mutation • EGFR-TKI • Gefitinib • Erlotinib • Afatinib

K. Kobayashi (✉) • H. Kagamu
Department of Respiratory Medicine, Saitama Medical University International Medical Center, 1397-1 Yamane, Hidaka, Saitama 350-1298, Japan
e-mail: care-notebook@nifty.com

© Springer Science+Business Media Singapore 2017

Y. Takiguchi (ed.), *Molecular Targeted Therapy of Lung Cancer*,
DOI 10.1007/978-981-10-2002-5_10

10.1 A Driver Mutation: *EGFR* Mutation

Dysregulation of protein kinases is frequently observed in cancer cells; therefore, protein kinases are attractive targets in the development of anticancer drugs. Small molecule inhibitors that block binding of adenosine-5-triphosphate (ATP) to the tyrosine kinase catalytic domain have been developed. The drugs gefitinib and erlotinib are the first generation of such agents and act as tyrosine kinase inhibitors (TKI) at the epidermal growth factor receptor (*EGFR*). In April 2004, three groups of researchers reported that activating mutations of *EGFR*, detected by direct sequencing, were present in a subset of non-small cell lung cancer (NSCLC) and that tumors with *EGFR* mutations were highly sensitive to EGFR-TKIs [1–3]. Namely, tumor harboring EGFR mutation of exon19 deletion or a point mutation of L858R on exon 21 is sensitive to gefitinib and erlotinib, which are called as the first generation of EGFR-TKI. It was found that T790M on exon 20 was inactive to these EGFR-TKIs.

Polymorphisms by DNA sequencing stimulated a search for protein kinase "driver" mutations, which contribute to the transformation of normal cells into proliferating cancer cells, while other protein kinase mutations ("passenger" mutations) have been found and are considered to be mutations that occur in the course of cancer cell replication and proliferation. After making a discovery of the first driver gene, EGFR mutations, in NSCLCs, new driver mutations have been identified, the fusion gene between echinoderm microtubule-associated protein-like 4 (*EML4*) and the anaplastic lymphoma kinase (*ALK*) [4], the fusion genes with ret proto-oncogene (*RET*) [5–7], and ROS proto-oncogene 1 (*ROS1*) [8], which is the human homolog of the avian sarcoma virus UR2 transforming gene v-ros and encodes a receptor tyrosine kinase of the insulin receptor family.

The era of targeted therapy in oncology commenced with the identification of cancer genetic "driver mutations" that define unique molecular subsets of NSCLC. The first instance in the field of lung cancer was *EGFR* mutations and EGFR tyrosine kinase inhibitor (TKI).

10.1.1 EGFR *Mutation Tests*

10.1.1.1 Direct Sequencing Versus PCR

Direct sequencing for *EGFR* mutations requires histology of tissue samples obtained surgically. Tumor tissue is considered the definitive sample type for *EGFR* mutation analysis; however, for many patients, this sample type is not available. An alternative approach was developed by Hagiwara and colleagues in August 2004. This new method was termed the "peptide nucleic acid-locked nucleic acid polymerase chain reaction clamp" (PNA-LNA PCR clamp) [9] and makes use of both tissue-based assessment and cytology-based analysis for *EGFR* mutations. Briefly, genomic DNA fragments surrounding mutation hot spots in the *EGFR* gene are amplified by PCR in the presence of a clamp primer synthesized from PNA with a wild-type sequence. This leads to preferential amplification of the mutant sequence, which is detected by a fluorescent primer that incorporates LNA to increase specificity. As a result, a mutant *EGFR* sequence can be detected in the presence of a 100-fold wild-type sequence. Using the PNA-LNA PCR clamp, *EGFR* mutation from small numbers of *EGFR* mutation-positive cancer cells can be detected within 3 h. The sensitivity and specificity of the PNA-LNA PCR clamp were reported to be 97 % and 100 %, respectively [10]. Therefore, *EGFR* testing by the PNA-LNA PCR clamp method is possible in patients with extremely poor performance status (PS) and of advanced age. The studies cited later, all of the NEJ001, NEJ002, and NEJ003 series, used the PNA-LNA PCR clamp [11–13].

In 2012, a comparative assessment of the performance, sensitivity, and concordance among five *EGFR* tests using PCR-Invader, PNA-LNA PCR clamp, direct sequencing, Cycleave, and Scorpion amplification refractory mutation system (ARMS) was reported [14]. With the exception of direct sequencing, the tests detected mutations at ≥1 % mutant DNA. The success rates of analysis were 91.4–100 %, and the inter-assay concordance rates of successfully analyzed samples were 94.3–100 %. Thus, cytology-derived DNA is a viable alternative to formalin-fixed paraffin-embedded (FFPE) tissue samples for analyzing *EGFR* mutations.

EGFR-mutated NSCLCs occur in approximately 31 % of patients in Japan and 16.6 % in Europe [10, 15]. In Japan, about 50,000 patients are newly diagnosed with NSCLC per year. In 2011, around 48,000 tests for EGFR mutations were carried out under national health insurance indicating that most patients with NSCLC were screened in Japan.

10.1.1.2 Circulating Cell-Free Tumor DNA

The problems associated with re-biopsies during cancer progression are a major concern and have stimulated recent attempts to develop noninvasive approaches, based on plasma or serum samples, showing a great potential for monitoring EGFR-TKI treatments. Testing for circulating cell-free tumor DNA (ctDNA), which is released from tumor cells to circulating blood, could potentially be an alternative approach.

It was pointed out [16] that there had been several reports describing a wide range of sensitivities for ctDNA, ranging from 17.1 to 100 %. Several improvements in sensitivity have been tried, for example, different sample types (plasma versus serum), different DNA extraction methods, and different mutation detection methods. And two recent meta-analyses [17, 18] indicated that the pooled sensitivities for ctDNA were reported to be 67.4 % and 62 %, respectively. While, the two meta-analyses also revealed that ctDNA had high diagnostic accuracy with tumor tissues for the detection of EGFR mutations in NSCLC. And the specificities for ctDNA were 93.5 % and 95.9 %, respectively.

The improved sensitivity and the high concordance demonstrate that *EGFR* mutation status can be accurately assessed using ctDNA and also indicate that ctDNA from plasma could be used to identify changes in EGFR mutation types, especially in T790M [19], and to determine subsequent treatments after use of EGFR-TKIs.

10.2 History of Precision Medicine by First-Generation EGFR-TKIs: Gefitinib and Erlotinib

The pivotal studies presented in 2004 [1–3] clarified the relationship between *EGFR* mutation and efficacy of EGFR-TKI treatment. It indicated that there were two subpopulations with or without *EGFR* mutations in NSCLC, resulted in the possibility of treating NSCLC patients individually. From 2004 to 2010, two types of clinical studies have been undertaken (Table 10.1). First, the clinical efficacy of EGFR-TKIs, such as gefitinib or erlotinib, has been investigated initially in *unselected* patients [20–24] and subsequently on the basis of clinical characteristics [25]. Second, an approach of precision medicine to NSCLC treatment has been developed by investigating the clinical efficacy of EGFR-TKIs following molecular selection of patients for phase II studies [26–32], followed by phase III trials [12, 33–35]. The latter studies provided the first evidence for the value of characterizing NSLC patients for *EGFR* mutations and open precision medicine for advanced NSCLC.

10.2.1 Unselected Patients

Gefitinib was evaluated in a phase III study (Iressa Survival Evaluation in Advanced Lung Cancer (ISEL)) on NSCLC patients unselected by *EGFR* status [21]. In total, 1692 patients who were refractory to chemotherapy or intolerant of chemotherapy were randomly allocated to gefitinib treatment (250 mg/day) or placebo plus best supportive care. The primary endpoint, median survival time (MST), was 5.1 months in the placebo group and 5.6 months in the gefitinib group, with no significant differences between the groups (p = 0.087). Therefore, efficacy of gefitinib in unselected NSCLC patients was not indicated. In a preplanned subgroup analysis of the ISEL trial [36], gefitinib was shown to extend survival in Asian patients (MST,

Table 10.1 EGFR-TKI studies for patients with good PS

	First-line series	Second-line series
Unselected patients	INTACT-1, INTACT-2	IDEAL 1 and 2
		BR.21
		Iressa Survival Evaluation in Lung Cancer
		V-15-32
		INTEREST
Selection by background	IPASS	
Selection by EGFR mutation	NEJ 002	
	WJTOG 3405	
	EURTAC-SLCG GECP06/01	
	OPTIMAL (CTONG 0802)	

9.5 months vs. 5.5 months; HR = 0.66, $p = 0.01$). In addition, a covariate analysis of demographic subsets of patients of Asian origin who had been treated with gefitinib showed a survival advantage for never smokers (HR, 0.37; $p = 0.0004$) and adenocarcinoma patients (HR, 0.54; $p = 0.0028$).

10.2.2 Selection by Background

In March 2006, the Iressa Pan-Asia Study (IPASS) was initiated to investigate the effectiveness of gefitinib as a first-line therapy for previously untreated patients in East Asia who had advanced pulmonary adenocarcinoma and who were light- or non-smokers [25]. A total of 1217 NSCLC patients were selected by background and received either gefitinib or carboplatin (CBDCA) plus paclitaxel (PTX). For progression-free survival (PFS), the primary endpoint of this study, an HR of 0.741 was reported (95 % CI, 0.651–0.845) for the gefitinib group. However, since the survival curves for the two groups crossed each other, it was difficult to interpret the HR (Fig. 10.1a). Cox proportional hazards model should be applied where there is a constant relationship between HR and time [37], which is not the case when crossing curves. For example, PFS of gefitinib was variously better, the same, or worse than that of CDBCA + PTX at 12, 6, and 3 months, respectively, indicating changing the HR values in time.

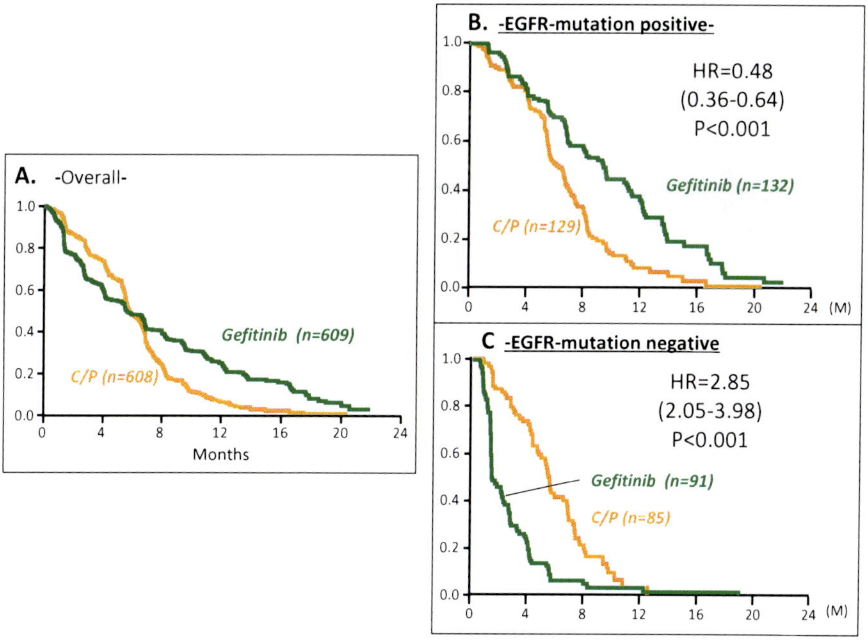

Fig. 10.1 Survival curves in IPASS study. In (**a**) the survival curves for the two groups crossed each other, indicating no indication for applying Cox model, while in (**b**, **c**) the crossing of the survival curves for the two treatment groups disappeared

Although the results regarding the primary endpoint was inconclusive, the importance of the IPASS report is demonstrated in its subset analyses. [25] An *EGFR* mutation test (amplification mutation refractory system) was performed on tumor samples from 437 of the 1217 patients (36 %). *EGFR* mutations were identified in 261 patients; PFS in these patients was significantly longer for those who received gefitinib compared to CBDCA+ PTX (HR = 0.48; $P < 0.001$); by contrast, in the 176 patients who were negative for *EGFR* mutation, PFS was significantly longer among those who received CBDCA plus PTX (HR = 2.85; $P < 0.001$). Thus, the *subset* analyses of IPASS study indicated the *possibility* of indication for gefitinib treatment in patients who were positive for *EGFR* mutations.

In addition, the biomarkers were compared in the IPASS study for *EGFR* gene copy number (fluorescent in situ hybridization), EGFR protein expression (immunohistochemistry), and EGFR mutations (direct sequencing) [38]. PFS after gefitinib treatment was significantly longer in patients whose tumors had *EGFR* mutations (HR, 0.48) and was significantly shorter in patients whose tumors had no *EGFR* mutations (HR, 3.85). Among the three biomarkers, *EGFR* mutation was the strongest predictor of PFS and tumor response to gefitinib versus CBDCA plus PTX (Fig. 10.2), while selection of patients who were of Asian origin, had an adenocarcinoma histology,

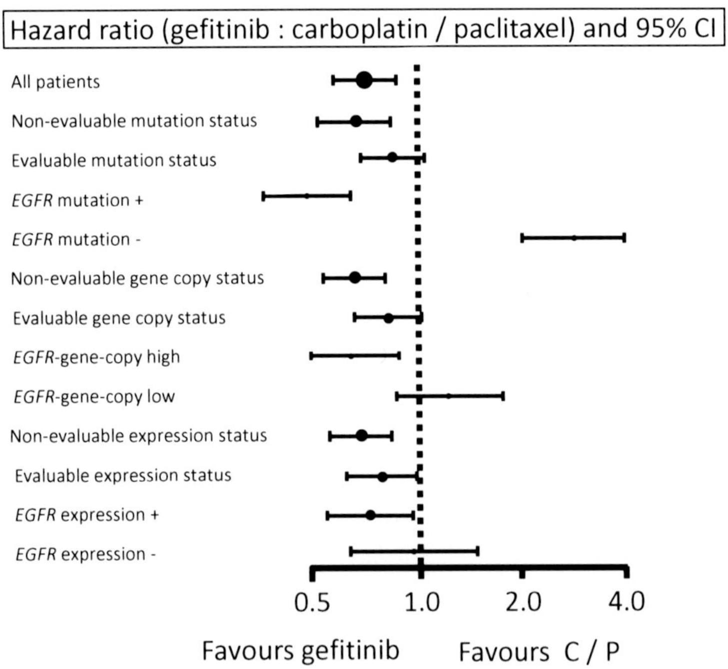

Fig. 10.2 Biomarkers for EGFR-TKI. Forest plot of progression-free survival (PFS) by epidermal growth factor receptor (EGFR) mutation status, gene copy number, and protein expression status. Hazard ratio 1 implies a lower risk of progression or death for patients treated with gefitinib. The size of the point estimate reflects the number of events in the subgroup, with a larger circle indicating more events

and were light- or non-smokers resulted in producing an *EGFR* mutation-rich population (60 %, 261 EGFR mutations/437 patients evaluated). If a strategy of selection by background is employed, there is an approximately 40 % risk of including patients for TKI treatment who do not carry *EGFR* mutations.

10.2.3 Selection by EGFR Mutation

Since the pivotal studies of 2004 reported on the relationship between EGFR mutations and TKI sensitivity, a number of phase II studies in Japan have confirmed the striking response to EGFR-TKIs in patients harboring sensitive *EGFR* mutations [26–32]. A combined analysis employing these phase II studies, named I-CAMP (IRESSA Combined Analysis of the Mutation Positives), indicated longer PFS with gefitinib than with standard chemotherapies [39]. In March 2006, at the same time that the IPASS study started, two phase III trials, the North East Japan (NEJ) 002 Study [12] and the West Japan Thoracic Oncology Group (WJTOG) 3405 [33], were initiated to compare gefitinib with standard chemotherapies as a first-line treatment for *EGFR*-mutated NSCLC (Table 10.1). NEJ 002 confirmed that PFS in the gefitinib group was significantly longer than in the CBDCA plus PTX group (10.8 months versus 5.4 months, HR = 0.30, $P < 0.001$) (Fig. 10.3). The gefitinib-treated group in WJTOG3405 also had a significantly longer PFS compared to a cisplatin plus docetaxel group, with a median PFS of 9.2 months versus 6.3 months (HR, 0.489, $p < 0.0001$).

The efficacy of another EGFR-TKI, erlotinib, was investigated in phase III studies of OPTIMAL [34] initiated in August 2008 and EURTAC [35] started in February 2007. OPTIMAL compared the PFS of erlotinib with that by gemcitabine plus carboplatin as the first-line treatment for Chinese patients with advanced *EGFR*-mutated NSCLC. Median PFS was significantly longer in erlotinib-treated patients than those on chemotherapy (13.1 vs. 4.6 months; HR = 0.16; $p < 0.0001$). In EURTAC, PFS was also used as the endpoint in a comparison of erlotinib with standard chemotherapy for first-line treatment of European patients with advanced *EGFR*-mutated NSCLC. The preplanned interim analysis showed that the median PFS was 9.7 months in the erlotinib group, compared with 5.2 months in the standard chemotherapy group (HR = 0·37; $p < 0.0001$).

A retrospective study by the National Cancer Center Hospital of Japan found that OS became significantly longer among the *EGFR*-mutant patients treated after gefitinib approval compared to those treated before gefitinib approval (MST, 27.2 vs. 13.6 months, respectively; $P < 0.001$) [40], indicating the importance of using EGFR-TKIs for *EGFR*-mutant patients. NEJ 002 [41] and WJTOG 3405 [42] showed identical OS between gefitinib and chemotherapy in first-line treatment of NSCLC patients harboring sensitive *EGFR* mutations (Fig. 10.3). Almost all patients who were treated with first-line chemotherapy in NEJ 002 and WJTOG 3405 were subsequently given a crossover treatment with gefitinib. Therefore, from the viewpoint of OS, the effect of gefitinib is additive to that of the chemotherapy.

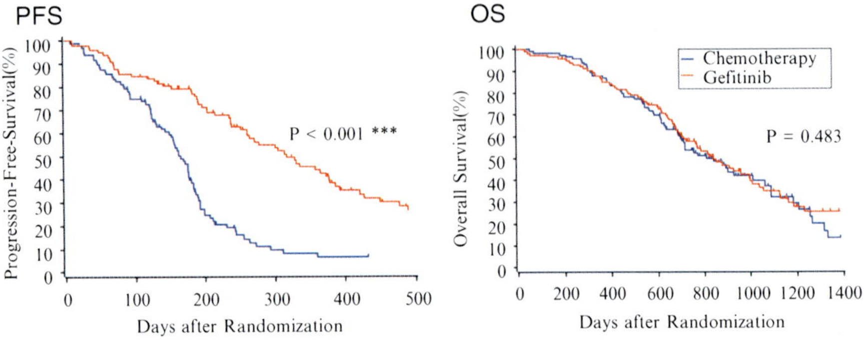

Fig. 10.3 NEJ 002 – chemotherapy vs. gefitinib. Kaplan-Meier curves for progression-free survival (PFS) (Maemondo et al. [12]) and overall survival (OS) (Inoue et al. [41]) of patients treated with gefitinib (*red line*) and those treated with standard chemotherapy (*blue line*) are shown

With regard to the timing of use of an EGFR-TKI, it is indicated that both first-line and second-line gefitinib are acceptable.

When OS is identical between two arms, then improvement in quality of life (QoL) is the key goal of treatment of NSCLC. NEJ 002 and OPTIMAL presented QoL results [43, 44]. In NEJ 002, QoL of patients was assessed weekly using the Care Notebook [45]; the primary endpoint of the QoL analysis was time to defined deterioration from a baseline on physical, mental, and life well-being scales. Kaplan-Meier probability curves and log-rank tests showed that time to 9.1 % or 27.3 % deterioration in daily functioning significantly favored gefitinib over chemotherapy (HR = 0.43, $p < 0.0001$, and HR, 0.32, $p < 0.0001$, respectively) (Fig. 10.4). It was indicated that QoL was maintained much longer in patients treated with gefitinib than in those treated with standard chemotherapy. In OPTIMAL, the Functional Assessment of Cancer Therapy (FACT) measuring system showed that, in comparison to the gemcitabine plus carboplatin group, the erlotinib group had a clinically relevant improvement in QoL, as assessed by scores on the FACT-L (73 % vs. 29.6 %; odds ratio (OR) = 6.9; $p < 0.0001$), the LCSS (75.7 % vs. 31.5 %; OR = 6.77; $p < 0.0001$), and the TOI (71.6 % vs. 24.1 %; OR = 7.79; $p < 0.0001$). These QoL results conclusively indicate that EGFR-TKIs should be considered as the standard first-line therapy for advanced EGFR-mutated NSCLC despite the lack of any survival advantage.

10.2.4 EGFR-Mutated Patients with Poor Performance Status and Advanced Age

The multicenter phase II NEJ001 study was undertaken to investigate the efficacy and feasibility of gefitinib treatment for advanced NSCLC patients harboring *EGFR* mutations, but who were ineligible for chemotherapy due to poor performance status (PS) [11]. The overall response rate in this patient group was 66 %, and median

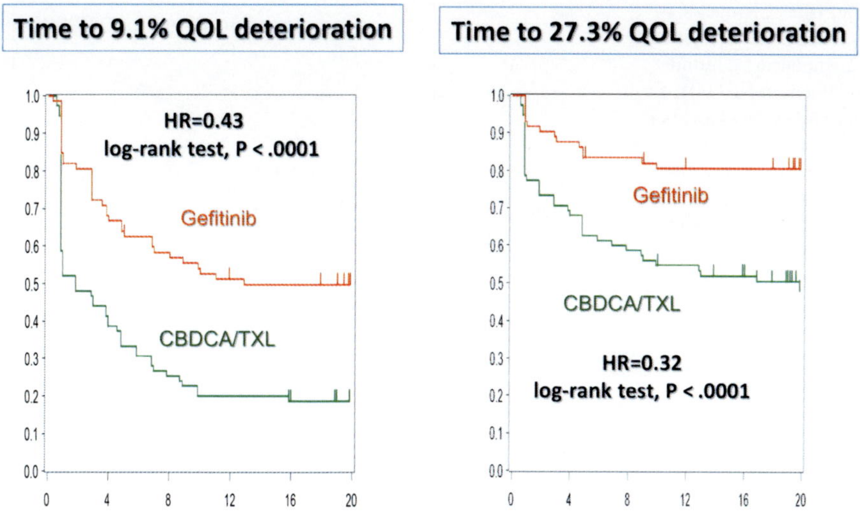

Fig. 10.4 Daily functioning in NEJ 002. Time to 9.1 and 27.3 % QoL deterioration for daily functioning are shown (Oizumi et al. [43])

PFS and MST were 6.5 months and 17.8 months, respectively. PS improved by 79 % ($p < 0.00005$); in particular, 14 (68 %) of 22 patients improved from PS ≥ 3 at baseline to PS 0 or 1 (Fig. 10.5). Thus, the "Lazarus Response" (in which Jesus restored Lazarus to life 4 days after his death) was observed in treatment-naïve, poor PS patients with NSCLC and *EGFR* mutations [46]. In patients with sensitive *EGFR* mutations but with *extremely poor PS* (suspected MST less than 4 months with best supportive care (BSC)), the difference in benefit with or without gefitinib treatment was so marked that a randomized phase III study to compare gefitinib to BSC alone may not be justified. This was the first occasion on which changes in treatment guidelines were provoked by a phase II study of NSCLC. Since previously there has been no standard treatment for these patients with short life expectancy other than BSC, examination of *EGFR* mutations as a biomarker is also strongly recommended in this patient population.

With regard to so-called "fit" elderly patients harboring *EGFR* mutations, the NEJ003 phase II study [13] investigated patients with a chemotherapy-naïve history, a median age of 80 years (range, 75–87 years), and *a* PS 0–1, who were treated with gefitinib as a first-line treatment. The response rate was 74 %, and the median PFS and OS were 12.3 months and 33.8 months, respectively. Considering its strong antitumor activity and mild toxicity, first-line gefitinib may be preferable to standard chemotherapy in this population.

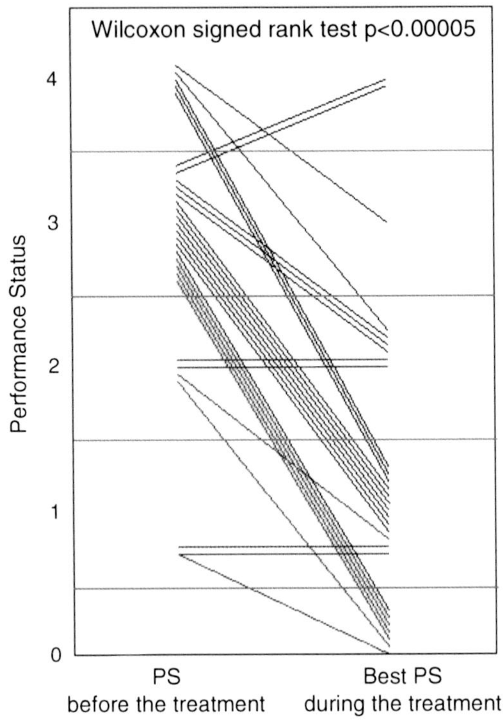

Fig. 10.5 Change of performance status of each patient during gefitinib treatment in NEJ 001. Each line shows the change of performance status (PS) of a patient from baseline to best status during the treatment. A clinically valuable improvement in 68 % of patients was observed, i.e., they improved from PS 3–4 at baseline to PS 0–1

10.2.5 Interstitial Lung Disease (ILD) After Gefitinib and Erlotinib Treatment

Common adverse events associated with EGFR-TKI treatments are diarrhea, skin rashes, and nausea; these are mild in severity and manageable. However, EGFR-TKI agents can induce ILD, which has the potential to be fatal. The incidence of ILD during EGFR-TKI treatments was reported to range from 1 to 5.4 %. The US Food and Drug Administration (FDA) reported 1 % worldwide incidence of ILD in 50,000 patients who received gefitinib in 2003 [47]. The incidence of ILD in Japanese populations was reported to be 4.0 % (95 % confidence interval, 3.0–5.1 %) [48]. These suggest that the incidence of ILD during EGFR-TKI treatments varies among Japanese and non-Japanese populations.

Risk factors for ILD have been identified in several Japanese studies: preexisting pulmonary fibrosis, poor PS, prior thoracic irradiation, male, smoking, older age (>55 years), recent NSCLC diagnosis, reduced normal lung on computed tomography scan, and concurrent cardiac disease [48]. Furthermore, a Korean study identified lower albumin levels, which might be related to poor PS, as a risk factor [49]. However, the mechanism by which EGFR-TKI may cause ILD is still unclear. Hagiwara K et al. focused on MUC4, a mucin protein encoded by the *MUC4* gene, and reported that specific polymorphisms might be associated with the risk of EGFR-TK-induced ILD (patented).

EGFR-TKI-induced ILD is often life-threatening. The main treatments for EGFR-TKI-induced ILD are immediate discontinuation of the EGFR-TKI and systemic corticosteroids; however, there have been no controlled trials to evaluate the efficacy of these strategies. Other supportive treatments include supplemental oxygen, empirical antibiotics, and mechanical ventilation. However, the mortality for gefitinib-induced ILD is approximately 30–40 % [48].

10.3 Second- and Third-Generation EGFR-TKI

10.3.1 Second-Generation EGFR-TKI: Afatinib

Second-Generation EGFR-TKIs

First-generation EGFR-TKIs were designed to inhibit ATP binding to wild-type EGFR tyrosine kinase via reversible competitive binding. By contrast, second-generation EGFR-TKIs, such as afatinib, neratinib, and dacomitinib, were designed to covalently bind to ERBB receptor family members and irreversibly block their enzymatic activity. Second-generation EGFR-TKIs inhibit tyrosine kinase activity in EGFR/ERBB1 and also in HER2/ERBB2 and HER4/ERBB4, which contain an electrophilic group capable of a Michael addition to conserved cysteine residues within the catalytic domains of EGFR (Cys797), HER2 (Cys805), and HER4 (Cys803). The precise role of HER2 and HER4 in lung cancer remains unclear; the conformation of HER2 resembles a ligand-activated state and may favor pathogenic EGFR signaling after formation of a heterodimer with EGFR [50]. The K_d of afatinib for wild-type EGFR, exon19 EGFR, and L858R EGFR is 0.25 nM, 0.11 nM, and 0.2 nM, respectively (Table 10.2) [51]. Afatinib shows comparable ability to inhibit EGFR tyrosine kinases with uncommon mutations, such as G719X or L861Q. Notably, the K_d of afatinib for EGFR with both L858R and T790M is 1.1 nM, and antitumor activity was estimated for EGFR acquire T790M by these enzymatic analyses.

Antitumor Activity of Afatinib in Preclinical Analyses

Cell-based assays showed that afatinib suppresses EGFR tyrosine kinase activity for a longer time than the reversible first-generation EGFR-TKIs; this effect is due to the irreversible suppression of kinase activity that continues until the cancer cells synthesize new EGFR [52]. The effective concentration of afatinib is one to two orders of magnitude below those needed for inhibition of colony formation by erlotinib in soft agar assays of tumor cells harboring EGFR mutations [53]. Although in vitro assays have shown the potency of afatinib (IC_{50} of 9–10 nM) against cancer cells carrying the EGFR L858R/T790M double mutation [54], a xenograft model showed that afatinib alone did not exhibit enough therapeutic efficacy against tumor cells harboring the exon20 T790M mutation. An additional anti-EGFR mAb, cetuximab, was required to overcome EGFR-TKI resistance in cells with T790M [55].

Table 10.2 A quantitative dissociation constant (K_d (μM)) of afatinib, erlotinib, and gefitinib for ERBBB family kinases

Kinases	Afatinib	Erlotinib	Gefitiib
EGFR	0.25	0.67	1
EGFR(E746 = A750del)	0.11	0.48	0.54
EGFR(G719C)	0.1	0.85	2
EGFR(G719S)	0.19	0.52	1.1
EGFR(L747 = E749del,A750P)	0.14	0.52	0.57
EGFR(L747 = S752del,P753S)	0.12	0.47	0.57
EGFR(L747 = T751del,Sins)	0.12	0.35	0.52
EGFR(L858R)	0.2	0.97	0.94
EGFR(L858R,T790M)	1.1	190	140
EGFR(L861Q)	0.23	1.2	1.4
EGFR(S752 = I759del)	0.14	1.6	0.98
EGFR(T790M)	0.61	140	40
ERBB2	5	2900	3500
ERBB3	4500	1100	790
ERBB4	6.3	230	410

Mindy I Davis, et al. Nature biotechnology, 2011

Clinical Benefits of Afatinib for Common EGFR Mutations

Two phase III trials, LUX-Lung 3 and LUX-Lung 6, have been conducted to evaluate the efficacy of first-line afatinib on overall survival of patients who have advanced lung adenocarcinoma harboring EGFR-activating mutations [56, 57, 58]. It is noteworthy that overall survival was significantly longer for patients with EGFR exon19 del in the afatinib group than in the chemotherapy group in both trials (Fig. 10.6). By contrast, there were no significant differences between treatments for patients with EGFR L858R tumors. Four randomized phase III studies conducted to evaluate first-generation EGFR-TKIs, such as gefitinib and erlotinib, for advanced lung cancer patients harboring EGFR-activating mutations could not demonstrate superiority of EGFR-TKI to platinum doublet chemotherapy in overall survival of patients with either exon19 del or L858R; in these studies, most patients who were assigned to the chemotherapy group received first-generation EGFR-TKIs as a second-line treatment. It is likely that EGFR-TKI and platinum doublet chemotherapy are mutually non-cross resistant and that the sequence of the two treatments makes no differences to antitumor efficacy. In contrast, in LUX-Lung 3 and LUX-Lung 6, most patients who were assigned to chemotherapy received first-generation EGFR-TKIs but not second-generation EGFR-TKIs as a second-line treatment. Kato et al. analyzed the Japanese subgroup in LUX-Lung 3 and demonstrated that first-line afatinib produced a consistent and significant improvement in overall survival results in patients with exon19 del but not in patients with L858R, even though most patients (93.5 %) received subsequent first-generation EGFR-TKI therapy [59]. Thus, LUX-Lung 3 and LUX-Lung 6 indicated a possible superiority of afatinib to first-generation EGFR-TKIs in EGFR exon19 del patients and showed for the first time biologically significant differences between two common activating EGFR mutations, exon19 del and L858R.

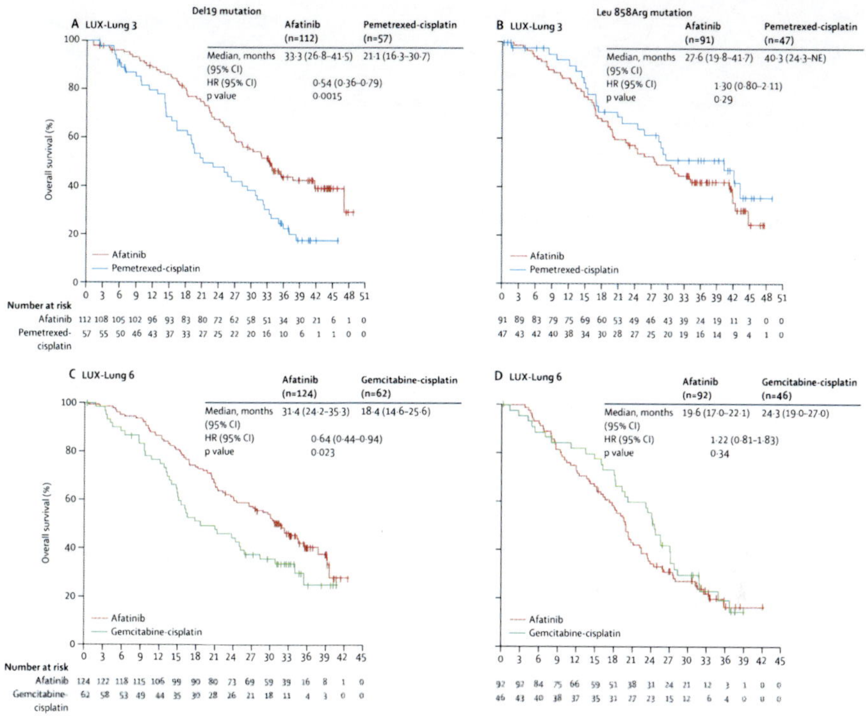

Fig. 10.6 Overall survival in lung cancer patients harboring EGFR exon19 del or L858R who were treated in LUX-Lung 3 and LUX-Lung 6. Overall survival of EGFR exon19 del-positive disease patients in (**a**) LUX-Lung 3 and (**c**) LUX-Lung 6. L858R-positive disease patients in (**b**) in LUX-Lung 3 and (**d**) LUX-Lung 6. HR, hazard ratio. *NE* not estimable

Differences Between Exon19 Del and L858R

It has been established that wild-type EGFR is primarily driven by ligand-binding-induced extracellular domain dimerization. After extracellular dimerization, a critical step in EGFR activation is the formation of an asymmetric dimer of kinase domains, in which the C-terminal lobe of one kinase domain (donor) and the N-terminal lobe of another kinase domain (acceptor) associate (Fig. 10.7a). EGFR with L858R preferentially assumes the acceptor role and requires a wild-type EGFR donor for superacceptor activity (Fig. 10.7b) [60, 61]. By contrast, EGFR with an exon19 del is active as either acceptor or donor and is oncogenic even in the absence of dimerization (Fig. 10.7c) [62]. Overall, the oncogenic signaling of exon19 del is dimerization independent and that of L858R is dependent on heterodimer formation with wild-type EGFR. It is still uncertain why afatinib and first-generation EGFR-TKI show different antitumor activities for lung cancer patients harboring exon19 del EGFR. It is likely that EGFR with exon19 del has a higher oncogenic potential and requires stronger inhibition by EGFR-TKI for optimal antitumor efficacy.

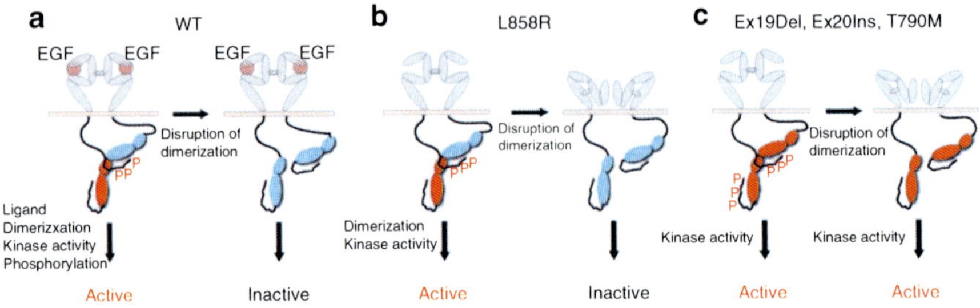

Fig. 10.7 Proposed model of cellular transformation by wild-type (**a**) EGFR and oncogenic mutations (**b**, **c**)

Clinical Benefits of Afatinib for Uncommon Mutations

Approximately 10 % of lung cancer patients with mutated EGFR harbor uncommon mutations, such as G719X, L861Q, or rare mutations. However, there is a paucity of data regarding the sensitivity of these tumors to EGFR-TKI. Yang et al. [[63]]collected data from LUX-Lung 2, LUX-Lung 3, and LUX-Lung 6 and conducted a post hoc analysis to assess antitumor efficacy of afatinib for uncommon mutations. As estimated from the affinity data for the inhibition of enzymatic activity of mutated EGFR, afatinib demonstrated significant antitumor activity against lung cancer with uncommon EGFR mutations, especially G719X, L861Q, and S768I.

Antitumor Effect of Afatinib for EGFR T790M in Clinical Trials

Although afatinib showed antitumor activity in cell-based assays, it had to be combined in vivo with anti-EGFR mAb to exhibit antitumor efficacy against cancer harboring EGFR T790M [64, 65]. One of the reasons why the in vitro antitumor activity could not be translated into in vivo antitumor efficacy is that a plasma concentration sufficient to inhibit EGFR T790M could not be obtained because nonselective inhibition of wild-type EGFR by afatinib resulted in intolerable adverse events.

Adverse Events with Afatinib

The most common grade 3 or 4 adverse events related to afatinib treatment were rash or acne, diarrhea, paronychia, and stomatitis or mucositis. Compared to first-generation EGFR-TKIs, afatinib induced grade 3 or 4 skin rashes, acne, stomatitis, and diarrhea more frequently because of its higher affinity to wild-type EGFR. The severity of these common adverse events depends on afatinib plasma concentration. By contrast, the frequency of interstitial lung disease in patients treated with afatinib is similar to that in patients treated with first-generation EGFR-TKIs [56, 57]. Grade 3 or 4 transaminase elevation was less frequent in patients treated with afatinib than in patients treated with gefitinib, because afatinib is not metabolized by cytochrome enzymes in the liver.

10.3.2 Third-Generation EGFR-TKI

Third-Generation EGFR-TKI

Although second-generation EGFR-TKIs show stronger activity for inhibiting mutated EGFR tyrosine kinases by forming irreversible covalent bonds, the nonselective inhibition of wild-type EGFR tyrosine kinase results in adverse events such as skin rashes, acne, stomatitis, and diarrhea. Furthermore, the acquisition of a T790M mutation lowers the affinity of EGFR-TKIs for mutant EGFR, and the EGFR-TKI concentrations needed to inhibit tyrosine kinase activity are not achievable due to toxicity related to nonselective inhibition of wild-type EGFR. Third-generation EGFR-TKIs are designed to inhibit EGFR tyrosine kinase harboring activating mutations and T790M through the formation of irreversible bonds while sparing the activity of wild-type EGFR tyrosine kinase. WZ4002 was the first agent to be published. Rociletinib (CO1686), which is closely related to WZ4002, is in clinical trials. Osimertinib (AZD9291) and HM61713 are other third-generation EGFR-TKIs that have progressed to clinical trials [66].

Antitumor Activity of Third-Generation EGFR-TKI in Preclinical Analysis

Rociletinib is a potent 2,4-disubstituted pyrimidine molecule that covalently modifies the conserved Cys797 in the ATP-binding pocket of the EGFR tyrosine kinase domain [67]. Enzymatic assays show that rociletinib is a potent inhibitor of EGFR tyrosine kinase with L858R and T790M and is approximately 22-fold more selective than wild-type EGFR tyrosine kinase. Cell-based assays and in vivo xenograft model assays indicated that rociletinib potently inhibits proliferation of lung cancer cells harboring EGFR with L858R or exon19 del in the presence or absence of an additional T790M mutation, while it has minimal antitumor activity against wild-type EGFR cells.

Osimertinib is a mono-anilino-pyrimidine compound that selectively inhibits mutated EGFR tyrosine kinase [68]. Osimertinib also binds to EGFR kinase irreversibly by targeting Cys797. In enzyme assays, it exhibits nearly 200 times greater potency against EGFR with L858R and T790M than wild-type EGFR. Cell-based assays, xenograft models, and transgenic models demonstrated that osimertinib shows potent antitumor activity against lung cancer with EGFR harboring L858R, exon19 del, L858R + T790M, or exon19 del + T790M, with lower activity against wild-type EGFR.

Antitumor Efficacy of Third-Generation EGFR-TKI in Clinical Trials

Phase 1–2 studies were conducted to assess the value and antitumor efficacy of rociletinib and osimertinib in lung cancer patients with an EGFR mutation who had shown disease progression during an existing EGFR-TKI treatment [69, 70]. Both rociletinib and osimertinib demonstrated potent antitumor efficacy against progressive disease in cases with a T790M mutation.

Adverse Events of Third-Generation EGFR-TKIs

The predominant grade 3 adverse event in the rociletinib phase 1–2 study was hyperglycemia at therapeutic doses [69]. Preclinical studies suggest that hyperglycemia is caused by a rociletinib metabolite that inhibits type I insulin-like growth factor and, to a lesser extent, insulin receptor kinases [71, 72]. Grade 3 prolongation of the QTc was observed but caused no symptoms. As expected, adverse events related to wild-type EGFR inhibition, such as acne, rash, and diarrhea, were infrequent and less severe with mutant EGFR-specific rociletinib.

In the osimertinib phase 1–2 study, the most common adverse events were diarrhea, rash, nausea, and decreased appetite. In contrast to rociletinib, hyperglycemia was rare [70]. Six cases of potential pneumonitis-like events were noted.

Resistant Mechanisms of Third-Generation EGFR-TKIs

Preclinical studies addressed acquired resistance mechanisms to rociletinib and osimertinib. Lung cancer cells harboring L858R and T790M mutations were exposed to increasing doses of rociletinib until resistance developed [67]. A comparison of cell morphology with parental cells indicated that rociletinib-resistant cells seemed to acquire a spindle-like morphology; no additional mutations in EGFR or any of the other oncogenes tested, including *MET*, *BRAF*, *ERBB2*, *HRAS*, *NRAS*, *KRAS*, and *PIK3CA*, were identified. RNA expression analysis of the parental and rociletinib-resistant cells demonstrated a significant enrichment of genes involved in epithelial-mesenchymal transition. Consistent with a mesenchymal cell signature, vimentin was upregulated and E-cadherin downregulated in rociletinib-resistant clones. Analysis of biopsy samples obtained from EGFR-mutant lung cancer patients with T790M who developed resistance to rociletinib showed that the lung cancer cells lost T790M upon progression in half of the patients [73]. By contrast, serial cell-free plasma DNA (cfDNA) and biopsy specimens collected from lung cancer patients who developed resistance to osimertinib indicated that acquisition of a C797S mutation or the loss of T790M contributed to the resistance mechanism [74, 75, 76]. Mutation of the EGFR C797 codon is a predicted resistance mechanism to irreversible EGFR-TKIs that form covalent bonds targeting Cys797 [77]. Collectively, it seems that tumor heterogeneity, additional C797S mutation, and the loss of T790M play a role in developing resistance to third-generation EGFR-TKIs.

10.4 Combination Therapy with EGFR-TKIs

10.4.1 Combination Therapy with Bevacizumab

The beta phase III study assessed the efficacy of erlotinib and bevacizumab combination therapy in patients with recurrent or refractory NSCLC after failure of first-line treatment [78]. Although a subgroup analysis of overall survival times seemed to favor a combination treatment of erlotinib and bevacizumab in patients with EGFR-mutated tumors compared with those with EGFR wild-type tumors, the difference did not achieve significance. Seto et al. [79] conducted a phase II study to assess the efficacy and safety of the combination of erlotinib and bevacizumab in patients with non-squamous lung cancer harboring activating EGFR mutations. The combination therapy significantly prolonged PFS. Phase III studies to evaluate the superiority of the combination of erlotinib and bevacizumab over erlotinib alone are still ongoing. The mechanisms by which bevacizumab and EGFR-TKIs exhibit synergistic effects remain to be elucidated.

10.4.2 Combination Therapy with Chemotherapy

For unselected NSCLC patients, no clinical benefits of adding EGFR-TKI to platinum-based doublet chemotherapy have been shown in phase III trials [80–82]. In the CALGB30406 trial, PFS and OS favored EGFR-mutated patients who were treated with erlotinib in combination with carboplatin + paclitaxel as a first-line chemotherapy [83]. The NEJ005/TCOG0902 study compared concurrent versus sequential alternating gefitinib and chemotherapy in NSCLC patients harboring sensitive EGFR mutations [84]. Phase III studies to assess the role of concurrent EGFR-TKI treatment to cytotoxic agents in EGFR-mutant patients have been conducted.

10.5 Treatment Modalities with EGFR-TKIs

10.5.1 Surgery with EGFR-TKIs

Antitumor therapeutic efficacy in an adjuvant setting has been examined in phase III trials (BR19, RADIANT) [85, 86]. The BR19 study assessed disease-free survival and OS of completely resected (stage IB, II, or IIIA) NSCLC patients who were randomly assigned to receive gefitinib or a placebo. No clinical benefit of gefitinib treatment has been demonstrated in patients with either EGFR wild-type tumors or EGFR mutation-positive tumors. The RADIANT study employed a randomized, double-blind, placebo-controlled protocol in patients with completely resected IB to IIIA NSCLC; the tumors in these patients were shown to express EGFR protein by immunohistochemistry or to have EGFR amplification by FISH. OS in the erlotinib and placebo groups appeared identical even in patients with EGFR mutations. Thus, the efficacy of EGFR-TKI for lung cancer in an adjuvant setting was not indicated.

10.5.2 Radiation Treatment and EGFR-TKIs

A preclinical study using cell-based assays and xenograft models indicated that erlotinib enhances the induction of apoptosis following radiation exposure and promotes an increase in radiosensitivity. Moreover, erlotinib appeared to cause a profound inhibition of tumor growth when combined with radiation. However, no clinical studies have yet demonstrated any synergistic antitumor efficacy of EGFR-TKI in patients with locally advanced NSCLC who receive curative thoracic radiation therapy.

10.5.3 Immunotherapy with EGFR-TKIs

Immune checkpoint inhibitors, such as anti-PD-1 Abs and anti-PD-L1 Abs, that promote antitumor T-cell immunity exhibit potent antitumor efficacy against both non-squamous and squamous lung cancer in phase III trials [87]. However, not all patients treated with immune checkpoint inhibitors develop effective antitumor immune reactions. The CheckMate 057 phase III clinical trial indicated that nivolumab may be less effective for lung cancer harboring EGFR mutations.

A cell-based study demonstrated that EGFR pathway activation is correlated with upregulation of PD-1, PD-L1, and CTLA-4; moreover, PD-L1 expression on tumor cells is attenuated with EGFR-TKI treatment [88]. Analysis of surgically resected lung cancer samples showed PD-L1 expression is up-regulated in patients whose cancer cells harbor activating EGFR mutations [89]. Usually, PD-L1 expression on tumor cells indicates an immune reaction in the tumor microenvironment resulting in IFNγ production by T cells, because most tumor cells express PD-L1 upon IFNγ exposure. However, EGFR signal-dependent PD-L1 expressed on EGFR-mutant tumor cells may abrogate T-cell immunity without IFNγ secretion at tumor sites. Thus, it will be necessary to investigate that inhibition of EGFR signal-dependent expression of PD-L1 by EGFR-TKI may have a synergistic effect to promote antitumor T-cell immunity and/or immune-related adverse events with immune checkpoint inhibitors.

References

1. Lynch TJ, Bell DW, Sordella R et al (2004) Activating mutations in the epidermal growth factor receptor underlying responsiveness of non-small-cell lung cancer to gefitinib. N Engl J Med 350(21):2129–2139
2. Paez JG, Janne PA, Lee JC et al (2004) EGFR mutations in lung cancer: correlation with clinical response to gefitinib therapy. Science 304(5676):1497–1500
3. Pao W, Miller V, Zakowski M et al (2004) EGF receptor gene mutations are common in lung cancers from "never smokers" and are associated with sensitivity of tumors to gefitinib and erlotinib. Proc Natl Acad Sci U S A 101(36):13306–13311
4. Soda M, Choi YL, Enomoto M et al (2007) Identification of the transforming EML4-ALK fusion gene in non-small-cell lung cancer. Nature 448(7153):561–566
5. Kohno T, Ichikawa H, Totoki Y et al (2012) KIF5B-RET fusions in lung adenocarcinoma. Nat Med 18(3):375–377
6. Takeuchi K, Soda M, Togashi Y (2012) ROS1 and ALK fusions in lung cancer. Nat Med 18(3):378–381
7. Lipson D, Capelletti M, Yelensky R et al (2012) Identification of new ALK and RET gene fusions from colorectal and lung cancer biopsies. Nat Med 18(3):382–384
8. Rikova K, Guo A, Zeng Q et al (2007) Global survey of phosphotyrosine signaling identifies oncogenic kinases in lung cancer. Cell 131:1190–1203
9. Nagai Y, Miyazawa H, Huqun et al (2005) Genetic heterogeneity of the epidermal growth factor receptor in non-small cell lung cancer cell lines revealed by a rapid and sensitive detection system, the peptide nucleic acid-locked nucleic acid PCR clamp. Cancer Res 65(16):7276–7282
10. Tanaka T, Matsuoka M, Sutani A et al (2010) Frequency of and variables associated with the EGFR mutation and its subtypes. Int J Cancer 126(3):651–655
11. Inoue A, Kobayashi K, Usui K et al (2009) First-line gefitinib for patients with advanced non-small-cell lung cancer harboring epidermal growth factor receptor mutations without indication for chemotherapy. J Clin Oncol 27(9):1394–1400
12. Maemondo M, Inoue A, Kobayashi K et al (2010) Gefitinib or chemotherapy for non-small-cell lung cancer with mutated EGFR. N Engl J Med 362(25):2380–2388
13. Maemondo M, Minegishi Y, Inoue A et al (2012) First-line gefitinib in patients aged 75 or older with advanced non-small cell lung cancer harboring epidermal growth factor receptor mutations: NEJ 003 study. J Thorac Oncol 7(9):1417–1422
14. Goto K, Satouchi M, Ishii G et al (2012) An evaluation study of EGFR mutation tests utilized for non-small-cell lung cancer in the diagnostic setting. Ann Oncol 23(11):2914–2919
15. Rosell R, Moran T, Queralt C et al (2009) Screening for epidermal growth factor receptor mutations in lung cancer. N Engl J Med 361(10):958–967
16. Sun W, Yuan X, Tian Y et al (2015) Non-invasive approaches to monitor EGFR-TKI treatment in non-small-cell lung cancer. J Hematol Oncol 8:95
17. Qiu M, Wang J, Xu Y et al (2015) Circulating tumor DNA is effective for the detection of EGFR mutation in non-small cell lung cancer: a meta-analysis. Cancer Epidemiol Biomark Prev 24(1):206–212
18. Luo J, Shen L, Zheng D (2014) Diagnostic value of circulating free DNA for the detection of EGFR mutation status in NSCLC: a systematic review and meta-analysis. Sci Rep 4:6269
19. Maheswaran S, Sequist LV, Nagrath S et al (2008) Detection of mutations in EGFR in circulating lung-cancer cells. N Engl J Med 359(4):366–377
20. Shepherd FA, Rodrigues Pereira J et al (2005) Erlotinib in previously treated non-small-cell lung cancer. N Engl J Med 353(2):123–132
21. Thatcher N, Chang A, Parikh P et al (2005) Gefitinib plus best supportive care in previously treated patients with refractory advanced non-small-cell lung cancer: results from a randomised, placebo controlled, multicentre study (Iressa Survival Evaluation in Lung Cancer). Lancet 366:1527–1537

22. Kim ES, Hirsh V, Mok T et al (2008) Gefitinib versus docetaxel in previously treated non-small-cell lung cancer (INTEREST): a randomised phase 3 trial. Lancet 372(9652):1809–1818
23. Maruyama R, Nishiwaki Y, Tamura T et al (2008) Phase III study, V-15-32, of gefitinib versus docetaxel in previously treated Japanese patients with non-small-cell lung cancer. J Clin Oncol 26(26):4244–4252
24. Lee DH, Park K, Kim JH et al (2010) Randomized Phase III trial of gefitinib versus docetaxel in non-small cell lung cancer patients who have previously received platinum-based chemotherapy. Clin Cancer Res 16(4):1307–1314
25. Mok TS, Wu YL, Thongprasert S et al (2009) Gefitinib or carboplatin-paclitaxel in pulmonary adenocarcinoma. N Engl J Med 361(10):947–957
26. Inoue A, Suzuki T, Fukuhara T et al (2006) Prospective phase II study of gefitinib for chemotherapy-naive patients with advanced non-small-cell lung cancer with epidermal growth factor receptor gene mutations. J Clin Oncol 24:3340–3346
27. Asahina H, Yamazaki K, Kinoshita I et al (2006) A phase II trial of gefitinib as first-line therapy for advanced non-small cell lung cancer with epidermal growth factor receptor mutations. Br J Cancer 95:998–1004
28. Sutani A, Nagai Y, Udagawa K et al (2006) Gefitinib for non-small-cell lung cancer patients with epidermal growth factor receptor gene mutations screened by peptide nucleic acid-locked nucleic acid PCR clamp. Br J Cancer 95:1483–1489
29. Yoshida K, Yatabe Y, Park JY et al (2007) Prospective validation for prediction of gefitinib sensitivity by epidermal growth factor receptor gene mutation in patients with non-small cell lung cancer. J Thorac Oncol 2:22–28
30. Sunaga N, Tomizawa Y, Yanagitani N et al (2007) Phase II prospective study of the efficacy of gefitinib for the treatment of stage III/IV non-small cell lung cancer with EGFR mutations, irrespective of previous chemotherapy. Lung Cancer 56:383–389
31. Tamura K, Okamoto I, Kashii T et al (2008) Multicentre prospective phase II trial of gefitinib for advanced non-small cell lung cancer with epidermal growth factor receptor mutations: results of the West Japan Thoracic Oncology Group trial (WJTOG0403). Br J Cancer 98:907–914
32. Sugio K, Uramoto H, Onitsuka T et al (2009) Prospective phase II study of gefitinib in non-small cell lung cancer with epidermal growth factor receptor gene mutations. Lung Cancer 64(3):314–318
33. Mitsudomi T, Morita S, Yatabe Y et al (2010) Gefitinib versus cisplatin plus docetaxel in patients with non-small-cell lung cancer harbouring mutations of the epidermal growth factor receptor (WJTOG3405): an open label, randomised phase 3 trial. Lancet Oncol 11(2):121–128
34. Zhou C, Wu YL, Chen G et al (2011) Erlotinib versus chemotherapy as first-line treatment for patients with advanced EGFR mutation-positive non-small-cell lung cancer (OPTIMAL, CTONG-0802): a multicentre, open-label, randomised, phase 3 study. Lancet Oncol 12:735–742
35. Rosell R, Carcereny E, Gervais R et al (2012) Erlotinib versus standard chemotherapy as first-line treatment for European patients with advanced EGFR mutation-positive non-small-cell lung cancer (EURTAC): a multicentre, open-label, randomised phase 3 trial. Lancet Oncol 13:239–246
36. Chang A, Parikh P, Thongprasert S et al (2006) Gefitinib (IRESSA) in patients of Asian origin with refractory advanced non-small cell lung cancer: subset analysis from the ISEL study. J Thorac Oncol 1(8):847–855
37. Cox DR (1972) Regression models and life-tables. J R Stat Soc Ser B Methodol 34(2):187–220
38. Fukuoka M, Wu YL, Thongprasert S et al (2011) Biomarker analyses and final overall survival results from a phase III, randomized, open-label, first-line study of gefitinib versus carboplatin/paclitaxel in clinically selected patients with advanced non-small-cell lung cancer in Asia (IPASS). J Clin Oncol 29(21):2866–2874

39. Morita S, Okamoto I, Kobayashi K et al (2009) Combined survival analysis of prospective clinical trials of gefitinib for non-small cell lung cancer with EGFR mutations. Clin Cancer Res 15(13):4493–4498
40. Takano T, Fukui T, Ohe Y et al (2008) EGFR mutations predict survival benefit from gefitinib in patients with advanced lung adenocarcinoma: a historical comparison of patients treated before and after gefitinib approval in Japan. J Clin Oncol 26:5589–5595
41. Inoue A, Kobayashi K, Maemondo M et al (2013) Updated overall survival results from a randomized phase III trial comparing gefitinib with carboplatin–paclitaxel for chemo-Naïve non-small 5 cell lung cancer with sensitive EGFR gene mutations (NEJ002). Ann Oncol 24(1):54–59
42. Mitsudomi T, Morita S, Yatabe Y, et al (2012) Updated overall survival results of WJTOG 3405, a randomized phase III trial comparing gefitinib with cisplatin plus docetaxel as the first-line treatment for patients with non-small cell lung cancer harboring mutations of the epidermal growth factor receptor (EGFR). J Clin Oncol 30 suppl; ASCO 2012 abstr 7521
43. Oizumi S, Kobayashi K, Inoue A et al (2012) Quality of life with gefitinib in patients with EGFR-mutated non-small cell lung cancer: quality of life analysis of North East Japan Study Group 002 Trial. Oncologist 17(6):863–870
44. ZhouC, Wu YL, Chen G, et al (2011) Updated efficacy and quality-of-life (QoL) analyses in OPTIMAL, a phase III, randomized, open-label study of first-line erlotinib versus gemcitabine/carboplatin in patients with *EGFR*-activating mutation-positive (*EGFR* Act Mut+) advanced non-small cell lung cancer (NSCLC). J Clin Oncol 29 suppl; ASCO 2011 abstr 7520
45. Kobayashi K, Green J, Shimonagayoshi M et al (2005) Validation of the care notebook for measuring physical, mental and life well-being of patients with cancer. Qual Life Res 14(4):1035–1043
46. Langer CJ (2009) The "lazarus response" in treatment-naive, poor performance status patients with non-small-cell lung cancer and epidermal growth factor receptor mutation. J Clin Oncol 27(9):1350–1354
47. Cohen MH, Williams GA, Sridhara R et al (2003) FDA drug approval summary: gefitinib (ZD1839) (Iressa) tablets. Oncologist 8:303–306
48. Kudoh S, Kato H, Nishiwaki Y et al Interstitial lung disease in Japanese patients with lung cancer: a cohort and nested case-control study. Am J Respir Crit Care Med 177:1348–1357
49. Beom SH, Kim DW, Sim SH (2015) Gefitinib-induced interstitial lung disease in Korean lung cancer patients. Cancer Res Treat. doi:10.4143/crt.2014.201 [Epub ahead of print]
50. Cho HS, Mason K, Ramyar KX et al (2003) Structure of the extracellular region of HER2 alone and in complex with the Herceptin Fab. Nature 421(6924):756–760
51. Davis MI, Hunt JP, Herrgard S et al (2011) Comprehensive analysis of kinase inhibitor selectivity. Nat Biotechnol 29(11):1046–1051
52. Spicer JF, Rudman SM (2010) EGFR inhibitors in non-small cell lung cancer (NSCLC): the emerging role of the dual irreversible EGFR/HER2 inhibitor BIBW 2992. Target Oncol 5(4):245–255
53. Li D, Ambrogio L, Shimamura T et al (2008) BIBW2992, an irreversible EGFR/HER2 inhibitor highly effective in preclinical lung cancer models. Oncogene 27(34):4702–4711
54. Solca F, Dahl G, Zoephel A et al (2012) Target binding properties and cellular activity of afatinib (BIBW 2992), an irreversible ErbB family blocker. J Pharmacol Exp Ther 343(2):342–350
55. Regales L, Gong Y, Shen R et al (2009) Dual targeting of EGFR can overcome a major drug resistance mutation in mouse models of EGFR mutant lung cancer. J Clin Invest 119(10):3000–3010
56. Sequist LV, Yang JC, Yamamoto N et al (2013) Phase III study of afatinib or cisplatin plus pemetrexed in patients with metastatic lung adenocarcinoma with EGFR mutations. J Clin Oncol 31(27):3327–3334
57. Wu Y-L, Zhou C, Hu C-P et al (2014) Afatinib versus cisplatin plus gemcitabine for first-line treatment of Asian patients with advanced non-small-cell lung cancer harbouring EGFR mutations (LUX-Lung 6): an open-label, randomised phase 3 trial. Lancet Oncol 15(2):213–222

58. Yang JC-H, Wu Y-L, Schuler M et al (2015) Afatinib versus cisplatin-based chemotherapy for EGFR mutation-positive lung adenocarcinoma (LUX-Lung 3 and LUX-Lung 6): analysis of overall survival data from two randomised, phase 3 trials. Lancet Oncol 16(2):141–151
59. Kato T, Yoshioka H, Okamoto I et al (2015) Afatinib versus cisplatin plus pemetrexed in Japanese patients with advanced non-small cell lung cancer harboring activating EGFR mutations: subgroup analysis of LUX-Lung 3. Cancer Sci 106(9):1202–1211
60. Red Brewer M, Yun CH, Lai D, Lemmon MA, Eck MJ, Pao W (2013) Mechanism for activation of mutated epidermal growth factor receptors in lung cancer. Proc Natl Acad Sci U S A 110(38):E3595–E3604
61. Shan Y, Eastwood MP, Zhang X et al (2012) Oncogenic mutations counteract intrinsic disorder in the EGFR kinase and promote receptor dimerization. Cell 149(4):860–870
62. Cho J, Chen L, Sangji N et al (2013) Cetuximab response of lung cancer-derived EGF receptor mutants is associated with asymmetric dimerization. Cancer Res 73(22):6770–6779
63. Yang JCH, Sequist LV, Geater SL et al (2015) Clinical activity of afatinib in patients with advanced non-small-cell lung cancer harbouring uncommon EGFR mutations: a combined post-hoc analysis of LUX-Lung 2, LUX-Lung 3, and LUX-Lung 6. Lancet Oncol 16(7):830–838
64. Katakami N, Atagi S, Goto K et al (2013) LUX-Lung 4: a phase II trial of afatinib in patients with advanced non-small-cell lung cancer who progressed during prior treatment with erlotinib, gefitinib, or both. J Clin Oncol 31(27):3335–3341
65. Janjigian YY, Smit EF, Groen HJ et al (2014) Dual inhibition of EGFR with afatinib and cetuximab in kinase inhibitor-resistant EGFR-mutant lung cancer with and without T790M mutations. Cancer Discov 4(9):1036–1045
66. Riely GJ, Yu HA (2015) EGFR: the paradigm of an oncogene-driven lung cancer. Clin Cancer Res 21(10):2221–2226
67. Walter AO, Sjin RT, Haringsma HJ et al (2013) Discovery of a mutant-selective covalent inhibitor of EGFR that overcomes T790M-mediated resistance in NSCLC. Cancer Discov 3(12):1404–1415
68. Cross DA, Ashton SE, Ghiorghiu S et al (2014) AZD9291, an irreversible EGFR TKI, overcomes T790M-mediated resistance to EGFR inhibitors in lung cancer. Cancer Discov 4(9):1046–1061
69. Sequist LV, Soria JC, Goldman JW et al (2015) Rociletinib in EGFR-mutated non-small-cell lung cancer. N Engl J Med 372(18):1700–1709
70. Janne PA, Yang JC, Kim DW et al (2015) AZD9291 in EGFR inhibitor-resistant non-small-cell lung cancer. N Engl J Med 372(18):1689–1699
71. Vazquez-Martin A, Cufi S, Oliveras-Ferraros C et al (2013) IGF-1R/epithelial-to-mesenchymal transition (EMT) crosstalk suppresses the erlotinib-sensitizing effect of EGFR exon 19 deletion mutations. Sci Rep 3:2560
72. Cortot AB, Repellin CE, Shimamura T et al (2013) Resistance to irreversible EGF receptor tyrosine kinase inhibitors through a multistep mechanism involving the IGF1R pathway. Cancer Res 73(2):834–843
73. Piotrowska Z, Niederst MJ, Karlovich CA et al (2015) Heterogeneity underlies the emergence of EGFRT790 wild-type clones following treatment of T790M-positive cancers with a third-generation EGFR inhibitor. Cancer Discov 5(7):713–722
74. Thress KS, Paweletz CP, Felip E et al (2015) Acquired EGFR C797S mutation mediates resistance to AZD9291 in non-small cell lung cancer harboring EGFR T790M. Nat Med 21(6):560–562
75. Planchard D, Loriot Y, Andre F et al (2015) EGFR-independent mechanisms of acquired resistance to AZD9291 in EGFR T790M-positive NSCLC patients. Ann Oncol 26(10):2073–2078
76. Yu HA, Tian SK, Drilon AE et al (2015) Acquired resistance of EGFR-Mutant Lung Cancer to a T790M-specific EGFR inhibitor: emergence of a third mutation (C797S) in the EGFR Tyrosine Kinase Domain. JAMA Oncol 1(7):982–984

77. Yu Z, Boggon TJ, Kobayashi S et al (2007) Resistance to an irreversible epidermal growth factor receptor (EGFR) inhibitor in EGFR-mutant lung cancer reveals novel treatment strategies. Cancer Res 67(21):10417–10427
78. Herbst RS, Ansari R, Bustin F et al (2011) Efficacy of bevacizumab plus erlotinib versus erlotinib alone in advanced non-small-cell lung cancer after failure of standard first-line chemotherapy (BeTa): a double-blind, placebo-controlled, phase 3 trial. Lancet 377(9780):1846–1854
79. Seto T, Kato T, Nishio M et al (2014) Erlotinib alone or with bevacizumab as first-line therapy in patients with advanced non-squamous non-small-cell lung cancer harbouring EGFR mutations (JO25567): an open-label, randomised, multicentre, phase 2 study. Lancet Oncol 15(11):1236–1244
80. Giaccone G, Herbst RS, Manegold C et al (2004) Gefitinib in combination with gemcitabine and cisplatin in advanced non-small-cell lung cancer: a phase III trial--INTACT 1. J Clin Oncol 22(5):777–784
81. Herbst RS, Giaccone G, Schiller JH et al (2004) Gefitinib in combination with paclitaxel and carboplatin in advanced non-small-cell lung cancer: a phase III trial—INTACT 2. J Clin Oncol 22(5):785–794
82. Herbst RS, Prager D, Hermann R et al (2005) TRIBUTE: a phase III trial of erlotinib hydrochloride (OSI-774) combined with carboplatin and paclitaxel chemotherapy in advanced non-small-cell lung cancer. J Clin Oncol 23(25):5892–5899
83. Janne PA, Wang X, Socinski MA et al (2012) Randomized phase II trial of erlotinib alone or with carboplatin and paclitaxel in patients who were never or light former smokers with advanced lung adenocarcinoma: CALGB 30406 trial. J Clin Oncol 30(17):2063–2069
84. Sugawara S, Oizumi S, Minato K et al (2015) Randomized phase II study of concurrent versus sequential alternating gefitinib and chemotherapy in previously untreated non-small cell lung cancer with sensitive EGFR mutations: NEJ005/TCOG0902. Ann Oncol 26(5):888–894
85. Goss GD, O'Callaghan C, Lorimer I et al (2013) Gefitinib versus placebo in completely resected non-small-cell lung cancer: results of the NCIC CTG BR19 study. J Clin Oncol 31(27):3320–3326
86. Kelly K, Altorki NK, Eberhardt WE et al (2015) Adjuvant erlotinib versus placebo in patients with stage IB-IIIA non-small-cell lung cancer (RADIANT): a randomized, double-blind, Phase III trial. J Clin Oncol 33:4007–4014
87. Brahmer J, Reckamp KL, Baas P et al (2015) Nivolumab versus docetaxel in advanced squamous-cell non-small-cell lung cancer. N Engl J Med 373(2):123–135
88. Akbay EA, Koyama S, Carretero J et al (2013) Activation of the PD-1 pathway contributes to immune escape in EGFR-driven lung tumors. Cancer Discov 3(12):1355–1363
89. Azuma K, Ota K, Kawahara A et al (2014) Association of PD-L1 overexpression with activating EGFR mutations in surgically resected nonsmall-cell lung cancer. Ann Oncol 25(10):1935–1940

Chapter 11
ALK Mutant

Akihiko Gemma

Abstract Anaplastic lymphoma tyrosine kinase (ALK) inhibitors have strong antitumor effects in patients with non-small cell lung cancer (NSCLC) with ALK fusion gene. The ALK inhibitors crizotinib, ceritinib, and alectinib were developed. Companion diagnostic and therapeutic agents for specific ALK inhibition have been simultaneously approved, but this is causing severe inconvenience in clinical practice for diagnosing ALK-positive lung cancer.

The therapeutic strategy for the patients is mainly by molecularly targeted therapy. The current status of ALK inhibitors and specificity of biomarkers in ALK-positive lung cancer are reviewed in this study. In summary, there are many arguments relating to the appropriate use of crizotinib, ceritinib, or alectinib as the situation demands and regarding which agent to use first. Many clinicians question the limitations of companion diagnostics and therapeutic agents; a more flexible response will be expected in order to accurately diagnose and provide proper treatment of ALK-positive lung cancer.

Keywords Alk inhibitor • Molecular diagnosis • Molecular targeting

11.1 Introduction

Anaplastic lymphoma tyrosine kinase (ALK) inhibitors produce strong antitumor effects in patients with non-small cell lung cancer (NSCLC) who test positive for the ALK fusion gene [1–3]. Development of the ALK inhibitors crizotinib, ceritinib, and alectinib has raised issues on the future use of ALK inhibitors in general.

The use of fluorescence in situ hybridization (FISH), immunohistochemistry (IHC), and reverse transcription-polymerase chain reaction (RT-PCR) as laboratory procedures for diagnosing ALK-positive lung cancer requires an organized diagnostic algorithm. In addition, companion diagnostic and therapeutic agents for specific

A. Gemma (✉)
Division of Pulmonary Medicine, Graduate School of Medicine, Nippon Medical School, 1-1-5 Sendagi, Bunkyo-ku, Tokyo 113-8602, Japan
e-mail: agemma@nms.ac.jp

© Springer Science+Business Media Singapore 2017
Y. Takiguchi (ed.), *Molecular Targeted Therapy of Lung Cancer*,
DOI 10.1007/978-981-10-2002-5_11

ALK inhibition have been simultaneously approved, but this is causing severe inconvenience in clinical practice.

The therapeutic strategy whereby patients with a driver mutation receive the corresponding molecularly targeted therapy is the same as that used in epidermal growth factor receptor (EGFR) mutation-positive lung cancer. However, because of the difficulty of fusion gene detection, the relationship between companion diagnostics and ALK inhibitors, etc., there are specificities and issues not shared by EGFR-positive lung cancer in terms of ALK-positive lung cancer treatment.

The current status of ALK inhibitors and specificity of biomarkers in ALK-positive lung cancer are reviewed in this study.

11.2 The Guidelines

In 2011, the US Food and Drug Administration (FDA) granted approval for crizotinib as a therapeutic agent for ALK-positive NSCLC, and the Japanese government approved the manufacture and sale of this drug in 2012. In the PROFILE 1007 trial, progression-free survival (PFS) and response rate (RR) in patients previously treated for ALK fusion-positive NSCLC were 7.7 months and 65 %, respectively [1]. PFS and RR in treatment-naïve patients with ALK fusion-positive NSCLC were 10.9 months and 74 %, respectively, in the PROFILE 1014 trial. The ALK/MET/ROS1 inhibitor crizotinib showed a significant improvement in PFS compared with that of standard chemotherapy in the same trial [2]. Reported adverse events were visual disorders, diarrhea, nausea, an increased aspartate aminotransferase/alanine aminotransferase (AST/ALT) ratio, etc. These results indicate that crizotinib should be given in the early stage of treatment, at least up to the point of second-line treatment for patients with ALK-positive advanced NSCLC.

The second-generation ALK inhibitor alectinib selectively functions by fusing ALK gene, and efficacy of this agent has been demonstrated in crizotinib-resistant tumors harboring the L1196M and C1156Y gatekeeper mutations in vitro. When a phase I/II trial (AF-001JP) with alectinib in treatment-naïve patients with ALK-positive advanced NSCLC was conducted in Japan, the RR was 93.5 % [3]. Observed grade 3 or higher adverse reactions were neutropenia (4 %) and increased levels of serum bilirubin (2 %), ALT (2 %), and creatine phosphokinase (CPK, 2 %), indicating a mild incidence of adverse events [3]. The Japanese government approved the manufacture and sale of this drug in July 2014 based on results from this clinical trial. Ceritinib (also called LDK378), another second-generation ALK inhibitor, has more potent ALK inhibitory activity against crizotinib-resistant ALK-positive NSCLC, also confirming an antitumor effect (RR: 56 %). An application for ceritinib was approved by the Japanese government after the FDA approval in April 2014 [4].

The challenge with a diagnostic approach to the ALK fusion gene is the difficulty with detection, because the site of mutation is not confined as with EGFR gene mutation. Attention must be paid to the diagnosis due to advantages and disadvantages of FISH, IHC, or RT-PCR. The Biomarker Committee of the Japan Lung Cancer Society has issued guidelines for genetic testing for the ALK fusion gene in December 2011, and it is currently recommended that ALK-positive results should be detectable by as many laboratory procedures (two or more) as possible [5]. However, it is difficult to perform such procedures in every lung cancer patient whose tissue specimens may be difficult to obtain. Achievement of a reliable diagnosis in every patient is contributing to the establishment of a robust methodology. In verification of ALK testing results from 2337 specimens performed by Pfizer Inc., inconsistency has been reported between FISH and IHC results, particularly many FISH-positive and IHC-negative patterns with low RR. Yatabe performed reanalysis with 14 patients who showed inconsistency between FISH and highly sensitive IHC results, and as a result, RRs in FISH-positive ALK and IHC-positive ALK were 20 % and 50 %, respectively [6]. A previous clinical trial with crizotinib reported that RR was approximately 60 % in patients assessed on the basis of FISH positivity alone, but when RR was assessed along with IHC and RT-PCR in FISH-positive patients, RR reportedly exceeded 80 % [7]. On the other hand, the AF-001JP trial with alectinib was conducted in Japan by considering "FISH positive and IHC positive" or "RT-PCR positive" to indicate ALK fusion-positive results and showed an extremely good RR (93.5 %) [3].

FISH has become established as a diagnostic technique. The above clinical trial with crizotinib was conducted in FISH-positive patients, but problems using FISH as a false-positive diagnosis still remained. Because RT-PCR is a highly sensitive and specific technique, there can be little doubt that RT-PCR diagnoses ALK RT-PCR-positive patients, but high-quality RNA is required for such diagnosis, and false-negative results may occur when fusion genes involve translocation partners. Furthermore, in present circumstances, therapeutic agents have not been commercially available.

ALK is usually rarely expressed in lung tissues. If IHC detects these proteins as ALK positive, they are most likely to be ALK fusion positive. In this sense the IHC technique is highly effective. It has been difficult to detect a very small amount of ALK fusion protein expressed in lung tissues using conventional IHC methods. However, in recent years highly sensitive IHC has allowed us to amplify highly sensitive antibodies (clone 5A4 and D5F3) and signals using a high-sensitivity visualization system (iAEP, EnVision FLEX), resulting in high sensitivity and specificity in detection. In the present circumstances, we conducted screening using a highly sensitive IHC, followed by FISH methods for verification of IHC. IHC- and FISH-positive patients are considered to be cases which respond to ALK inhibitors. Even when only one of these tests is positive, the patient is considered to be responsive to ALK inhibitors; therefore, we need to decide whether or not to administer ALK inhibitors on the basis of a balance between clinical benefit and clinical harm.

11.3 Related Clinical Trials

11.3.1 Crizotinib Therapy Versus Chemotherapy for Advanced ALK Fusion-Positive Lung Cancer (PROFILE 1007)

Purpose Conduct a prospective study to compare the effectiveness of crizotinib therapy versus chemotherapy in patients previously treated for ALK fusion-positive lung cancer.

Methods ALK FISH fusion-positive tumors showed platinum-based chemotherapy-related exacerbation of ALK-positive locally advanced or metastatic NSCLC confirmed as median values.

Patients were randomly assigned in a crizotinib:pemetrexed/docetaxel 1:1 ratio.

Primary endpoint: PFS (central decision)

Results Three hundred and forty-seven patients were enrolled between February 2010 and February 2012 (crizotinib therapy group 173 patients, chemotherapy group 174 patients). PFSs were 7.7 months in the crizotinib therapy group and 3.0 months in the chemotherapy group [hazard ratio (HR): 0.49, $p < 0.001$]; RRs were 65 % and 20 % in the crizotinib and chemotherapy groups, respectively ($p < 0.001$). There was no significant difference in overall survival (OS) time at the data cutoff point (HR: 1.02, $p = 0.54$).

Conclusion Crizotinib therapy was more effective than chemotherapy in patients previously treated for ALK fusion-positive NSCLC.

11.3.2 Crizotinib Therapy Versus Chemotherapy in the Early Stage of Treatment of ALK-Positive Lung Cancer (PROFILE 1014)

Purpose To examine and compare the effectiveness of crizotinib therapy versus chemotherapy in the early stage of treatment of ALK-positive NSCLC

Methods A phase III clinical trial was conducted by comparing crizotinib therapy and standard chemotherapy in 343 treatment-naïve patients with ALK-positive non-squamous NSCLC. Patients in the crizotinib therapy group were administered crizotinib 250 mg twice a day, and those in the chemotherapy group were administered pemetrexed 500 mg/m^2 + cisplatin (CDDP) 75 mg/m^2 or carboplatin (CBDCA) area under the curve (AUC) = 5–6, every 3 weeks for 6 cycles.

Primary endpoint: PFS.

Results Median PFS was significantly longer in the crizotinib therapy group compared with the chemotherapy group (10.9 months versus 7.9 months, HR: 0.45,

$p < 0.001$). RRs were 74 % in the crizotinib therapy group and 45 % in the chemotherapy group ($p < 0.001$). Median OS was not achieved in either group. One-year survival rates in the crizotinib therapy group and chemotherapy group were 84 % and 79 %, respectively. Adverse events on crizotinib therapy were mainly visual disorders, diarrhea, nausea, and edema and in chemotherapy were mainly nausea, vomiting, malaise, and decreased appetite. There was a positive correlation between decline of symptoms of lung cancer and improvement of quality of life in the crizotinib therapy group.

Conclusion Crizotinib therapy was more effective than standard chemotherapy (pemetrexed + platinum-based chemotherapy) in treatment-naïve patients with ALK-positive NSCLC.

11.3.3 Phase I/II Trial with Alectinib in Advanced ALK-Positive NSCLC (AF-001JP)

Purpose Safety and efficacy study of alectinib, a new selective oral ALK inhibitor

Methods A phase I trial was conducted to determine dose-limiting toxicity (DLT), maximum tolerated dose (MTD), and recommended dose (RD) of alectinib, and also a phase II trial based on the RD was conducted in patients with advanced ALK fusion-positive NSCLC who were previously untreated with ALK inhibitors.

Primary endpoint (phase II trial): RR

Results Twenty-four patients (phase I trial) and forty-six patients (phase II trial) were enrolled between September 2010 and August 2012. Because DLT of alectinib was not determined in the phase I trial, alectinib MTD (300 mg twice a day) was administered to patients as RD in the phase II trial.

In the phase II trial, RR (a primary endpoint) was 93.5 %; no grade 4 or higher adverse events were observed; and grade 3 or higher adverse reactions included neutropenia (4 %), increased levels of CPK (2 %), etc.

Conclusion Alectinib showed strong tolerability and effectiveness in advanced ALK fusion-positive NSCLC.

11.4 Limitations and Challenges of Clinical Trials

PFS was significantly longer in the crizotinib therapy group in both the PROFILE 1007 and 1014 trials, but OS did not show a difference compared with chemotherapy because of the number of events and crossover designs. Furthermore, these clinical trials were conducted in FISH-positive patients. Compared with RR in FISH-positive and IHC-positive/RT-PCR-positive patients, RR tended to be lower in FISH-positive only patients. Therefore, false-positive results may be included in ALK FISH-positive patents in both clinical trials.

Although the manufacture and sale of alectinib has been approved by the Japanese government on the basis of the results of the AF-001JP trial, only a small number of patients (46 cases) have been treated with alectinib, and insufficient information about adverse events and side effects management is available. It is necessary to await the reports of case studies and results of the phase III trial (J-ALEX trial).

11.5 Words of Caution

The PROFILE 1007 and 1014 trials were conducted in FISH-positive patients, but currently in Japan confirmation of a positive test result using FISH after highly sensitive IHC screening is the standard for ALK mutation testing. A specific companion diagnostic for each ALK inhibitor is basically required (e.g., Vysis ALK Break Apart FISH for crizotinib and ALK iAEP® IHC kit and Vysis FISH kit for alectinib). Because there are some inconsistencies in clinical practice, the Biomarker Committee of the Japan Lung Cancer Society has requested flexible responses in order to prevent patient disadvantages that could accompany these inconsistencies.

11.6 Comments

1. There are many arguments relating to the appropriate use of crizotinib, ceritinib, or alectinib as the situation demands and regarding which agent to use first. Future directions will be indicated based on the results of the phase III trial (J-ALEX) comparing crizotinib and alectinib.
2. In Japan, high detection sensitivity and specificity are achieved with highly sensitive IHC resulting from use of the combination of a high-affinity monoclonal antibody and sensitization with high sensitivity, and it is important to broaden the use of highly sensitive IHC screening.
3. Many clinicians question the limitations of companion diagnostics and therapeutic agents, and patients may suffer from disadvantages that accompany these limitations. A more flexible response will be expected in order to accurately diagnose and provide proper treatment of ALK-positive lung cancer, along with reduction of associated medical expenses.

References

1. Shaw AT, Kim DW, Nakagawa K et al (2013) Crizotinib versus chemotherapy in advanced ALK-positive lung cancer. N Engl J Med 368(25):2385–2394
2. Solomon BJ, Mok T, Kim DW et al (2014) First-line crizotinib versus chemotherapy in ALK-positive lung cancer. N Engl J Med 371(23):2167–2177
3. Seto T, Kiura K, Nishio M et al (2013) CH5424802 (RO5424802) for patients with ALK-rearranged advanced non-small-cell lung cancer (AF-001JP study): a single-arm, open-label, phase 1-2 study. Lancet Oncol 14(7):590–598
4. Shaw AT, Kim DW, Mehra R et al (2014) Ceritinib in ALK-rearranged non-small-cell lung cancer. N Engl J Med 370(13):1189–1197
5. Biomarker Committee, the Japan Lung Cancer Society 2011 Guidance for ALK gene testing in lung cancer patients
6. Biomarker Committee, the Japan Lung Cancer Society 2013 General information and handling inconsistency between highly sensitive IHC and FISH testing for ALK gene (the second report)
7. Chihara D, Suzuki R (2011) More on crizotinib. N Engl J Med 364(8):776–777

Chapter 12
Minor-Driver Mutant

Akihiko Miyanaga

Abstract Representative driver oncogenes such as epidermal growth factor receptor (*EGFR*), anaplastic lymphoma kinase (*ALK*), *KRAS*, and B-rapidly accelerated fibrosarcoma (*BRAF*) have recently been identified as new genetic aberrations in patients with non-small cell lung cancer (NSCLC). Additionally, rearranged during transfection (*RET*) and c-ros oncogene 1 (*ROS1*) fusion genes, which are minor-driver oncogenes, are each found in 1–2 % of NSCLC and represent distinct molecular subsets. Studies based on preclinical and clinical studies of several fusion-positive patients indicate that inhibiting the kinase activity of the *RET* and *ROS1* fusion proteins is a promising therapeutic strategy. Therefore, there are several ongoing clinical trials aimed at examining the efficacy of tyrosine kinase inhibitors (TKIs) against fusion proteins in patients with fusion-positive NSCLC. Other minor gene mutations (*HER2/ERBB2*, *NTRK1*, *NRG1*, *FGFR1/FGFR3*, *DDR2*, and *PIK3CA*) that are targetable by existing TKIs have also been identified in patients with NSCLCs. It is necessary to establish systematic genomic testing algorithms to identify defined subsets of patients with NSCLC for whom effective drug therapies are available either commercially or through clinical trials.

Keywords Minor mutation • Driver oncogene • Lung cancer • RET fusion • ROS1 fusion

12.1 Introduction

Non-small cell lung cancer (NSCLC) is the leading cause of death from cancer. The use of cytotoxic chemotherapy is associated with a response rate of 20–35 % and a median survival time of 10–12 months among patients with advanced NSCLC [1]. However, the discovery of recurrent driver mutations, such as the epidermal growth factor receptor (*EGFR*) kinase and anaplastic lymphoma kinase (*ALK*) fusions, has led to a marked change in the treatment of patients with NSCLC and specifically

A. Miyanaga (✉)
Department of Pulmonary Medicine and Oncology, Graduate School of Medicine, Nippon Medical School, 1-1-5 Sendagi, Bunkyo-ku, Tokyo 113-8602, Japan
e-mail: a-miyanaga@nms.ac.jp

© Springer Science+Business Media Singapore 2017

Y. Takiguchi (ed.), *Molecular Targeted Therapy of Lung Cancer*,
DOI 10.1007/978-981-10-2002-5_12

Table 12.1 The characteristic of minor-driver mutations in lung cancer

Gene	Smoking status	Histology	Genetic alteration
ROS1	Primarily never smokers	AD	ROS1 translocations
RET	Primarily never smokers	AD	RET translocations
HER2	Primarily never smokers	AD	HER2 mutations
			HER2 amplification
BRAF	Primarily smokers	AD/SCC	BRAF mutations
cMET	N/A	AD/SCC	cMET overexpression, amplification, and mutation
PI3K pathway	More in smokers	SCC/AD	PTEN loss
			PI3K amplification
			PI3K mutations
			AKT mutations
DDR2	–	SCC	DDR2 mutation
FGFR1	Primarily smokers	SCC	FGFR1 amplification

AD adenocarcinoma, *SCC* squamous cell carcinoma

lung adenocarcinoma (LADC). Treatment with EGFR kinase inhibitors (gefitinib, erlotinib, and afatinib) and ALK inhibitors (crizotinib, ceritinib, and alectinib) is superior to standard chemotherapy in patients with lung cancers that have *EGFR* mutations or *ALK* fusions, respectively [2–4]. More recently, it has become evident that subsets of NSCLC can be further defined targeting mutations in B-rapidly accelerated fibrosarcoma (*BRAF*), *HER2/ERBB2*, neuroblastoma rat sarcoma (*NRAS*), and *PIK3CA*; in addition, fusions that involve c-ros oncogene 1 (*ROS1*), rearranged during transfection (*RET*), and focal fibroblast growth factor receptor 1/3 (*FGFR1/FGFR3*) demonstrate great potential for therapeutic intervention [5–8]. The oncogene addiction model proposes that cancers harboring such gene amplifications, rearrangements, or mutations rely on the protein encoded by the gene, which dictates their malignant phenotype and can be thus referred to as driver alterations. Driver mutations lead to constitutive activation of mutant signaling proteins that induce and sustain tumorigenesis. Mutations can be found in all NSCLC histologies (including LADC, squamous cell carcinoma (SCC), and large cell carcinoma) and in current, former, and never smokers. In particular, never smokers with LADC have the highest incidence of *EGFR*, *ALK*, and *HER2* mutations or *RET* and *ROS1* fusion genes (Table 12.1). In a recent study, minor-driver mutations including *FGFR1/FGFR3*, *DDR2*, and *PIK3CA* that may be linked to outcomes with targeted therapies in SCC are emerging. In addition, results from a recent large genomic study in lung SCC have added a variety of potential therapeutic targets that await validation in prospective clinical trials [9]. Among NSCLCs, rearrangements in *ALK* and *ROS1* are present in approximately 5 % of LADC [10–13]. *RET* rearrangements have been identified in 1–2 % of Asian patients with LADC [5, 6, 14] (Fig. 12.1). The *HER2* mutation has been found in 2–3 %, *BRAF* in 0.5–1 %, *MET* amplification in <1 %, and *NRAS* in <1 % of cases [15]. Another multi-arm phase II trial that screened 427 NSCLC patients for at least one gene mutation found a frequency

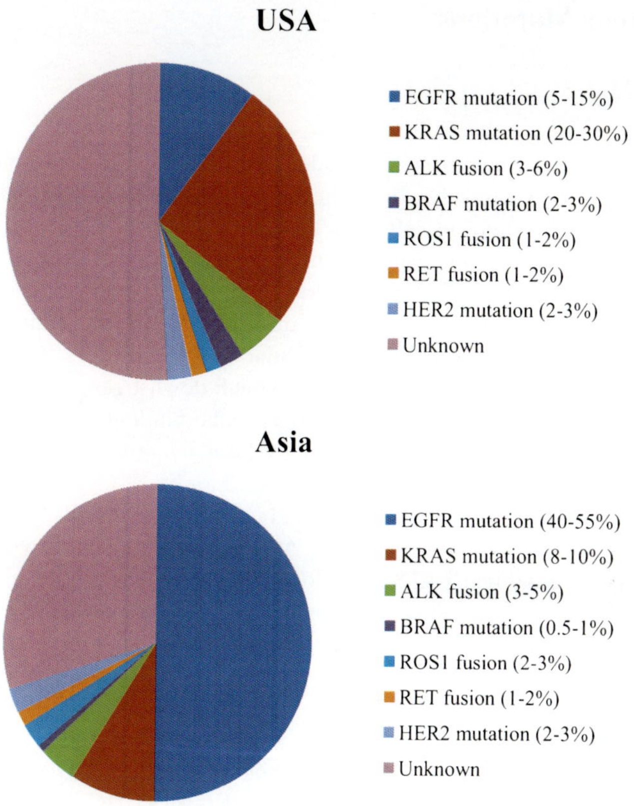

Fig. 12.1 The frequencies of oncogenic driver mutations in *LADC* (Data from patients in the USA and Asia were taken by previous reports (13–14). *LADC* lung adenocarcinoma)

of 3.0 % for MET pathway, 2.8 % for *BRAF*, and 2.8 % for *HER2/ERBB2* mutations [16].

Kris et al. have reported a median survival of 3.5 years (interquartile range [IQR], 1.96–7.70) for 260 patients with an oncogenic driver and genotype-directed therapy compared with 2.4 years (IQR, 0.88–6.20) for 318 patients with any oncogenic drivers who did not receive genotype-directed therapy [15]. This showed that those individuals with drivers receiving a matched targeted agent lived longer and that randomized trials are required to determine whether targeting therapy based on oncogenic drivers improves survival. Therefore, it will be necessary to develop systematic genomic testing algorithms to identify defined subsets of patients with NSCLC for whom effective drug therapies are available either commercially or through clinical trials.

In this article, we focus on these minor-driver mutations in NSCLC, regarding their characteristics, frequencies, mechanism whereby agents target them, and clinical evidence relating to the use of these agents.

12.2 Minor Mutations

12.2.1 *RET*

The *RET* receptor tyrosine kinase is encoded by the RET proto-oncogene localized on human chromosome 10q11.2. It is believed to be required for development of the kidneys and enteric system, as well as for the differentiation and survival of neurons [17, 18]. RET is the signaling receptor for the glial cell-derived neurotrophic factor (GDNF) family of ligands: GDNF, neurturin (NRTN), persephin (PSPN), and artemin (ARTN) [18]. After ligand binding, the intracellular kinase domain is activated, followed by autophosphorylation of intracellular tyrosine residues. These phosphotyrosine residues then serve as a platform to which downstream signaling proteins carrying SRC homology 2 (SH2) or phosphotyrosine-binding (PTB) domains bind and transmit signals into the cell, leading to the activation of RAS/ERK1/2 and PI3K/AKT pathways [19].

RET fusion-positive cases have been found in about 1–2 % of LADC patients in Asian and European populations [5–7, 20]. All reported lung tumors with *RET* rearrangements have exhibited LADC histology. Where overlap was evaluated, *RET* rearrangements have been shown to occur in tumors without other common oncogenic drivers such as *EGFR*, *ALK*, and *ROS1*. The most commonly reported fusion gene was *KIF5B-RET* [6]. Other commonly observed fusion genes are *CCDC6-RET*, *NCOA4-RET*, and *TRIM33-RET*.

RET rearrangements may be identified by fluorescence in situ hybridization (FISH), reverse transcription-polymerase chain reaction (RT-PCR), or next-generation sequencing. The use of immunohistochemistry (IHC) for RET visualization has had variable results and is not popular for identification of *RET* fusion genes [21].

RET rearrangements result in the formation of fusion products, which are capable of undergoing constitutive dimerization, leading to subsequent ligand-independent kinase activation, potentially resulting in neoplastic transformation. *RET* rearrangements are oncogenic in vitro and in vivo. Preclinical studies have demonstrated that lung cancer cell lines with *RET* fusions may be sensitive to multi-kinase inhibitors such as vandetanib, sunitinib, and sorafenib [5, 20]. Other preclinical evidences have demonstrated that cell lines with *RET* fusion genes had transforming capacity and enhanced sensitivity to vandetanib and other RET inhibitors [22, 23].

A variety of agents targeted against RET pathways have been studied in other malignancies. These include vandetanib, cabozantinib, sunitinib, sorafenib, fostamatinib, and ponatinib which have shown good responses, especially in *RET* rearrangement tumors [19]. However, studies with *RET* inhibitors in NSCLC are lacking, because no inhibitor specific for RET is available. There are case reports of vandetanib leading to remission 4 weeks after starting therapy in LADC; *KIF5B-RET* positive fusions following chemotherapy, surgery, and radiation [24]; and partial response (PR) for 4 months in *CCDC6-RET* fusion-positive LADC [25]. A prospective phase II trial studying the role of the multi-kinase inhibitor cabozantinib in three *RET* fusion-positive NSCLC patients showed PR to treatment, and the third patient had a stable disease (SD). All three patients were progression-free at the

Table 12.2 Phase II clinical trials of RET-targeting therapies with RET fusion gene-positive NSCLC

Trial number	Treatment	Location	Study design	Primary end point	Enrollment
NCT01639508	Cabozantinib	USA	Open-label, single arm	Response rate	26
UMIN000010095	Vandetanib	Japan	Open-label, single arm	Response rate	19
NCT01823068	Vandetanib	Korea	Open-label, single arm	Response rate	18
NCT01877083	Lenvatinib	Global	Open-label, single arm	Response rate	20 or more
NCT01813734	Ponatinib	USA	Open-label, single arm	Response rate	20

NSCLC non-small cell lung cancer

time of reporting (4–8 months) [7]. It has been reported that the use of sunitinib in a case of *KIF5B-RET* fusion-positive NSCLC provided clinically significant disease control for 10 weeks until sunitinib was discontinued [26]. To date, five independent, open-label, single-arm, phase II studies have been conducted to evaluate the efficacy of sunitinib (NCT01829217), cabozantinib (NCT01639508), ponatinib (NCT01813734), vandetanib (NCT01823068), and lenvatinib (NCT01877083) in NSCLC with *RET* rearrangements (Table 12.2). In Japan, the Lung Cancer with RET Rearrangement (LURET) study (UMIN000010095) has evaluated the efficacy of vandetanib in 19 patients with *RET* fusion gene-positive NSCLC (Table 12.2). Among 17 eligible patients, nine (53 % [95 % CI 28–77]) achieved an objective response, with a median progression-free survival of 4.7 months (95 % CI 2.8–8.5) [27]. Similar impressive clinical activity has been observed in phase II study of carbozantinib in RET-positive NSCLC. The overall response was 28 % (95 % CI 12–49), which met the primary endpoint, and the median progression-free survival was 5.5 months (95 % CI 3.8–8.4) [28].

12.2.2 ROS1

ROS1 is a human receptor tyrosine kinase, encoded by the *ROS1* gene which is closely related to the *ALK* gene [29]. ROS1 has considerable amino acid homology with ALK [30]. Oncogenic activation of ROS1 as a result of different chromosomal rearrangements has been identified in a variety of human tumors including LADC, where it has been reported in approximately 5 % of such cases [11, 31]. *ROS1* rearrangements tend to occur in LADC with solid, papillary, cribriform, or signet ring cell histologic patterns, tend to produce mucin, and tend to arise in patients who are younger and never smokers [11]. *ALK* and *ROS1* fusions are nonoverlapping with other known drivers in lung cancer, such as *KRAS* and *EGFR* mutations [11]. Several fusion genes have been identified including *CD74-ROS1*, *SLC34A2-ROS1*, *EZR-ROS1*, *TPM3-ROS1*, and *SDC4-ROS1*; the *ROS1* kinase gene is retained in all of these fusion events, and the expressed fusion genes are believed to play a role in carcinogenesis [11, 31, 32]. The mechanisms of oncogenic transformation provoked by these fusion genes are believed to involve upregulation of the phosphatase SHP-2, the PI3K/AKT/mTOR pathway, the JAK/STAT pathway, and the MAPK/ERK pathway [31].

The presence of *ROS1* rearrangements may be detected by FISH with the *ROS1* break-apart probe and by RT-PCR [5, 7, 21]. FISH testing is not dependent on the specific fusion partner. Specific fusion partners are detected by RT-PCR [11]. IHC can be used to screen for positive ROS1 which can then be confirmed by FISH. IHC has been reported to be highly sensitive for ROS1-positive LADC confirmed by FISH and RT-PCR with strong diffuse expression; however, false-positive IHC has been reported to occur in some ROS1-negative LADC with results confirmed by FISH as required [33].

Preclinical studies have reported that *ROS1* fusions are associated with sensitivity to tyrosine kinase inhibitors that have off-target activity against crizotinib [11]. Clinically, two patients with tumors harboring *ROS1* fusions have shown partial responses to crizotinib [11, 31]. A retrospective analysis of 1073 NSCLC cases demonstrated no difference in OS between *ROS1* rearrangement positive and negative subgroups, but suggested sensitivity of *ROS1* rearrangement positive cell lines [11].

A phase I study of the efficacy of crizotinib in *ROS1*-rearranged NSCLC showed marked antitumor activity, with an objective response rate of 72 % (95 % CI, 58–84) and median progression-free survival (PFS) of 19.2 months (95 % CI, 14.4 to not reached). Although several types of *ROS1* rearrangement have been described, this study did not find any correlation between the type of rearrangement and response to therapy [34]. In a retrospective European case study, 31 *ROS1*-positive NSCLC cases treated with crizotinib were retrospectively reviewed, and an objective response rate of 80 % and 9.1-month median PFS were calculated in this cohort [35].

Acquired resistance to crizotinib has been observed in LADC with *ROS1* fusion genes, as with *ALK*-positive LADC. The mechanism of acquired resistance to *ROS1* inhibition has been reported to involve an acquired mutation in *CD74-ROS1* fusion, whereby a glycine-to-arginine substitution occurs at codon 2032 in the ROS1 kinase domain [36]; in addition, a different mechanism may involve activation of alternative signaling via the EGFR pathway [37]. Furthermore, preclinical studies have reported that foretinib is a potent *ROS1* inhibitor [38].

12.2.3 BRAF

The oncogene *BRAF* encodes a serine/threonine kinase that lies downstream of RAS protein in the RAS-RAF-MEK-ERK signaling pathway [39]. *BRAF* mutations are seen in about 50 % of melanomas, where *BRAF V600E* is a driver mutation that can be effectively targeted with selective *BRAF* and/or *MEK* inhibitors [40]. *BRAF* mutations have been found in about 1–5 % of patients with NSCLC [41–43]. In contrast to melanoma, the *BRAF* mutations found in NSCLC were *V600E* (50 %), G469A (39 %), and D594G (11 %) [42]. *V600E BRAF* mutations occurred more frequently in women and never smokers, whereas non-*V600E* mutations occurred more commonly in current and former smokers. The *BRAF V600E* genotype has been associated with more aggressive tumor histology and poorer prognosis compared to non-V600E genotypes [44].

BRAF V600E mutations can be detected with targeted next-generation sequencing [45]. IHC using the VE1 antibody has also been reported as a successful screening tool for *BRAF V600E* mutation in LADC [46].

Many of these non-*V600E* mutations show only intermediate or low kinase activity, and preclinical studies have suggested that non-*V600E BRAF*-mutated kinases were resistant to *BRAF*-targeted therapy, although some may be sensitive to downstream pathway inhibitors such as MEK inhibitors [47]. In a histology-independent, biomarker-selected, early phase II basket study of vemurafenib, this agent showed modest antitumor activity in cancers that sporadically express *BRAF V600* mutations [48], whereas the *BRAF G469L* mutant LADC did not respond to vemurafenib [49]. These studies suggested that *BFAF V600E* mutations were predictive biomarkers for therapy of LADC with vemurafenib. The selective *BRAF V600E* mutant kinase inhibitor dabrafenib has increased PFS in *BRAF V600E*-mutated metastatic melanoma compared to conventional therapy [50]. Two patients with *BRAF V600* mutant NSCLC are reported to have had a PR to dabrafenib [51]. Currently, agents targeting BRAF or downstream effectors in ongoing clinical trials include dabrafenib for patients with NSCLC and prospectively identified *BRAF V600E* mutations: the MEK inhibitor, trametinib, for patients with non-*V600E BRAF* mutations; and dasatinib for patients with NSCLC and inactivating or uncharacterized *BRAF* mutations (NCT01336634, NCT01362296, and NCT01514864, respectively).

Multiple mechanisms of acquired resistance to targeted *BRAF V600* inhibitors have been described in melanoma, including upregulation of receptor tyrosine kinases, activation of the AKT pathway, and acquired mutation in *NRAS* [52]. Recent data suggest that inhibition of *BRAF V600E* can activate feedback leading to increased activity of and dependence on RAS [53]. In a case of LADC, it was reported that the acquired *KRAS G12D* mutation was primarily responsible for acquired dabrafenib resistance in this patient [54].

12.2.4 *HER2*

HER2 mutations have been reported in approximately 2 % of NSCLC patients, with these generally lacking other *EGFR*, *ALK*, and *KRAS* mutations. *HER2* mutations are more prevalent in LADC from patients who are never smokers and are more common in Asians and females. Similarly, *HER2* gene amplification has been found in approximately 2 % of NSCLC patients identified by FISH using criteria for HER2 amplification in breast cancer [55, 56].

HER2 mutations mostly occur in exon 20 as in-frame insertions, leading to constitutive activation of the receptor and downstream AKT and MEK pathways [57, 58]. *HER2* mutations respond to the genetic driver definition, and preclinical models have proved the concept of the transforming property of such a genetic alteration [59]. *HER2* mutations may be more relevant in lung carcinogenesis than HER2 amplification or overexpression. Some ongoing clinical trials are enrolling *HER2*-mutated NSCLC patients, mixed together with *HER2*-amplified or *EGFR*-mutated NSCLC patients. Clinical trials with *HER2* mutants in NSCLC have shown promising preliminary results for therapy with afatinib [55, 60], trastuzumab [55], dacomitinib [61], and neratinib plus temsirolimus [62].

12.2.5 *NTRK1*

Chromosomal rearrangements involving neurotrophic tyrosine kinase 1 (*NTRK1*) occur in a subset of NSCLC, leading to expression of an oncogenic tropomyosin-related kinase (TrkA) fusion protein. *NTRK1* fusion genes in NSCLC were first described among a population of LADC patients in the absence of detectable *EGFR* or *KRAS* mutation, or *ALK* or *ROS1* fusion genes [63]. Two different *NTRK1* fusions have been described in NSCLC using next-generation sequencing, namely, *MPRIP-NTRK1* and *CD74-NTRK1* [63]. Preclinical evidence supports the role of these fusions in TrkA autophosphorylation leading to oncogenic processes [63]. It was indicated that entrectinib, an orally available small molecule inhibitor of TrkA, TrkB, TrkC, ROS1, and ALK, showed significant antitumor activity in a patient with NSCLC harboring an *SQSTM1-NTRK1* gene rearrangement [64].

12.2.6 *NRG1*

The *CD74-NRG1* fusion genes which are a chimeric transcript fusing the first six exons of *CD74* to the exons encoding the EGF-like domain of the neuregulin-1 (NRG1) III-β3 isoform have been shown to occur specifically in invasive mucinous adenocarcinomas (IMAs) of never smokers, a tumor type that is otherwise associated with *KRAS* mutations [65]. It has been reported that *NRG1* fusions were present in approximately 1.7 % of lung adenocarcinomas and in 17.6 % of *KRAS*-negative IMAs [65, 66]. The *CD74-NRG1* fusion activated HER2/HER3 signaling, whereas the *EZR-ERBB4* and *TRIM24-BRAF* fusions constitutively activated the ERBB4 and BRAF kinases, respectively.

12.2.7 FGFR1/FGFR3

Lung SCCs have fewer treatment options than those with non-SCC NSCLC. Thus, driver mutations are emerging that may be linked to outcomes with targeted therapies in SCC. Altered genes include *FGFR1* and *DDR2* as well as *PIK3CA*. In addition, results from a recent large-scale genomic study in lung SCC have added a variety of potential therapeutic targets that await validation in prospective clinical trials.

The fibroblast growth factor receptor (*FGFR*) *1/FGFR3* genes encode one member of the FGFR TK family, which includes four kinases: FGFR1, FGFR2, FGFR3, and FGFR4. FGFR TKs belong to the immunoglobulin superfamily and act as receptors for the various fibroblast growth factors (FGFs). Amplification or activation of *FGFR1* has been reported in lung SCC from former/current smokers [67, 68]. The Cancer Genome Atlas (TCGA) reports demonstrated *FGFR3*_missense mutations (3 %), amplifications (0.6 %), fusions (2.2 %), and deletions (1.7 %) in lung SCC and *FGFR3* amplifications (1.3 %) and a single mutation event to S779R (0.4 %) in LADC [9, 69]. *FGFR3-TACC3* fusions were identified in 0.5 % of cases in lung adenocarcinoma [70, 71]. In other reports, lung SCCs were most notable for their 9 % frequency of *FGFR1* amplification, which is in contrast to only 4 % of LADC harboring any FGFR abnormality [72]. Preclinical studies have shown that cancers harboring *FGFR3-TACC3* fusions and other *FGFR3* fusions are sensitive to pan-FGFR inhibitors and FGFR-selective agents [8, 70]. In addition, such studies indicated that cancer cells with amplified *FGFR1* could display addiction to FGFR signaling [69]. Clinical trials with FGFR inhibitors are currently under way.

12.2.8 DDR2

Discoidin death receptor 2 (DDR2) is a member of the DDR family of receptor tyrosine kinases that are stimulated by collagen rather than peptide growth factors. The precise mechanism by which *DDR2* mutations promote cellular transformation remains unclear, but ectopic expression of *DDR2* has been shown to correlate with STAT5 and Src phosphorylation [73]. *DDR2* mutations have been observed at a rate of 4 % in SCC and have been associated with sensitivity to dasatinib [73]. Preclinical evidence has suggested a synergistic effect of potential RTK-driven adaptive resistant mechanisms on DDR2 targeting and dasatinib combined with MET and insulin-like growth factor receptor (IGF1R) inhibitors; furthermore, ligand stimulation of

EGFR and MET rescued DDR2 mutant lung SCC cells from dasatinib-induced loss of cell viability [74].

12.2.9 PIK3CA

Phosphatidylinositol 3-kinases (PI3K) are a family of lipid kinases involved in many cellular processes, including cell growth, proliferation, differentiation, motility, and survival. *PIK3CA* mutations have been found in 1–3 % of NSCLC, appear to be more common in squamous cell histology compared to adenocarcinoma (AD), and occur in both never smokers and ever smokers [75]. *PIK3CA* mutations can co-occur with *EGFR* mutations [75, 76]. Preclinical data have shown that introduction of activating *PIK3CA* mutations into *EGFR*-mutated lung cancer cell lines confers resistance to EGFR-TKIs, and *PIK3CA* mutations also have been detected in 5 % of *EGFR*-mutated lung cancer patients with acquired resistance to EGFR-TKI [77, 78]. Preclinical studies have shown that inhibition of multiple PI3K pathway components blocks the growth of PI3K-dependent NSCLC cell lines and induces tumor regression in mouse xenograft models of *PIK3CA* mutant lung cancer [79].

12.3 Challenges and Conclusions

We have reviewed the minor oncogenic gene mutations associated with NSCLCs. Ongoing developments in high-throughput sequencing analysis and systematic genomic technologies have led to the identification of novel molecular events that characterize NSCLC transformation and may represent critical oncogenic drivers. Recently, genetic events such as *EGFR* mutations and *ALK* fusions have become targetable with currently available molecular agents. Preclinical and clinical trials of other minor genetic oncogenes are ongoing. Such a potential paradigm change toward personalized targeted therapy has raised several new challenges. First, patients with NSCLC are well recognized to be a variable population, based on interpatient tumor heterogeneity. Second, the dynamic change within the cancer genome during the disease course is now being recognized as an additional challenge because the tumor genetic change may undergo substantial alteration during disease progression or in response to the treatment. Third, both quantity and quality of tumor tissues are essential for genomic testing. When only very small amounts of material can be obtained from biopsies, it is difficult to develop diagnostic systems that enable simultaneous examination of multiple gene alterations in routine formalin-fixed and paraffin-embedded (FFPE) clinical specimens. Fourth, although high-throughput genome sequencing holds revolutionary potential for personalized cancer therapy, analysis of the vast amount of genomic data available to permit identification of clinically relevant drug targets and genomic variants constitutes a significant challenge. These reforms in our understanding of NSCLC biology emphasize the importance of an individualized therapeutic approach based on molecular profiles.

References

1. Ohe Y, Ohashi Y, Kubota K et al (2007) Randomized phase III study of cisplatin plus irinotecan versus carboplatin plus paclitaxel, cisplatin plus gemcitabine, and cisplatin plus vinorelbine for advanced non-small-cell lung cancer: Four-Arm Cooperative Study in Japan. Ann Oncol 18:317–323
2. Maemondo M, Inoue A, Kobayashi K et al (2010) Gefitinib or chemotherapy for non-small-cell lung cancer with mutated EGFR. N Engl J Med 362:2380–2388
3. Shaw AT, Kim DW, Nakagawa K et al (2013) Crizotinib versus chemotherapy in advanced ALK-positive lung cancer. N Engl J Med 368:2385–2394
4. Mok TS, Wu YL, Thongprasert S et al (2009) Gefitinib or carboplatin-paclitaxel in pulmonary adenocarcinoma. N Engl J Med 361:947–957
5. Kohno T, Ichikawa H, Totoki Y et al (2012) KIF5B-RET fusions in lung adenocarcinoma. Nat Med 18:375–377
6. Takeuchi K, Soda M, Togashi Y et al (2012) RET, ROS1 and ALK fusions in lung cancer. Nat Med 18:378–381
7. Drilon A, Wang L, Hasanovic A et al (2013) Response to Cabozantinib in patients with RET fusion-positive lung adenocarcinomas. Cancer Discov 3:630–635
8. Wu YM, Su F, Kalyana-Sundaram S et al (2013) Identification of targetable FGFR gene fusions in diverse cancers. Cancer Discov 3:636–647
9. Cancer Genome Atlas Research Network (2012) Comprehensive genomic characterization of squamous cell lung cancers. Nature 489:519–525
10. Koivunen JP, Mermel C, Zejnullahu K et al (2008) EML4-ALK fusion gene and efficacy of an ALK kinase inhibitor in lung cancer. Clin Cancer Res 14:4275–4283
11. Bergethon K, Shaw AT, Ou SH et al (2012) ROS1 rearrangements define a unique molecular class of lung cancers. J Clin Oncol 30:863–870
12. Pao W, Hutchinson KE (2012) Chipping away at the lung cancer genome. Nat Med 18:349–351
13. Li T, Kung HJ, Mack PC, Gandara DR (2013) Genotyping and genomic profiling of non-small-cell lung cancer: implications for current and future therapies. J Clin Oncol 31:1039–1049
14. Kohno T, Tsuta K, Tsuchihara K et al (2013) RET fusion gene: translation to personalized lung cancer therapy. Cancer Sci 104:1396–1400
15. Kris MG, Johnson BE, Berry LD et al (2014) Using multiplexed assays of oncogenic drivers in lung cancers to select targeted drugs. JAMA 311:1998–2006
16. Lopez-Chavez A, Thomas A, Rajan A et al (2015) Molecular profiling and targeted therapy for advanced thoracic malignancies: a biomarker-derived, multiarm, multihistology phase II basket trial. J Clin Oncol 33:1000–1007
17. Eng C (1999) RET proto-oncogene in the development of human cancer. J Clin Oncol 17:380–393
18. Blume-Jensen P, Hunter T (2001) Oncogenic kinase signalling. Nature 411:355–365
19. Plaza-Menacho I, Mologni L, McDonald NQ (2014) Mechanisms of RET signaling in cancer: current and future implications for targeted therapy. Cell Signal 26:1743–1752
20. Lipson D, Capelletti M, Yelensky R et al (2012) Identification of new ALK and RET gene fusions from colorectal and lung cancer biopsies. Nat Med 18:382–384
21. Tsuta K, Kohno T, Yoshida A et al (2014) RET-rearranged non-small-cell lung carcinoma: a clinicopathological and molecular analysis. Br J Cancer 110:1571–1578

22. Matsubara D, Kanai Y, Ishikawa S et al (2012) Identification of CCDC6-RET fusion in the human lung adenocarcinoma cell line, LC-2/ad. J Thorac Oncol 7:1872–1876
23. Chao BH, Briesewitz R, Villalona-Calero MA (2012) RET fusion genes in non-small-cell lung cancer. J Clin Oncol 30:4439–4441
24. Gautschi O, Zander T, Keller FA et al (2013) A patient with lung adenocarcinoma and RET fusion treated with vandetanib. J Thorac Oncol 8:e43–e44
25. Falchook GS, Ordonez NG, Bastida CC et al (2016) Effect of the RET inhibitor vandetanib in a patient with RET fusion-positive metastatic non-small-cell lung cancer. J Clin Oncol 34:e141–e144
26. Wu H, Shih JY, Yang JC (2015) Rapid response to sunitinib in a patient with lung adenocarcinoma harboring KIF5B-RET fusion gene. J Thorac Oncol 10:e95–e96
27. Yoh K, Seto T, Satouchi M et al (2016) Vandetanib in patients with previously treated RET-rearranged advanced non-small-cell lung cancer (LURET): an open-label, multicentre phase 2 trial. Lancet Respir Med
28. Drilon A, Rekhtman N, Arcila M et al (2016) Cabozantinib in patients with advanced RET-rearranged non-small-cell lung cancer: an openlabel, single-centre, phase 2, single-arm trial. Lancet Oncol
29. Robinson DR, Wu YM, Lin SF (2000) The protein tyrosine kinase family of the human genome. Oncogene 19:5548–5557
30. Chin LP, Soo RA, Soong R, Ou SH (2012) Targeting ROS1 with anaplastic lymphoma kinase inhibitors: a promising therapeutic strategy for a newly defined molecular subset of non-small-cell lung cancer. J Thorac Oncol 7:1625–1630
31. Davies KD, Le AT, Theodoro MF et al (2012) Identifying and targeting ROS1 gene fusions in non-small cell lung cancer. Clin Cancer Res 18:4570–4579
32. Ou SH, Bartlett CH, Mino-Kenudson M et al (2012) Crizotinib for the treatment of ALK-rearranged non-small cell lung cancer: a success story to usher in the second decade of molecular targeted therapy in oncology. Oncologist 17:1351–1375
33. Sholl LM, Sun H, Butaney M et al (2013) ROS1 immunohistochemistry for detection of ROS1-rearranged lung adenocarcinomas. Am J Surg Pathol 37:1441–1449
34. Shaw AT, Ou SH, Bang YJ et al (2014) Crizotinib in ROS1-rearranged non-small-cell lung cancer. N Engl J Med 371:1963–1971
35. Mazieres J, Zalcman G, Crino L et al (2015) Crizotinib therapy for advanced lung adenocarcinoma and a ROS1 rearrangement: results from the EUROS1 cohort. J Clin Oncol 33:992–999
36. Awad MM, Katayama R, McTigue M et al (2013) Acquired resistance to crizotinib from a mutation in CD74-ROS1. N Engl J Med 368:2395–2401
37. Davies KD, Mahale S, Astling DP et al (2013) Resistance to ROS1 inhibition mediated by EGFR pathway activation in non-small cell lung cancer. PLoS One 8:e82236
38. Davare MA, Saborowski A, Eide CA et al (2013) Foretinib is a potent inhibitor of oncogenic ROS1 fusion proteins. Proc Natl Acad Sci U S A 110:19519–19524
39. Roberts PJ, Der CJ (2007) Targeting the Raf-MEK-ERK mitogen-activated protein kinase cascade for the treatment of cancer. Oncogene 26:3291–3310
40. Flaherty KT, Infante JR, Daud A et al (2012) Combined BRAF and MEK inhibition in melanoma with BRAF V600 mutations. N Engl J Med 367:1694–1703
41. Davies H, Bignell GR, Cox C et al (2002) Mutations of the BRAF gene in human cancer. Nature 417:949–954
42. Paik PK, Arcila ME, Fara M et al (2011) Clinical characteristics of patients with lung adenocarcinomas harboring BRAF mutations. J Clin Oncol 29:2046–2051
43. Cardarella S, Ogino A, Nishino M et al (2013) Clinical, pathologic, and biologic features associated with BRAF mutations in non-small cell lung cancer. Clin Cancer Res 19:4532–4540
44. Marchetti A, Felicioni L, Malatesta S et al (2011) Clinical features and outcome of patients with non-small-cell lung cancer harboring BRAF mutations. J Clin Oncol 29:3574–3579

45. McCourt CM, McArt DG, Mills K et al (2013) Validation of next generation sequencing technologies in comparison to current diagnostic gold standards for BRAF, EGFR and KRAS mutational analysis. PLoS One 8:e69604
46. Ilie M, Long E, Hofman V et al (2013) Diagnostic value of immunohistochemistry for the detection of the BRAFV600E mutation in primary lung adenocarcinoma Caucasian patients. Ann Oncol 24:742–748
47. Pratilas CA, Hanrahan AJ, Halilovic E et al (2008) Genetic predictors of MEK dependence in non-small cell lung cancer. Cancer Res 68:9375–9383
48. Hyman DM, Puzanov I, Subbiah V et al (2015) Vemurafenib in multiple nonmelanoma cancers with BRAF V600 mutations. N Engl J Med 373:726–736
49. Gautschi O, Peters S, Zoete V et al (2013) Lung adenocarcinoma with BRAF G469 L mutation refractory to vemurafenib. Lung Cancer 82:365–367
50. Hauschild A, Grob JJ, Demidov LV et al (2012) Dabrafenib in BRAF-mutated metastatic melanoma: a multicentre, open-label, phase 3 randomised controlled trial. Lancet 380:358–365
51. Falchook GS, Long GV, Kurzrock R et al (2012) Dabrafenib in patients with melanoma, untreated brain metastases, and other solid tumours: a phase 1 dose-escalation trial. Lancet 379:1893–1901
52. Sullivan RJ, Flaherty KT (2013) Resistance to BRAF-targeted therapy in melanoma. Eur J Cancer 49:1297–1304
53. Lito P, Pratilas CA, Joseph EW et al (2012) Relief of profound feedback inhibition of mitogenic signaling by RAF inhibitors attenuates their activity in BRAFV600E melanomas. Cancer Cell 22:668–682
54. Rudin CM, Hong K, Streit M (2013) Molecular characterization of acquired resistance to the BRAF inhibitor dabrafenib in a patient with BRAF-mutant non-small-cell lung cancer. J Thorac Oncol 8:e41–e42
55. Mazieres J, Peters S, Lepage B et al (2013) Lung cancer that harbors an HER2 mutation: epidemiologic characteristics and therapeutic perspectives. J Clin Oncol 31:1997–2003
56. Heinmoller P, Gross C, Beyser K et al (2003) HER2 status in non-small cell lung cancer: results from patient screening for enrollment to a phase II study of herceptin. Clin Cancer Res 9:5238–5243
57. Stephens P, Hunter C, Bignell G et al (2004) Lung cancer: intragenic ERBB2 kinase mutations in tumours. Nature 431:525–526
58. Perera SA, Li D, Shimamura T et al (2009) HER2YVMA drives rapid development of adenosquamous lung tumors in mice that are sensitive to BIBW2992 and rapamycin combination therapy. Proc Natl Acad Sci U S A 106:474–479
59. Shimamura T, Ji H, Minami Y et al (2006) Non-small-cell lung cancer and Ba/F3 transformed cells harboring the ERBB2 G776insV_G/C mutation are sensitive to the dual-specific epidermal growth factor receptor and ERBB2 inhibitor HKI-272. Cancer Res 66:6487–6491
60. De Greve J, Teugels E, Geers C et al (2012) Clinical activity of afatinib (BIBW 2992) in patients with lung adenocarcinoma with mutations in the kinase domain of HER2/neu. Lung Cancer 76:123–127
61. Janne PA, Ou SH, Kim DW et al (2014) Dacomitinib as first-line treatment in patients with clinically or molecularly selected advanced non-small-cell lung cancer: a multicentre, open-label, phase 2 trial. Lancet Oncol 15:1433–1441
62. Gandhi L, Bahleda R, Tolaney SM et al (2014) Phase I study of neratinib in combination with temsirolimus in patients with human epidermal growth factor receptor 2-dependent and other solid tumors. J Clin Oncol 32:68–75
63. Vaishnavi A, Capelletti M, Le AT et al (2013) Oncogenic and drug-sensitive NTRK1 rearrangements in lung cancer. Nat Med 19:1469–1472
64. Farago AF, Le LP, Zheng Z et al (2015) Durable clinical response to entrectinib in NTRK1-rearranged non-small cell lung cancer. J Thorac Oncol 10:1670–1674
65. Fernandez-Cuesta L, Plenker D, Osada H et al (2014) CD74-NRG1 fusions in lung adenocarcinoma. Cancer Discov 4:415–422

66. Nakaoku T, Tsuta K, Ichikawa H et al (2014) Druggable oncogene fusions in invasive mucinous lung adenocarcinoma. Clin Cancer Res 20:3087–3093
67. Dutt A, Ramos AH, Hammerman PS et al (2011) Inhibitor-sensitive FGFR1 amplification in human non-small cell lung cancer. PLoS One 6:e20351
68. Weiss J, Sos ML, Seidel D et al (2010) Frequent and focal FGFR1 amplification associates with therapeutically tractable FGFR1 dependency in squamous cell lung cancer. Sci Transl Med 2:62ra93
69. Cancer Genome Atlas Research Network (2014) Comprehensive molecular profiling of lung adenocarcinoma. Nature 511:543–550
70. Capelletti M, Dodge ME, Ercan D et al (2014) Identification of recurrent FGFR3-TACC3 fusion oncogenes from lung adenocarcinoma. Clin Cancer Res 20:6551–6558
71. Majewski IJ, Mittempergher L, Davidson NM et al (2013) Identification of recurrent FGFR3 fusion genes in lung cancer through kinome-centred RNA sequencing. J Pathol 230:270–276
72. Helsten T, Elkin S, Arthur E et al (2016) The FGFR landscape in cancer: analysis of 4,853 tumors by next-generation sequencing. Clin Cancer Res 22:259–267
73. Hammerman PS, Sos ML, Ramos AH et al (2011) Mutations in the DDR2 kinase gene identify a novel therapeutic target in squamous cell lung cancer. Cancer Discov 1:78–89
74. Bai Y, Kim JY, Watters JM et al (2014) Adaptive responses to dasatinib-treated lung squamous cell cancer cells harboring DDR2 mutations. Cancer Res 74:7217–7228
75. Sun Y, Ren Y, Fang Z et al (2010) Lung adenocarcinoma from East Asian never-smokers is a disease largely defined by targetable oncogenic mutant kinases. J Clin Oncol 28:4616–4620
76. Kawano O, Sasaki H, Endo K et al (2006) PIK3CA mutation status in Japanese lung cancer patients. Lung Cancer 54:209–215
77. Sequist LV, Waltman BA, Dias-Santagata D et al (2011) Genotypic and histological evolution of lung cancers acquiring resistance to EGFR inhibitors. Sci Transl Med 3:75ra26
78. Engelman JA, Mukohara T, Zejnullahu K et al (2006) Allelic dilution obscures detection of a biologically significant resistance mutation in EGFR-amplified lung cancer. J Clin Invest 116:2695–2706
79. Engelman JA, Chen L, Tan X et al (2008) Effective use of PI3K and MEK inhibitors to treat mutant Kras G12D and PIK3CA H1047R murine lung cancers. Nat Med 14:1351–1356

Chapter 13
Mechanism of Resistance to Targeted Molecular Therapy

Masahiro Seike

Abstract Patients with non-small cell lung cancer (NSCLC) who have activating mutations of the epidermal growth factor receptor (EGFR) gene and anaplastic lymphoma kinase (ALK) fusion gene have shown a strong response to EGFR tyrosine kinase inhibitors (EGFR-TKIs) and ALK-TKIs. However, the emergence of acquired resistance is virtually inevitable, thereby limiting improvement in patient outcomes. Several mechanisms of acquired resistance to EGFR-TKI have been identified, including an exon 20 T790M secondary mutation, HGF/MET signal activation, epithelial-mesenchymal transition (EMT), and conversion to small cell lung cancer (SCLC). Two major mechanisms of resistance to ALK-TKIs have been demonstrated. Half of the resistant tumors exhibit ALK signal-dependent activation, such as ALK secondary mutations and/or amplification. Another common mechanism of resistance is the activation of alternate survival pathways, such as those mediated by EGFR, KRAS, or IGF-1R. Further studies should identify additional mechanisms associated with acquired resistance to EGFR-TKIs and ALK-TKIs. Understanding the mechanisms of acquired resistance to EGFR-TKIs and ALK-TKIs should facilitate the development of targeted molecular therapies to overcome this resistance.

In this review, we summarize the mechanisms of resistance to targeted molecular therapies to EGFR-TKIs and ALK-TKIs and therapeutic strategies aimed at overcoming this resistance.

Keywords Lung cancer • Drug resistance • Targeted molecular therapy • EGFR-TKI • ALK-TKI

M. Seike (✉)
Department of Pulmonary Medicine and Oncology, Graduate School of Medicine, Nippon Medical School, 1-1-5 Sendagi, Bunkyo-ku, Tokyo 113-8603, Japan
e-mail: mseike@nms.ac.jp

© Springer Science+Business Media Singapore 2017
Y. Takiguchi (ed.), *Molecular Targeted Therapy of Lung Cancer*,
DOI 10.1007/978-981-10-2002-5_13

13.1 Oncogenic Driver Mutations and Molecular-Targeted Therapy in Non-small Cell Lung Cancer

Recently, oncogenic driver mutations have been identified in non-small cell lung cancer (NSCLC) patients, such as epidermal growth factor receptor (EGFR) mutations and the anaplastic lymphoma kinase (ALK) fusion gene [1–3]. Several tyrosine kinase inhibitors (TKIs) are currently approved for the treatment of NSCLCs with oncogenic driver mutations. Recent randomized phase III trials have shown that treatment of advanced NSCLC patients harboring EGFR mutations with first-line EGFR-TKIs including gefitinib, erlotinib, and afatinib improves progression-free survival (PFS) [4–7]. ALK fusion gene-positive NSCLC patients also showed a dramatic response to ALK-TKIs such as crizotinib, alectinib, and ceritinib [8–10]. Next-generation sequencing technologies have been used to identify novel driver mutations, including those in the RET, ROS1, NTRK1, and NRG1 genes in NSCLC patients [11–13]. Molecular-targeted therapies for NSCLC patients with oncogenic driver mutations in genes other than EGFR and ALK are under clinical development. Unfortunately, despite the initial marked response to TKIs, most NSCLC patients with oncogenic driver mutations eventually acquire resistance. Therefore, there is a clinical need to develop strategies to overcome the resistance of patients to EGFR-TKIs and ALK-TKIs.

13.2 Mechanisms of Resistance to EGFR-TKI

Two major mechanisms of acquired resistance to EGFR-TKIs were identified in patients with EGFR-mutant NSCLC [14–17] (Fig. 13.1). Approximately half of resistant tumors develop a secondary EGFR mutation in exon 20 T790M, which prevents inhibition by EGFR-TKIs due to steric hindrance or an increased binding affinity for ATP [14]. Mutations leading to activation of alternative signaling pathways, such as those inducing MET amplification, high-level hepatocyte growth factor (HGF) expression, or PTEN downregulation, have also been identified as mechanisms of resistance to EGFR-TKIs [15–17]. Morphological and phenotypic changes such as conversion to small cell lung cancer (SCLC) or the epithelial-mesenchymal transition (EMT) have also been identified as mechanisms of resistance to EGFR-TKIs [18, 19] (Fig. 13.1). However, the mechanism of resistance in approximately 20–30 % of cases remains unknown (Fig. 13.1).

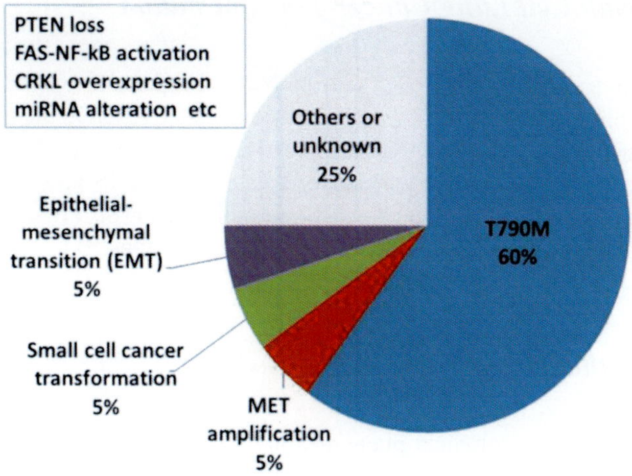

Fig. 13.1 Mechanisms of acquired resistance to EGFR-TKI

13.2.1 The T790M Secondary Mutation

Third-generation EGFR-TKIs such as osimertinib (AZD9291) and CO-1686 have shown promising activity in treatment-resistant EGFR mutation-positive NSCLCs containing the T790M mutation [20, 21]. However, acquisition of an EGFR C797S mutation has been identified as a novel mechanism of resistance to osimertinib (AZD9291) [22].

13.2.2 HGF-MET Activation

An additional 10–20 % of tumors from refractory EGFR-mutant NSCLC patients undergo MET gene amplification, which causes HER3-dependent activation of the signaling cascade downstream of EGFR despite its inhibition by TKIs [15]. Recently, MET inhibitors have been administered to NSCLC patients who are naïve or resistant to EGFR-TKIs [23]. This phase II study showed that PFS was extended in the group treated with erlotinib plus the MET inhibitor tivantinib than in the group treated with erlotinib alone, especially among patients harboring KRAS mutations [23]. Yano et al. also showed that HGF, a ligand of the MET oncoprotein, induces gefitinib resistance in EGFR-mutant lung adenocarcinoma cells by restoring PI3K/AKT signaling via phosphorylation of MET, but not EGFR or ErbB3 [16]. Inhibition of HGF-MET signaling may be a useful strategy to overcome resistance to EGFR-TKIs depending on the status of HGF-MET signaling.

13.2.3 Small Cell Lung Cancer Transformation

Morphological transformation to SCLC from NSCLC represents one of the mechanisms of acquired resistance to EGFR-TKIs. Sequist et al. reported the morphological transformation of five drug-resistant NSCLC tumors (14 % of a total of 37 patients) containing EGFR mutations into SCLC [18]. The existing EGFR mutation was maintained during SCLC transformation in all cases [18]. These transformed SCLC tumors were sensitive to standard SCLC chemotherapy.

13.2.4 Epithelial-Mesenchymal Transition

EMT is a progressive biological phenomenon that includes loss of epithelial cell adhesion and induction of a mesenchymal phenotype. Several studies have demonstrated that EMT is associated with reduced drug sensitivity and acquisition of resistance to EGFR-TKIs in NSCLC, whereas retention of an epithelial phenotype ensured a good response to EGFR-TKIs [24, 25]. Recent studies revealed that overexpression of AXL led to resistance to EGFR-TKI in NSCLC cells undergoing EMT and that AXL was a potential therapeutic target in patients with acquired resistance to EGFR-TKIs [19]. These reports suggested that EMT might be a mechanism of resistance to EGFR-TKI. However, the molecular mechanisms of the development of EMT-related resistance to EGFR-TKI are still not fully understood. Therapeutic strategies aiming to prevent EMT with a view to restoring sensitivity to EGFR-TKIs remain to be investigated.

13.2.5 Cancer Stem Cell Properties

Cancer stem cells (CSCs) are characterized by the capacity for pluripotency and self-renewal and are thought to represent a renewable source of cancer cells. The significance of CSC-like properties to the mechanism of resistance to EGFR-TKIs in NSCLCs has recently been investigated [26, 27]. Sharma et al. identified a drug-tolerant cancer cell subpopulation (DTP) that maintains viability under conditions where the vast majority of the cell population is rapidly killed following treatment with gefitinib [26]. The putative CSC marker CD133 was shown to be overexpressed in these DTPs, suggesting a CSC-like phonotype. Shien et al. established 13 gefitinib-resistant, EGFR-mutant NSCLC cell lines. Four of the latter lines showed an EMT phenotype and CSC-like properties, accompanied by overexpression of the CSC markers ALDHA1, ABCG2, and CD44 [27]. These findings may provide clues to overcoming resistance to EGFR-TKIs.

13.2.6 PTEN Loss, FAS/NF-κB Activation, and CRKL Overexpression

Loss of PTEN expression has been associated with decreased sensitivity to EGFR-TKIs owing to activation of PI3K-AKT signaling, impairment of ligand-induced ubiquitination, and degradation of activated EGFR in EGFR-mutant cells [17].

A recent study showed that knockdown of FAS and NF-κB enhanced cell death induced by the EGFR-TKI erlotinib in EGFR-mutant lung cancer cells [28]. Increased expression of the NF-κB inhibitor IκB predicted an improved response and PFS in EGFR-mutant NSCLC patients who received EGFR-TKI therapy [28]. These findings suggest that simultaneous inhibition of EGFR and NF-κB may be useful for the treatment of EGFR-mutant NSCLC.

CRKL overexpression induced resistance to EGFR-TKI mediated by ERK and AKT signaling in EGFR-mutant NSCLC cells [29]. CRKL amplification was found in a lung adenocarcinoma treated with an EGFR inhibitor [29]. These results suggest that CRKL is a therapeutic target for a subset of EGFR-mutant NSCLCs that harbor CRKL amplifications.

13.2.7 MicroRNA Alterations

MicroRNAs (miRNAs) can function as either tumor suppressors or oncogenes and are used as diagnostic, prognostic, and therapeutic biomarkers in lung cancer. Four miRNAs (miR-30b, miR-30c, miR-221, and miR-222) have been shown to play important roles in gefitinib-induced apoptosis and EMT in NSCLC cells in vitro [30]. Several reports demonstrated that EGFR-activated miR-21 is a potential therapeutic target in tumors with mutations in EGFR [31, 32]. Members of the miR-200 family, targeting the E-cadherin suppressors ZEB1 and ZEB2, have been recognized as key suppressors of EMT associated with the resistance to EGFR-TKIs [27, 32]. These findings suggest that miRNAs may be promising therapeutic targets in EGFR-mutant NSCLC patients.

13.2.8 Intrinsic Resistance

Recent studies have identified molecules associated with intrinsic resistance to EGFR-TKIs. BIM (BCL2L11) is a member of the Bcl-2 family encoding a pro-apoptotic protein. Upregulation of BIM is required for apoptosis induction by EGFR-TKIs in EGFR-mutant NSCLCs. Notably, a BIM deletion polymorphism occurs naturally in 13 % of East Asian individuals. This polymorphism can mediate intrinsic resistance to and reduced responses to EGFR-TKIs in EGFR-mutant NSCLC patients [33]. PFS in response to first-line EGFR-TKIs was significantly shorter in patients with the BIM deletion polymorphism than in those patients with wild-type BIM (8.6 and 4.6 months, respectively [$p = 0.004$]) [34]. An HDAC inhibitor could restore BIM function and resistance to EGFR-TKI [35]. Treatment with an HDAC inhibitor combined with an EGFR-TKI could represent an attractive strategy to treat EGFR-mutant NSCLC patients harboring a BIM deletion polymorphism.

Yamaguchi et al. demonstrated that thyroid transcription factor-1 (TTF-1)-induced ROR1 was required to sustain EGFR survival signaling as an "Achilles heel" in lung adenocarcinoma. ROR1 could activate kinase-dependent c-Src as well as kinase-independent EGFR-ErbB3, ErbB3 phosphorylation, and PIK3 signaling [36]. Inhibition of ROR1 expression could also restore the sensitivity to the EGFR-TKI. ROR1 expression was evaluated in erlotinib-pretreated tumor samples from 45 EGFR-mutant patients in the EURTAC trial to assess its potential as a predictive biomarker of PFS and overall survival (OS) [6, 37]. The PFS of patients with elevated ROR1 expression was significantly shorter than in those patients with low/intermediate ROR1 expression (11.8 months and 5.8 months, respectively) [37].

Noro et al. recently reported that MET gene amplification could predict short PFS and overall survival (OS) after gefitinib treatment in lung adenocarcinoma (LADC) harboring EGFR mutations [38]. MET-FISH-positive LADC patients defined by MET amplification and gene copy number gains (CNGs) exhibited significantly shorter PFS and OS than patients who were MET-FISH negative [38].

These findings suggest that BIM, ROR1, and MET gene status are involved in intrinsic resistance to EGFR-TKIs and may thus be predictive biomarkers for selecting patients who would benefit from EGFR-TKI therapy.

13.2.9 Future Therapeutic Strategies

Understanding mechanisms of intrinsic and acquired resistance to EGFR-TKI, followed by the development of drugs targeted to molecules that can overcome this resistance, could serve as an important advance for targeting EGFR which is activated in NSCLC. Third-generation EGFR-TKIs such as osimertinib (AZD9291) and CO-1686 showed promising activity in EGFR mutation-positive NSCLCs harboring the T790M mutation [20, 21]. However, an EGFR C797S mutation arose as a novel mechanism of resistance to osimertinib (AZD9291) [22]. EGFR-TKIs combined with platinum-doublet chemotherapy showed prolonged PFS in EGFR-mutant NSCLCs in several clinical trials [39, 40]. A recent study demonstrated that the combination of erlotinib plus bevacizumab could be an effective first-line regimen to treat EGFR mutation-positive NSCLC patients [41]. PFS was significantly prolonged in patients receiving erlotinib plus bevacizumab compared to those that received erlotinib alone (16.0 months and 9.7 months, respectively). There are various mechanisms of acquired resistance to EGFR-TKIs. Future studies should clarify whether there exist other as yet unidentified mechanisms associated with acquired resistance to EGFR-TKI. In addition, treatments of individual patients

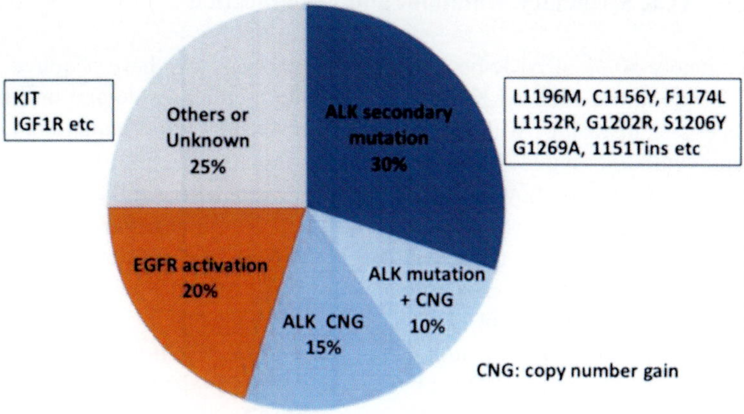

Fig. 13.2 Mechanisms of acquired resistance to crizotinib

should be based on assessment of the details of particular mechanism of resistance observed in the patient using re-biopsy or liquid biopsy samples.

13.3 Mechanisms of Resistance to ALK-TKIs

NSCLC patients harboring an ALK fusion gene have shown a strong response to ALK-TKIs such as crizotinib, alectinib, and ceritinib [8–10]. Based on the clinical data, first-line treatment with crizotinib has become the standard therapy for NSCLC patients with ALK fusion gene. However, acquired resistance to ALK-TKI remains virtually inevitable.

13.3.1 Crizotinib

Two major mechanisms of resistance to crizotinib have been demonstrated [42–47] (Fig. 13.2). ALK signal-dependent activation such as ALK secondary mutations and amplification has been reported in half of resistant tumors. The other mechanism is the activation of alternative survival signaling pathways including those mediated by EGFR, KRAS, cKIT, and/or IGF-1R. However, about 20–30 % of resistance mechanisms have yet to be identified.

13.3.1.1 ALK Secondary Mutations and Amplification

Approximately 30 % of ALK-positive NSCLC patients who have acquired resistance to crizotinib exhibit ALK secondary mutations both inside and outside the gatekeeper site, in analogy with the EGFR-TKI-resistant T790M mutation [42, 43]. Compared with the single T790M gatekeeper mutation of EGFR-TKIs, ALK-positive NSCLC tumors showed multiple gatekeeper mutations (L1196M, C1156Y, F1174L, L1152R, G1202R, S1206Y, G1269A, 1151Tins, etc.) after treatment with crizotinib. ALK gene fusion copy number gain (CNG) was also demonstrated as a mechanism of resistance to crizotinib in vitro [44]. Doebele et al. analyzed re-biopsied samples obtained from 11 ALK gene-rearranged NSCLC patients showing resistance to crizotinib and identified ALK secondary mutations in 36 % of the samples [43]. Two patients (18 %) exhibited ALK CNG [43]. One patient contained both an ALK secondary mutation (G1269A) and CNG [43]. Thus, ALK signal-dependent activation including ALK secondary mutations and amplification is recognized as a mechanism of resistance to crizotinib. Second-generation ALK inhibitors including alectinib and ceritinib exhibit substantial inhibitory potential against tumors with ALK secondary mutations and have been approved in Japan and the Food and Drug Administration (FDA) [9, 10].

13.3.1.2 EGFR Activation

Activation of alternate survival pathways is a major mechanism of resistance to ALK-TKIs. Sasaki et al. demonstrated that EGFR signaling and secretion of EGF and amphiregulin, which are EGFR ligands, are involved in the resistance of H3122 ALK-positive lung cancer cells to crizotinib [45]. EGFR activation could be induced by increased expression of several EGFR ligands, including TGF-α, HB-EGF, and NRG1 as well as EGF and amphiregulin [42, 46]. Almost all mechanisms of EGFR signal-dependent resistance retain ALK signaling. Previous studies reported that crizotinib therapy combined with EGFR-TKIs might be effective against tumors exhibiting EGFR signal-dependent resistance [42, 45].

13.3.1.3 KRAS Mutation, cKIT Amplification, and IGF-1R Activation

Activation of alternate survival signaling pathways such as those mediated by KRAS, cKIT, and/or IGF-1R has also been reported as a mechanism of resistance. KRAS mutations (G12C and G12V) were reported in 2 of 11 ALK-positive NSCLC patients using re-biopsied samples [43]. Two samples derived from crizotinib-resistant NSCLC patients showed amplification of the cKIT gene [42]. Treatment with the cKIT inhibitor imatinib restored the sensitivity to crizotinib. A recent study reported increased expression of IGF-1R and IRS-1 (an adaptor protein that binds to IGF-1R and ALK) in ALK fusion-positive NSCLC patients after crizotinib treatment [47]. Combined treatment with crizotinib and an IGF-1R inhibitor is being considered to overcome resistance to crizotinib in these patients.

13.3.2 Alectinib

Alectinib is a selective ALK inhibitor that was approved in Japan in 2014 [9]. Alectinib has substantial inhibitory potential against tumors with ALK gatekeeper secondary mutations (F1174L, L1196M, L1152R, C1156Y, 1151Tris, and G1269A). However, alectinib showed less efficacy against tumors harboring G1202R mutations. V1180L and I1171T ALK gatekeeper mutations were nevertheless reported as the mechanism of acquired resistance to alectinib [48, 49]. The second-generation ALK inhibitors, ceritinib and AP26113, were effective against ALK-positive NSCLC tumors harboring V1180L and I1171T mutations [48]. HGF/MET signal activation was found in alectinib-resistant NSCLC patients [50]. Crizotinib might be effective against alectinib-resistant NSCLC tumors exhibiting MET activation.

13.3.3 Ceritinib

Ceritinib is a second-generation ALK inhibitor that can block the activity of both ALK and IGF-1R. It was approved by the FDA in 2004 for the treatment of ALK fusion-positive NSCLC patients who failed to respond to crizotinib [10]. Ceritinib could suppress ALK-TKI-induced secondary mutations (L1196M, I1171T, S1206Y, and G1269A) and overcome the resistance to alectinib associated with I1171T and V1180L secondary mutations [51]. However, ALK secondary gatekeeper mutations (L1196M and G1269A) were also observed in ceritinib-resistant NSCLC cells [52, 53]. These resistances could be overcome by alectinib [52, 53].

13.3.4 Future Therapeutic Strategies

The response of ALK-positive NSCLC patients to ALK-TKIs is well documented, although the development of drug resistance is a challenge that must be overcome. Compared with the T790M gatekeeper mutation of EGFR-TKIs, ALK-positive NSCLC tumors developed multiple gatekeeper mutations following treatment with ALK-TKIs. The second-generation ALK-TKIs alectinib and ceritinib could overcome these gatekeeper mutations, whereas these TKIs might paradoxically induce these mutations. In addition, novel second-generation ALK-TKIs including AP26113, ASP3026, and TSR-001 are being developed and are expected to overcome drug resistance. Two randomized phase III trials (the J-ALEX and ALEX studies) are designed to evaluate the efficacy and safety of alectinib compared with crizotinib in ALK-positive NSCLC patients. The question of timing and the sequence of different ALK-TKIs treatments will be answered based on the results of these trials. Furthermore, ALK-TKIs combined with standard chemotherapy or molecular-targeted agents are being considered as a therapeutic strategy to overcome drug resistance. Evidence of chemotherapy with ALK-TKIs in ALK-positive patients has not yet been shown; however, ALK-TKI combination therapy is being planned for the future.

13.4 Mechanisms of Resistance to NSCLC Patients with Minor-Driver Mutation

RET and ROS1 rearrangements have recently been identified as oncogenic driver mutations in LADC [11–13]. Dovitinib can be used as a potential therapeutic agent for RET-rearranged LADC, and acquired resistance to dovitinib could be overcome by targeting Src [54]. Crizotinib showed remarkable responses in NSCLC patients harboring ROS1 fusions. Crizotinib resistance owing to a G2032R mutation in the ROS1 kinase domain was observed, and the cMET/RET/VEGFR inhibitor cabozantinib was able to overcome resistance to this secondary mutation in vitro [55]. The frequency of these gene rearrangements in LADC is rare, and dissection of the mechanism of resistance owing to minor-driver mutations is relatively difficult. However, identification of molecules associated with drug resistance should facilitate the development of treatments for LADC patients with minor-driver mutations.

References

1. Lynch TJ, Bell DW, Sordella R et al (2004) Activating mutations in the epidermal growth factor receptor. N Engl J Med 350:2129–2139
2. Paez JG, Janne PA, Lee JC et al (2004) EGFR mutations in lung cancer: correlation with clinical response to gefitinib therapy. Science 304:1497–1500
3. Soda M, Choi YL, Enomoto M et al (2007) Identification of the transforming EML4-ALK fusion gene in non-small-cell lung cancer. Nature 448(7153):561–566
4. Maemondo M, Inoue A, Kobayashi K et al (2010) Gefitinib or chemotherapy for non-small-cell lung cancer with mutated EGFR. N Engl J Med 362(25):2380–2388
5. Mitsudomi T, Morita S, Yatabe Y et al (2010) Gefitinib versus cisplatin plus docetaxel in patients with non-small-cell lung cancer harbouring mutations of the epidermal growth factor receptor (WJTOG3405): an open label, randomised phase 3 trial. Lancet Oncol 11(2):121–128
6. Rosell R, Carcereny E, Gervais R et al (2012) Erlotinib versus standard chemotherapy as first-line treatment for European patients with advanced EGFR mutation-positive non-small-cell lung cancer (EURTAC): a multicentre, open-label, randomised phase 3 trial. Lancet Oncol 13:239–246

7. Yang JC, Wu YL, Schuler M et al (2015) Afatinib versus cisplatin-based chemotherapy for EGFR mutation-positive lung adenocarcinoma (LUX-Lung 3 and LUX-Lung 6): analysis of overall survival data from two randomised, phase 3 trials. Lancet Oncol 16(2):141–151
8. Shaw AT, Kim DW, Nakagawa K et al (2013) Crizotinib versus chemotherapy in advanced ALK-positive lung cancer. N Engl J Med 368(25):2385–2394
9. Seto T, Kiura K, Nishio M et al (2013) CH5424802 (RO5424802) for patients with ALK-rearranged advanced non-small-cell lung cancer (AF-001JP study): a single-arm, open-label, phase 1–2 study. Lancet Oncol 14(7):590–598
10. Shaw AT, Kim DW, Mehra R et al (2014) Ceritinib in ALK-rearranged non-small-cell lung cancer. N Engl J Med 370(13):1189–1197
11. Kohno T, Ichikawa H, Totoki Y et al (2012) KIF5B-RET fusions in lung adenocarcinoma. Nat Med 18(3):375–377
12. Takeuchi K, Soda M, Togashi Y et al (2012) RET, ROS1 and ALK fusions in lung cancer. Nat Med 18(3):378–381
13. Lipson D, Capelletti M, Yelensky R et al (2012) Identification of new ALK and RET gene fusions from colorectal and lung cancer biopsies. Nat Med 18(3):382–384
14. Kobayashi S, Boggon TJ, Dayaram T et al (2005) EGFR mutation and resistance of non-small-cell lung cancer to gefitinib. N Engl J Med 352(8):786–792
15. Engelman JA, Zejnullahu K, Mitsudomi T et al (2007) MET amplification leads to gefitinib resistance in lung cancer by activating ERBB3 signaling. Science 18:1039–1043
16. Yano S, Wang W, Li Q et al (2008) Hepatocyte growth factor induces gefitinib resistance of lung adenocarcinoma with epidermal growth factor receptor-activating mutations. Cancer Res 68(22):9479–9487
17. Sos ML, Koker M, Weir BA et al (2009) PTEN loss contributes to erlotinib resistance in EGFR-mutant lung cancer by activation of akt and EGFR. Cancer Res 69(8):3256–3261
18. Sequist LV, Waltman BA, Dias-Santagata D et al (2011) Genotypic and histological evolution of lung cancers acquiring resistance to EGFR inhibitors. Sci Transl Med 3(75):75ra26
19. Zhang Z, Lee JC, Lin L et al (2012) Activation of the AXL kinase causes resistance to EGFR-targeted therapy in lung cancer. Nat Genet 44:852–860
20. Jänne PA, Yang JC, Kim DW et al (2015) AZD9291 in EGFR inhibitor-resistant non-small-cell lung cancer. N Engl J Med 372(18):1689–1699
21. Lecia V. Sequist, Jean-Charles Soria et al (2014) First-in-human evaluation of CO-1686, an irreversible, highly selective tyrosine kinase inhibitor of mutations of EGFR (activating and T790M). J Clin Oncol 32:5s, (suppl; abstr 8010)
22. Thress KS, Paweletz CP, Felip E et al (2015) Acquired EGFR C797S mutation mediates resistance to AZD9291 in non-small cell lung cancer harboring EGFR T790M. Nat Med 21(6):560–562
23. Sequist LV, von Pawel J, Garmey EG et al (2011) Randomized phase II study of erlotinib plus tivantinib versus erlotinib plus placebo in previously treated non-small-cell lung cancer. J Clin Oncol 29:3307–3315
24. Yauch RL, Januario T, Eberhard DA et al (2005) Epithelial versus mesenchymal phenotype determines in vitro sensitivity and predicts clinical activity of erlotinib in lung cancer patients. Clin Cancer Res 11:8686–8698
25. Thomson S, Buck E, Petti F et al (2005) Epithelial to mesenchymal transition is a determinant of sensitivity of non–small-cell lung carcinoma cell lines and xenografts to epidermal growth factor receptor inhibition. Cancer Res 65:9455–9462
26. Sharma SV, Lee DY, Li B et al (2010) A chromatin-mediated reversible drug-tolerant state in cancer cell subpopulations. Cell 141(1):69–80
27. Shien K, Toyooka S, Yamamoto H et al (2013) Acquired resistance to EGFR inhibitors is associated with a manifestation of stem cell-like properties in cancer cells. Cancer Res 73:3051–3061
28. Bivona TG, Hieronymus H, Parker J et al (2011) FAS and NF-κB signaling modulate dependence of lung cancers on mutant EGFR. Nature 471(7339):523–526

29. Cheung HW, Du J, Boehm JS et al (2011) Amplification of CRKL induces transformation and epidermal growth factor receptor inhibitor resistance in human non-small cell lung cancers. Cancer Discov 1(7):608–625
30. Garofalo M, Romano G, Di Leva G et al (2011) EGFR and MET receptor tyrosine kinase-altered microRNA expression induces tumorigenesis and gefitinib resistance in lung cancers. Nat Med 18(1):74–82
31. Seike M, Goto A, Okano T et al (2009) MiR-21 is an EGFR-regulated anti-apoptotic factor in lung cancer in never-smokers. Proc Natl Acad Sci U S A 106:12085–12090
32. Cuf S, Bonavia R, Vazquez-Martin A et al (2013) Silibinin suppresses EMT-driven erlotinib resistance by reversing the high miR-21/low miR-200c signature in vivo. Sci Rep 3:2459
33. Ng KP, Hillmer AM, Chuah CT et al (2012) A common BIM deletion polymorphism mediates intrinsic resistance and inferior responses to tyrosine kinase inhibitors in cancer. Nat Med 18(4):521–528
34. Lee JH, Lin YL, Hsu WH et al (2014) Bcl-2-like protein 11 deletion polymorphism predicts survival in advanced non-small-cell lung cancer. J Thorac Oncol 9(9):1385–1392
35. Nakagawa T, Takeuchi S, Yamada T et al (2013) EGFR-TKI resistance due to BIM polymorphism can be circumvented in combination with HDAC inhibition. Cancer Res 73(8):2428–2434
36. Yamaguchi T, Yanagisawa K, Sugiyama R et al (2012) NKX2-1/TITF1/TTF-1-Induced ROR1 is required to sustain EGFR survival signaling in lung adenocarcinoma. Cancer Cell 21(3):348–346
37. Karachaliou N, Gimenez-Capitan A, Drozdowskyj A et al (2014) ROR1 as a novel therapeutic target for EGFR-mutant non-small-cell lung cancer patients with the EGFR T790M mutation. Transl Lung Cancer Res 3(3):122–130
38. Noro R, Seike M, Zou F et al (2015) MET FISH-positive status predicts short progression-free survival and overall survival after gefitinib treatment in lung adenocarcinoma with EGFR mutation. BMC Cancer 15:31
39. Wu YL, Lee JS, Thongprasert S et al (2013) Intercalated combination of chemotherapy and erlotinib for patients with advanced stage non-small-cell lung cancer (FASTACT-2): a randomised, double-blind trial. Lancet Oncol 14(8):777–786
40. Sugawara S, Oizumi S, Minato K et al (2015) Randomized phase II study of concurrent versus sequential alternating gefitinib and chemotherapy in previously untreated non-small cell lung cancer with sensitive EGFR mutations: NEJ005/TCOG0902. Ann Oncol 26(5):888–894
41. Seto T, Kato T, Nishio M et al (2014) Erlotinib alone or with bevacizumab as first-line therapy in patients with advanced non-squamous non-small-cell lung cancer harbouring EGFR mutations (JO25567): an open-label, randomised, multicentre, phase 2 study. Lancet Oncol 15(11):1236–1244
42. Katayama R, Shaw AT, Khan TM et al (2012) Mechanisms of acquired crizotinib resistance in ALK-rearranged lung cancers. Sci Transl Med 4(120):120ra17
43. Doebele RC, Pilling AB, Aisner DL et al (2012) Mechanisms of resistance to crizotinib in patients with ALK gene rearranged non-small cell lung cancer. Clin Cancer Res 18(5):1472–1482
44. Katayama R, Khan TM, Benes C et al (2011) Therapeutic strategies to overcome crizotinib resistance in non-small cell lung cancers harboring the fusion oncogene EML4-ALK. Proc Natl Acad Sci U S A 108(18):7535–7540
45. Sasaki T, Koivunen J, Ogino A et al (2011) A novel ALK secondary mutation and EGFR signaling cause resistance to ALK kinase inhibitors. Cancer Res 71(18):6051–6060
46. Yamada T, Takeuchi S, Nakade J et al (2012) Paracrine receptor activation by microenvironment triggers bypass survival signals and ALK inhibitor resistance in EML4-ALK lung cancer cells. Clin Cancer Res 18(13):3592–3602
47. Lovly CM, McDonald NT, Chen H et al (2014) Rationale for co-targeting IGF-1R and ALK in ALK fusion-positive lung cancer. Nat Med 20(9):1027–1034

48. Katayama R, Friboulet L, Koike S et al (2014) Two novel ALK mutations mediate acquired resistance to the next-generation ALK inhibitor alectinib. Clin Cancer Res 20(22):5686–5696
49. Toyokawa G, Hirai F, Inamasu E et al (2014) Secondary mutations at I1171 in the ALK gene confer resistance to both crizotinib and alectinib. J Thorac Oncol 9(12):e86–e87
50. Isozaki H, Ichihara E, Ohashi K et al (2014) Acquired resistance to new ALK inhibitor, alectinib in lung cancer. Ann Oncol 25:iv546–iv563
51. Friboulet L, Li N, Katayama R, Lee CC et al (2014) The ALK inhibitor ceritinib overcomes crizotinib resistance in non-small cell lung cancer. Cancer Discov 4(6):662–673
52. Sakamoto H, Tsukaguchi T, Hiroshima S et al (2011) CH5424802, a selective ALK inhibitor capable of blocking the resistant gatekeeper mutant. Cancer Cell 19(5):679–690
53. Kodama T, Tsukaguchi T, Yoshida M et al (2014) Selective ALK inhibitor alectinib with potent antitumor activity in models of crizotinib resistance. Cancer Lett 351(2):215–221
54. Kang CW, Jang KW, Sohn J et al (2015) Antitumor activity and acquired resistance mechanism of dovitinib (TKI258) in RET-rearranged lung adenocarcinoma. Mol Cancer Ther 14(10):2238–2248
55. Katayama R, Kobayashi Y, Friboulet L et al (2015) Cabozantinib overcomes crizotinib resistance in ROS1 fusion-positive cancer. Clin Cancer Res 21(1):166–174

Chapter 14
Immunotherapy

Takahiro Ebata

Abstract The prognosis of advanced small cell lung cancer and non-small cell lung cancer was improved with development of chemotherapy and molecular target therapy but still remains poor. Recently, immunotherapy, especially immune checkpoint inhibitor that blocks negative co-stimulator of immune activation, showed promising efficacy.

Cytotoxic T-lymphocyte-associated protein 4 (CTLA-4) is a negative co-stimulator expressed in T cell. Interaction signal with CD80/86 and CTLA-4 between antigen-presenting cell and T cell leads to T-cell suppression. In malignant melanoma, ipilimumab, anti-CTLA-4 antibody, improved survival. On the other hand, programmed death-1 (PD-1) and its ligand PD-L1 are also negative co-stimulators between T cell and tumor cell. And anti-PD-1 antibody impedes the interaction with PD-1 and PD-L1 between T cell and tumor cell and leads to avoid apoptosis of T cell. Anti-PD-1 therapy showed survival benefit in malignant melanoma and non-small cell lung cancer in phase III study. However, it is unclear who can get the benefit. Expression of PD-L1 in tumor cell is expected as a predictive biomarker in non-small cell lung cancer. However, there remain many problems to estimate PD-L1 expression such as adequate antibody, cutoff value, and sample quality. Mutation burden in tumor cell is another promising predictive factor in immune checkpoint inhibitor. However, we need to investigate more validated marker. Further study to investigate the predictive factors and combination with other modalities or immunotherapies was warranted.

Keywords Lung cancer • Immunotherapy • Immune checkpoint inhibitor • PD-1 • CTLA-4

T. Ebata (✉)
Department of Clinical Oncology, Graduate School of Medicine, Chiba University, 1-8-1 Inohana, Chuo-ku, Chiba 260-8677, Japan
e-mail: tebata@chiba-u.jp

© Springer Science+Business Media Singapore 2017

Y. Takiguchi (ed.), *Molecular Targeted Therapy of Lung Cancer*,
DOI 10.1007/978-981-10-2002-5_14

14.1 Introduction

Lung cancer is the leading cause of death in the world [1]. It includes two groups, small cell lung cancer (SCLC) which accounts for 15–20 % of lung cancer and non-small cell lung cancer (NSCLC), which accounts for others. In advanced small cell lung cancer, platinum-based chemotherapy remains the standard treatment for this decade [2]. But the median overall survival was around 1 year. And second-line treatment such as amrubicin and topotecan has fewer benefits [3].

About 70 % of patients with NSCLC were diagnosed in advanced stage, and because it is difficult to cure advanced NSCLC, the main goal of treatment is to survive longer. Treatment of advanced NSCLC greatly developed in the past decade. Platinum-based chemotherapy with third-generation agent such as gemcitabine, paclitaxel, docetaxel (DOC), vinorelbine, and irinotecan showed efficacy as first-line chemotherapy for advanced NSCLC patients [4]. And especially in adenocarcinoma, using PEM with cisplatin followed by continuous maintenance PEM or adding bevacizumab is considered more survival benefit than them [5–7]. On the other hand, specific patients are able to receive great benefit of epidermal growth factor receptor tyrosine kinase inhibitor (EGFR-TKI) or anaplastic lymphoma kinase inhibitor (ALK inhibitor) [8–10]. However, prognosis of advanced NSCLC is still poor. Median progression-free survival (mPFS) of patients who received platinum-containing chemotherapy is about 4–6 months and that of specific patients who received molecular target drug for driver mutation is around 1 year. As for second-line treatment, DOC showed survival benefit for patients who received previous chemotherapy compared with best supportive care (BSC), vinorelbine, or ifosfamide [11, 12]. And in 2004, PEM was shown that its efficacy is comparable with DOC as second-line chemotherapy [13]. Separately from these results, erlotinib showed survival benefit compared with BSC for patients with advanced NSCLC who failed prior chemotherapy in 2005 [14]. Thus, the prognosis in advanced small cell and non-small cell lung cancer remains poor.

Immunotherapy is another strategy to improve survival of lung cancer. For a long time, the main strategy of immunotherapy for cancer is to enhance the immune reaction such as cancer vaccine and adaptive immunotherapy, which target tumor antigen. However, this treatment is not established as the standard treatment. On the other hand, recently, immune checkpoint, which enhances or suppresses immunoactivity, such as B7/cytotoxic T-lymphocyte-associated protein-4(CTLA-4) and programmed death-1(PD-1)/PD-L1, which reduce the immune response with interaction of this molecule, gathers attention for the target of cancer immunotherapy. Targeting this molecule is rapidly developing and becoming one of the most promising treatment strategies in cancer treatment.

In this session, we describe the mechanism of immune checkpoint and the efficacy of immune checkpoint inhibitor for cancer treatment.

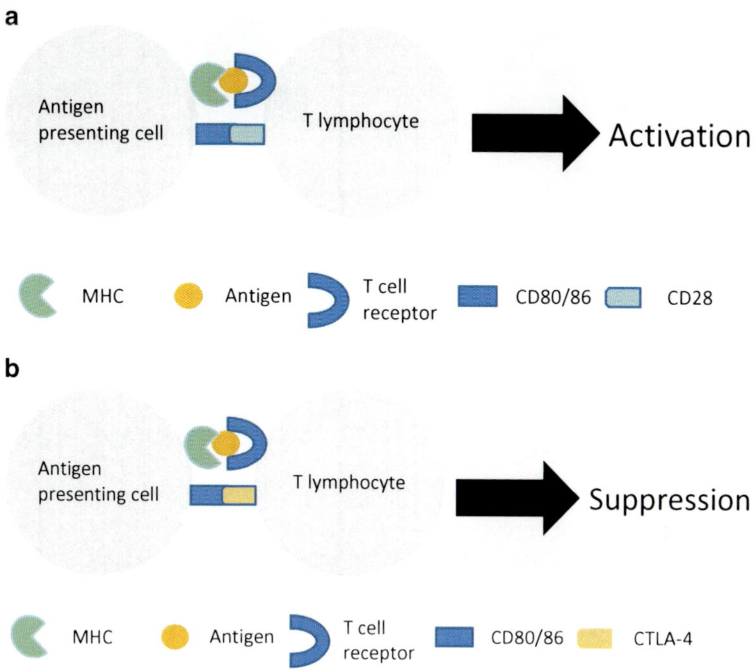

Fig. 14.1 (**a**) Interaction of TCR and antigen only is insufficient to T-cell activation. Co-stimulator signal with CD80/86 and CD28 leads to adequate activation. (**b**) CTLA-4 has higher affinity than CD28 and combines with CD80/86. This leads to suppression of T cell

14.2 B7/CTLA-4 Signal

Adaptive immune response is crucial for infection and cancer. At first, it requires antigen presentation between dendritic cell and T lymphocyte. T lymphocyte recognizes antigen by T-cell receptor. This interaction leads to activation of T lymphocyte which is specific to presented antigen and robust immune reaction. However, recognition of antigen alone is insufficient for adequate adaptive immune response. It also requires co-activation signal with interaction between B7 and CD28 (Fig. 14.1). Once this signal was activated, CTLA-4 was upregulated and combined with B7 because this has higher affinity than CD28. The interaction between B7 and CTLA-4 leads to suppressed immune reaction. This mechanism is to avoid excessive immune reaction, for example, CTLA-4 knockout mouse died within several months by autoimmune disease. In malignant melanoma, ipilimumab, a fully human IgG1 anti CTLA-4 antibody, combined with dacarbazine showed superior efficacy compared with dacarbazine alone [15]. The median overall survival of ipilimumab group and dacarbazine alone group was 11.2 and 9.1 months (hazard ratio for death, 0.72; $p < 0.001$). One of the impressive features of this agent is durable response. The 3-year survival rate was 20.8 and 12.2 % in each group. This result leads to eager clinical trial of immunotherapy for cancer.

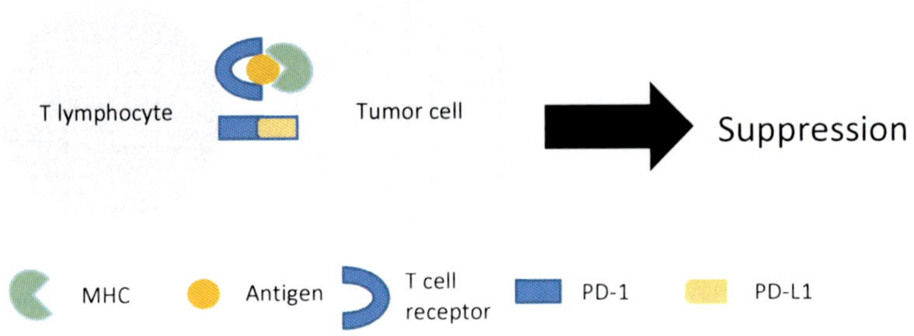

Fig. 14.2 Tumor cell evades immune surveillance by expressing PD-L1. Interaction of PD-1 and PD-L1 leads to T-cell suppression

14.3 PD-1/PD-L1 Signal

After the adequate antigen presentation followed by activation, T lymphocyte killed tumor cell in peripheral tissue. However, this process also includes suppression mechanism. Programmed death-1 (PD-1) is one of the CD28 family molecules discovered in 1992 [16]. T lymphocyte expresses this molecule and interaction with its ligand PD-L1 leads to suppressed immune reaction. The deficiency of PD-1 also leads to autoimmune disease. Its ligand, PD-L1, was expressed in antigen-presenting cell and tumor. Tumor cell escapes from immune reaction with PD-1/PD-L1 interaction, resulting in T-cell apoptosis (Fig. 14.2) [17].

Tumor sometimes expresses PD-L1, and this leads to poor prognosis in several cancers such as malignant melanoma, esophageal, ovarian, and lung cancer [18–21].

This might be because of immune suppression of tumor in peripheral tissue. Deficiency or inhibition of PD-1 leads to removal of the suppression of immune reaction to tumor. PD-1 knockout mouse or anti-PD-1 antibody showed antitumor efficacy. In malignant melanoma, nivolumab, a fully human IgG4 anti-PD-1 antibody, showed superior efficacy compared with dacarbazine [22]. Median survival time was not reached in nivolumab group and 10.8 months in dacarbazine group. Response rate was 40 % and duration of response was not reached in nivolumab group. This surprising result leads to the enthusiastic investigation of anti-PD-1 therapy.

Table 14.1 Efficacy of nivolumab compared with docetaxel in squamous cell lung carcinoma

	Nivolumab	Docetaxel	HR (95 % CI)	*P* value
Median overall survival (month) (95 % CI)	9.2 (7.3–13.3)	6.0 (5.1–7.3)	0.59 (0.44–0.79)	<0.001
1 year survival rate (%) (95 % CI)	42 (34–50)	24 (17–31)		
Median progression-free survival (month) (95 % CI)	3.5 (2.1–4.9)	2.8 (2.1–3.5)	0.62 (0.47–0.81)	<0.001
Median duration of response (month) (range)	Not reached (2.9–20.5)	8.4 (1.4–15.2)		
Response rate (%) (95 % CI)	20 (14–28)	9 (5–15)		0.008

14.4 Immunotherapy in Lung Cancer

14.4.1 Non-small Cell Lung Cancer

In phase I trial of anti-PD-1 and PD-L1 antibody in patients with advanced cancer, response was observed in 10–18 % of patients with non-small cell lung cancer [23, 24]. PD-1/PD-L1 pathway blockade was expected to have promising effect in non-small cell lung cancer.

In phase II trial for patients with advanced, refractory squamous non-small cell lung cancer, 117 patients received nivolumab [25]. Response rate was 14.5 % (95 % confidence interval [CI], 8.7–22.2). Median progression-free survival was only 1.9 months (95 % CI, 1.8–3.2), but median duration of response was not reached (95 % CI, 8.31; not reached).

In 2015, phase III trial proved the superiority in overall survival of nivolumab compared with docetaxel [26, 27]. In squamous cell carcinoma, 272 patients received nivolumab, at a dose of 3 mg/kg every 2 weeks or docetaxel at a dose of 75 mg/m^2 every 3 weeks. Median overall survival was 9.2 months (95% CI, 7.3–13.3) in nivolumab group and 6.0 months (95 % CI, 5.1–7.3) in docetaxel group (hazard ratio [HR] 0.59; 95 % CI, 0.44–0.79) (Table 14.1). The 1-year survival rate was 42 % (95 % CI, 34–50) and 24 % (95 % CI, 17–31). Expression of PD-L1 in tumor cell was not a predictive factor of nivolumab.

In non-squamous cell carcinoma, 582 patients received nivolumab or docetaxel at same schedule with squamous cell carcinoma. Median overall survival was 12.2 months (95 % confidence interval [CI], 9.7–15.0) in nivolumab group and 9.4 months (95 % CI, 8.1–10.7) in docetaxel group (hazard ratio [HR] 0.73; 95 % CI, 0.59–0.89) (Table 14.2). The 1-year survival rate was 51 % (95 % CI, 45–56) and 39 % (95 % CI, 33–45). In this report, although all subgroups favored nivolumab than docetaxel, PD-L1 expression was a strong predictive factor of nivolumab. On the other hand, pembrolizumab, a highly selective humanized monoclonal IgG4-kappa isotype PD-1 antibody, also showed efficacy [28]. In the phase I trial, 495 patients with advanced non-small cell lung cancer received pembrolizumab at a

Table 14.2 Efficacy of nivolumab compared with docetaxel in non-squamous cell lung carcinoma

	Nivolumab	Docetaxel	HR (95 % CI)	*P* value
Median overall survival (month) (95 % CI)	12.2 (9.7–15.0)	9.4 (8.1–10.7)	0.73 (0.59–0.89)	0.002
1 year survival rate (%) (95 % CI)	51 (45–56)	39 (33–45)		
Median progression-free survival (month) (95 % CI)	2.3 (2.2–3.3)	4.2 (3.5–4.9)	0.92 (0.77–1.11)	0.39
Median duration of response (month) (range)	17.2 (1.8–22.6)	5.6 (1.2–15.2)		
Response rate (%) (95 % CI)	19 (15–24)	12 (9–17)		0.02

dose of 2 or 10 mg/kg every 3 weeks or 10 mg/kg every 2 weeks. Objective response rate was 19.4 % and median progression-free survival was 3.7 months. In this study, PD-L1 expression seemed to be a predictive factor of response. Thus, immune checkpoint inhibitor caused a breakthrough in non-small cell lung cancer treatment. Further investigation to improve outcome is warranted.

14.4.2 Small Cell Lung Cancer

In small cell lung cancer (SCLC), the development of immune checkpoint inhibitor is delayed compared with non-small cell lung cancer. In ASCO 2015 in a phase I/II study, nivolumab with or without ipilimumab was reported [29]. Seventy-five patients received nivolumab 3 mg/kg every 2 weeks or nivolumab plus ipilimumab (1 + 1 mg/kg, 1 + 3 mg/kg, or 3 + 1 mg/kg) every 3 weeks for four cycles followed by nivolumab 3 mg/kg every 2 weeks. Overall response rate was 25 and 15 % in nivolumab with or without ipilimumab. On the other hand, in phase Ib trial, 16 PD-L1-positive SCLC patients received pembrolizumab [30]. Response rate was 25 % and durable response was observed.

14.4.3 Adverse Event

The feature of adverse event of immune checkpoint inhibitor is different from cytotoxic agent or molecular target therapy. Although the serious adverse event is rare, it causes immune-related adverse event such as endocrine system disorder. In a phase III trial, the frequency of a serious adverse event with nivolumab was lower than docetaxel (7 % vs. 55 %) [26]. Hematologic toxicity, which is one of the general adverse events in cytotoxic agent, is very low. Anemia and neutropenia occurred only 2 and 1 %. On the other hand, hypothyroidism and pneumonitis occurred in 4–5 % of patients. Anti-CTLA-4 antibody seems more toxic than anti-PD-1 antibody. In malignant melanoma, the frequency of immune-related adverse event of ipilimumab was 60 % [31]. Grade 3 or 4 event occurred in 10–15 % of patients.

The most common adverse event was diarrhea, which occurred in 27–31 % of patients. And endocrine disorder occurred in 3.9–7.6 % of patients. This adverse event occurs not only in the early phase of treatment but in the late phase and requires drug withdrawal or immunosuppressive therapy such as steroid and antitumor necrosis factor α-antibody.

14.4.4 The Feature of Response

The response of immune checkpoint inhibitor has some feature different from conventional chemotherapy. First, delayed response was often observed. Median time to response was 2.2 months (95 % CI, 1.6–11.8) [26]. This occurred even after the discontinuation of treatment. Second, tumor reduction after once tumor progression, called pseudo-progression, was observed. It is difficult to distinguish pseudo-progression from true progression. Then, new response criteria called irRECIST were put forward [32]. Validation of response criteria is warranted.

14.4.5 Predictive Factor

The feature of immunotherapy seemed to be that specific population showed efficacy and durable response. Then, it is crucial to investigate the predictive factor such as EGFR mutation for EGFR-TKI therapy.

The most expected simple answer for anti-PD-1 or anti-PD-L1 therapy is expression of PD-L1 in tumor cell. In non-squamous cell lung cancer treated with nivolumab and non-small cell lung cancer treated with pembrolizumab, expression of PD-L1 seemed to be a predictive biomarker [27, 28]. However, in squamous cell lung cancer treated with nivolumab, it was not a predictive factor [26]. And there was some problem for the estimation of expression of PD-L1. First, the most adequate antibody to estimate the expression of PD-L1 was unclear. In nivolumab study, anti-PD-L1 antibody clone 28-8 (Dako, North America) was used. And anti-PD-L1 antibody clone 22C3 (Merck) was used in pembrolizumab study. It is unknown which antibody is suitable for the estimation of PD-L1. Second, the cutoff value of PD-L1-positive tumor cell was also unclear. Third, expression of PD-L1 might be influenced by previous chemotherapy, molecular target therapy, and radiotherapy. It is also unknown how these treatments influence the expression of PD-L1. In nivolumab study, PD-L1 was estimated by archival samples in some cases. It might be suitable to estimate the samples obtained just before anti-PD-1 antibody treatment. However, there exists heterogeneity of PD-L1 expression. Ilie et al. reported that expression of PD-L1 might be underestimated in biopsy sample, and there was poor association between biopsy samples and surgically resected sample

Table 14.3 Difference of evaluation of PD-L1 expression in two studies

	Checkmate (nivolumab)	Keynote (pembrolizumab)
Antibody	Clone 28-8 (Dako, North America)	Clone 22C3 (Merck)
Cutoff value	1 %, 5 %, and 10 % or higher	1 %, 49 %, or higher
Material	Archival or recent	Contemporaneous
Predictive value of PD-L1	Non-squamous cell lung cancer only	Yes

[33]. These factors implicated whether PD-L1 expression in tumor cell is a predictive factor in anti-PD-1 therapy. It is desirable to establish the adequate method of estimation of PD-L1 expression. The difference of estimation of PD-L1 was shown in Table 14.3. And patients with PD-L1-negative tumor were also observed for response. It may be difficult to distinguish responder from nonresponder with PD-L1 expression alone. On the other hand, anti-PD-1 antibody is the most sensitive in several cancers such as malignant melanoma and non-small cell lung cancer which has higher mutational burden than other cancers. Then, other promising predictive factor is mutational burden of tumor. Rizvi et al. investigated the whole genome of tumor that received pembrolizumab [34]. They reported that tumors with higher nonsynonymous mutational burden, molecular smoking signature, and DNA repair pathway mutation showed good correlation with high response rate and progression-free survival of pembrolizumab. This result is consistent with ipilimumab in malignant melanoma [35]. In other cancers, mismatch repair deficiency predicted the response of immune checkpoint inhibitor [36]. Nonsynonymous mutational burden was correlated with neoantigen burden and might lead to good response of T cell. However, it might be difficult to identify crucial predictive factor such as EGFR mutation in EGFR-TKI therapy in the case of immune checkpoint inhibitor because immune reaction required a complicated process. Not only tumor factors but also host factors such as dendritic cell, T lymphocyte, and their activation process participate in the efficacy of immunotherapy.

14.5 Future Direction

Immune checkpoint inhibitor showed efficacy in non-small cell lung cancer. One of the next steps is combination of other immune therapies, molecular target therapies, or chemotherapies. Checkmate 223, a randomized trial which compares nivolumab with or without ipilimumab or platinum-doublet chemotherapy for advanced non-small cell lung cancer, is ongoing (NCT02477826). For ALK-positive advanced NSCLC, phase I study of the combination of crizotinib and pembrolizumab has been started (NCT02511184).

And clinical trial in another setting is also ongoing. As adjuvant setting, phase III trial which compared nivolumab for 1 year or observation for stage IB-IIIA non-small cell lung cancer after surgery followed by adjuvant chemotherapy

(NCT02595944). In unresectable stage III non-small cell lung cancer, phase II trial of concurrent chemoradiation followed by pembrolizumab is also ongoing (NCT02343952). The mechanism and efficacy of combination of immunotherapy and other modalities are unclear. This study gives the insight to this question. And not only CTLA-4 and PD-1 but also other immune checkpoint molecules such as LAG-3, TIM-3, and TIGIT are recognized [37–39]. The combination of stimulation of positive co-stimulator molecule and blockade of negative co-stimulator molecule may lead to further efficacy.

14.6 Conclusion

Immunotherapy in lung cancer, especially immune checkpoint inhibitor, is promising and rapidly developing. The efficacy of new agent or combination of other drugs will be discovered. On the other hand, further investigation of predictive factor is needed.

References

1. Siegel RL, Miller KD, Jemal A (2015) Cancer statistics. CA Cancer J Clin 65:5–29
2. Noda K, Nishiwaki Y, Kawahara M et al (2002) Irinotecan plus cisplatin compared with etoposide plus cisplatin for extensive small-cell lung cancer. N Engl J Med 346:85–91
3. von Pawel J, Jotte R, Spigel DR et al (2014) Randomized phase III trial of amrubicin versus topotecan as second-line treatment for patients with small-cell lung cancer. J Clin Oncol 32:4012–4019
4. Schiller JH, Harrington D, Belani CP et al (2002) Comparison of four chemotherapy regimens for advanced non-small-cell lung cancer. N Engl J Med 346:92–98
5. Scagliotti GV, Parikh P, von Pawel J et al (2008) Phase III study comparing cisplatin plus gemcitabine with cisplatin plus pemetrexed in chemotherapy-naive patients with advanced-stage non-small-cell lung cancer. J Clin Oncol 26:3543–3551
6. Paz-Ares LG, de Marinis F, Dediu M et al (2013) PARAMOUNT: final overall survival results of the phase III study of maintenance pemetrexed versus placebo immediately after induction treatment with pemetrexed plus cisplatin for advanced nonsquamous non-small-cell lung cancer. J Clin Oncol 31:2895–2902
7. Sandler A, Gray R, Perry MC et al (2006) Paclitaxel-carboplatin alone or with bevacizumab for non-small-cell lung cancer. N Engl J Med 355:2542–2550
8. Maemondo M, Inoue A, Kobayashi K et al (2010) Gefitinib or chemotherapy for non-small-cell lung cancer with mutated EGFR. N Engl J Med 362:2380–2388
9. Rosell R, Carcereny E, Gervais R et al (2012) Erlotinib versus standard chemotherapy as first-line treatment for European patients with advanced EGFR mutation-positive non-small-cell lung cancer (EURTAC): a multicentre, open-label, randomised phase 3 trial. Lancet Oncol 13:239–246
10. Solomon BJ, Mok T, Kim DW et al (2014) First-line crizotinib versus chemotherapy in ALK-positive lung cancer. N Engl J Med 371:2167–2177

11. Shepherd FA, Dancey J, Ramlau R et al (2000) Prospective randomized trial of docetaxel versus best supportive care in patients with non-small-cell lung cancer previously treated with platinum-based chemotherapy. J Clin Oncol 18(10):2095–2103
12. Fossella FV, DeVore R, Kerr RN et al (2000) Randomized phase III trial of docetaxel versus vinorelbine or ifosfamide in patients with advanced non-small-cell lung cancer previously treated with platinum-containing chemotherapy regimens. The TAX 320 Non-Small Cell Lung Cancer Study Group. J Clin Oncol 18(12):2354–2362
13. Hanna N, Shepherd FA, Fossella FV et al (2004) Randomized phase III trial of pemetrexed versus docetaxel in patients with non-small-cell lung cancer previously treated with chemotherapy. J Clin Oncol 22:1589–1597
14. Shepherd FA, Rodrigues Pereira J et al (2005) Erlotinib in previously treated non-small-cell lung cancer. N Engl J Med 353:123–132
15. Robert C, Thomas L, Bondarenko I et al (2011) Ipilimumab plus dacarbazine for previously untreated metastatic melanoma. N Engl J Med 364:2517–2526
16. Ishida Y, Agata Y, Shibahara K et al (1992) Induced expression of PD-1, a novel member of the immunoglobulin gene superfamily, upon programmed cell death. EMBO J 11:3887–3895
17. Dong H, Strome SE, Salomao DR et al (2002) Tumor-associated B7-H1 promotes T-cell apoptosis: a potential mechanism of immune evasion. Nat Med 8:793–800
18. Hino R, Kabashima K, Kato Y et al (2010) Tumor cell expression of programmed cell death-1 ligand 1 is a prognostic factor for malignant melanoma. Cancer 116:1757–1766
19. Ohigashi Y, Sho M, Yamada Y et al (2005) Clinical significance of programmed death-1 ligand-1 and programmed death-1 ligand-2 expression in human esophageal cancer. Clin Cancer Res 11:2947–2953
20. Hamanishi J, Mandai M, Iwasaki M et al (2007) Programmed cell death 1 ligand 1 and tumor-infiltrating CD8+ T lymphocytes are prognostic factors of human ovarian cancer. Proc Natl Acad Sci U S A 104:3360–3365
21. Mu CY, Huang JA, Chen Y et al (2011) High expression of PD-L1 in lung cancer may contribute to poor prognosis and tumor cells immune escape through suppressing tumor infiltrating dendritic cells maturation. Med Oncol 28:682–688
22. Robert C, Long GV, Brady B et al (2015) Nivolumab in previously untreated melanoma without BRAF mutation. N Engl J Med 372:320–330
23. Topalian SL, Hodi FS, Brahmer JR et al (2012) Safety, activity, and immune correlates of anti-PD-1 antibody in cancer. N Engl J Med 366(26):2443–2454
24. Brahmer JR, Tykodi SS, Chow LQ et al (2012) Safety and activity of anti-PD-L1 antibody in patients with advanced cancer. N Engl J Med 366(26):2455–2465
25. Rizvi NA, Mazieres J, Planchard D et al (2015) Activity and safety of nivolumab, an anti-PD-1 immune checkpoint inhibitor, for patients with advanced, refractory squamous non-small-cell lung cancer (CheckMate 063): a phase 2, single-arm trial. Lancet Oncol 16:257–265
26. Brahmer J, Reckamp KL, Baas P et al (2015) Nivolumab versus docetaxel in advanced squamous-cell non-small-cell lung cancer. N Engl J Med 373:123–135
27. Borghaei H, Paz-Ares L, Horn L et al (2015) Nivolumab versus docetaxel in advanced non-squamous non-small-cell lung cancer. N Engl J Med 373:1627–1639
28. Garon EB, Rizvi NA, Hui R et al (2015) Pembrolizumab for the treatment of non-small-cell lung cancer. N Engl J Med 372:2018–2028
29. Antonia SJ, Bendell JC, Taylor MH et al (2015) Phase I/II study of nivolumab with or without ipilimumab for treatment of recurrent small cell lung cancer (SCLC): CA209-032. ASCO Meet Abstr 33:7503
30. Ott PA, Fernandez MEE, Hiret S et al (2015) Pembrolizumab (MK-3475) in patients (pts) with extensive-stage small cell lung cancer (SCLC): preliminary safety and efficacy results from KEYNOTE-028. ASCO Meet Abstr 33:7502
31. Hodi FS, O'Day SJ, McDermott DF et al (2010) Improved survival with ipilimumab in patients with metastatic melanoma. N Engl J Med 363:711–723

32. Bohnsack O, Hoos A, Ludajic K (2014) 1070p adaptation of the immune related response criteria: irRECIST. Ann Oncol 25:iv369
33. Ilie M, Long-Mira E, Bence C et al (2015) Comparative study of the PD-L1 status between surgically resected specimens and matched biopsies of NSCLC patients reveal major discordances: a potential issue for anti-PD-L1 therapeutic strategies. Ann Oncol. doi:10.1093/annonc/mdv489
34. Rizvi NA, Hellmann MD, Snyder A et al (2015) Cancer immunology. Mutational landscape determines sensitivity to PD-1 blockade in non-small cell lung cancer. Science 348(6230):124–128
35. Snyder A, Makarov V, Merghoub T et al (2014) Genetic basis for clinical response to CTLA-4 blockade in melanoma. N Engl J Med 371:2189–2199
36. Le DT, Uram JN, Wang H et al (2015) PD-1 blockade in tumors with mismatch-repair deficiency. N Engl J Med 372:2509–2520
37. Shin DS, Ribas A (2015) The evolution of checkpoint blockade as a cancer therapy: what's here, what's next? Curr Opin Immunol 33:23–35
38. Anderson AC (2012) Tim-3, a negative regulator of anti-tumor immunity. Curr Opin Immunol 24(2):213–216
39. Chan CJ, Andrews DM, Smyth MJ (2012) Receptors that interact with nectin and nectin-like proteins in the immunosurveillance and immunotherapy of cancer. Curr Opin Immunol 24(2):246–251

Chapter 15
Lung Cancer Complicated with Interstitial Lung Diseases

Yuji Minegishi

Abstract Chronic interstitial lung diseases (ILDs) are one of the most common complications in patients with lung cancer. In the context of lung cancer patients with ILDs in Japan, the most serious toxicity is acute exacerbation of ILDs caused by anticancer treatment. Nevertheless, there is, so far, no consensus regarding optimal chemotherapy for advanced lung cancer patients with ILDs. On the other hand, introduction of molecular-targeted agents, in particular specific inhibitors targeting driver oncogene mutations, has dramatically changed the treatment of advanced non-small cell lung cancer (NSCLC). However, the application of molecular-targeted agents for lung cancer patients with preexisting ILDs should be carefully considered. In the case of patients treated with epithelial growth factor receptor (EGFR) tyrosine kinase inhibitors (TKI), some reports suggest that preexisting ILDs are significant risk factor for severe acute lung injury following treatment. In respect of molecular-targeted agents other than EGFR-TKIs, severe drug-induced ILDs have been reported. Further investigations will be needed to access the benefit and safety of molecular-targeted agents in lung cancer patients with chronic ILDs.

Keywords Lung cancer • Interstitial lung diseases • Acute exacerbation • Chemotherapy • Molecular-targeted therapy

15.1 Coexistence of Lung Cancer and Chronic Interstitial Lung Diseases

The co-occurrence of interstitial lung diseases (ILDs) is the most challenging complication in clinical practice for lung cancer. Evidence to date indicates that idiopathic pulmonary fibrosis (IPF), which is the most common subset of chronic ILDs, is associated with an increased risk of lung carcinogenesis. In Japan, it was reported that death due to lung cancer was the third most common cause of mortality (11 %)

Y. Minegishi (✉)
Department of Pulmonary Medicine and Oncology, Graduate School of Medicine, Nippon Medical School, 1-1-5 Sendagi, Bunkyo-ku, Tokyo 113-8602, Japan
e-mail: yminegis@nms.ac.jp

© Springer Science+Business Media Singapore 2017

Y. Takiguchi (ed.), *Molecular Targeted Therapy of Lung Cancer*,
DOI 10.1007/978-981-10-2002-5_15

during the course of IPF and death due to acute exacerbation (AE) of IPF and chronic respiratory failure were the first (40 %) and second (24 %) cause of mortality, respectively [1]. Turner-Warwick and colleagues found a high prevalence of lung cancer (9.8 %) in patients with IPF and an increased relative risk of lung cancer of 14.1 in patients with IPF compared to the general population [2]. A large population-based cohort study showed a significant increase in the incidence of lung cancer in IPF patients compared with control subjects (4.4 % versus 0.9 %, respectively). These investigators also concluded that IPF is an independent risk factor for lung carcinogenesis [3]. The cumulative incidence rate of lung cancer in patients with IPF increased in proportion of the duration of follow-up (3.3 %, 15.4 %, and 54.7 % at 1, 5, and 10 years, respectively) [4]. In a study including 83 autopsy cases of usual interstitial pneumonia (UIP), the prevalence of lung cancer in UIP (48.2 %) was reported as being significantly higher than that observed in the control population without UIP (9.1 %); moreover, the prevalence of multiple lung cancers in case with UIP (20.0 %) was markedly increased [5]. Kawasaki and colleagues reported evidence of UIP in background lung tissue in 53 (7.5 %) out of the 711 surgically resected lung cancer cases. Additionally, the incidence of multiple lung cancers in patients with UIP (17 %) was also significantly higher than seen in cases without UIP [6].

Thus, fibrotic lesions, especially those related to UIP, had a remarkable tendency to develop lung cancers. However, the carcinogenic mechanisms relating to this phenomenon have yet to be elucidated. UIP results in chronic injury to the bronchiolar and alveolar epithelia, which is characterized by fibrosis and remodeling of the peripheral lung tissue. This chronic inflammation may increase the risk of cancer developing through accumulated DNA damage. It has been reported that in the bronchiolar and alveolar epithelia within the honeycomb-like lesions of patients with IPF and/or lung cancer, several tumor suppressor genes such as p53, K-ras, and FHIT (fragile histidine triad) are frequently found to be mutated or deleted [7–12]. Those findings suggest that in patients with IPF, lung cancer might originate from these epithelial cells.

15.2 Acute Exacerbation of Chronic ILDs and Chemotherapy-Induced ILD

Research into lung cancer coexisting with, or complicated by, chronic ILDs is of particular consequence in Japan. Here, we will make reference to the current status of and issues associated with treatment of this patient population in Japan.

Idiopathic interstitial pneumonias (IIPs) are usually characterized by slowly progressive respiratory insufficiency. In particular, IPF is a relentlessly progressive and fatal disorder without an effective therapy. Moreover, some IIP patients experience acute exacerbations (AEs) generally characterized by suddenly progressive and severe respiratory failure, with new lung opacities and pathological lesions of diffuse alveolar damage (DAD). The concept of AE, which was first proposed in Japan, has recently come to be recognized globally. The criteria for identifying AE of IPF include an unexplained rapid worsening of dyspnea, severely impaired gas exchange, new radiographic diffused alveolar infiltrates, and the absence of alternate causes, such as infectious pneumonia, pulmonary embolism, pneumothorax, and heart failure [13]. This clinical condition is lethal and significantly affects the prognosis of patients with chronic ILDs, because there is no established treatment for AE.

Iatrogenic exacerbation of ILDs triggered by various anticancer treatments, such as surgical resection, thoracic radiotherapy, and chemotherapy, is the most common fatal complication. Moreover, it is considered that this AE of ILDs and severe drug-induced ILD (DILD), which is characterized pathologically by DAD, are more distinctive of Japanese populations than other races. Therefore, evidence of preexisting interstitial and/or fibrotic appearance on computed tomography (CT) scans of the chest poses significant concern in Japan. Nevertheless, there is no consensus as to the optimal treatment strategy in this context. In other words, the optimal approach for anticancer treatment and prophylaxis for treatment-related AE has not been elucidated. It has been suggested since the 1980s in Japan that acute respiratory disorder caused by anticancer treatment is associated with chronic ILDs such as IPF. The exacerbation of ILDs by chemotherapy agents has attracted attention following reports in Japan of ILD developing after treatment with gefitinib, an epidermal growth factor receptor (EGFR) tyrosine kinase inhibitor (TKI). Additionally, improvements in the resolution of CT scan have enabled better detection of ILDs.

The AE of chronic ILDs also seems to be frequent in people of the Mongolian race, as well as the Japanese. It has been suggested that racial etiologies could explain the higher incidence of some DILDs. For example, in 3166 Japanese patients with advanced/recurrent NSCLC who were enrolled in a cohort and nested case-control study, gefitinib-induced ILD was manifested in about 4.0 % of patients, which is about 13-fold higher compared to that observed in the USA, where the incidence is 0.3 % [FDA Approval Letter for Iressa]. Moreover, that study demonstrated that a predisposing background of preexisting interstitial pneumonias was an independent risk factor for developing acute ILD not only in the gefitinib cohort but also in the control cohort with conventional cytotoxic chemotherapies [14]. Similarly, the incidence of leflunomide-induced ILD in Japan is about 100-fold higher than that seen in the USA (1.8 % vs. 0.017 %) [Arava Periodic Safety Update Report]. According to the manufacturer of bleomycin, the incidence of bleomycin-induced ILD in Japan is about 60-fold higher than in other countries (0.66 % vs. 0.01 %).

15.2.1 Acute Exacerbation of ILDs Triggered by Anticancer Treatments

Pulmonary resection has been shown to trigger decompensating acute respiratory failure associated with high postoperative morbidity and mortality in lung cancer patients with ILDs. The AE of ILDs has been shown to be major cause of death for patients with lung cancer after pulmonary resection in the 2007 annual report of the Japanese Association for Thoracic Surgery. In 945 (3.6 %) out of 26,092 reported operations for lung cancer, complications with ILDs have been observed. The incidence of postoperative AE of ILDs was 16.2 % and the mortality rate due to postoperative AE was 42 %. The risk associated with surgery is significantly higher in these patients as compared to the 30-day operative and hospital mortality rates in all patients (0.46 % and 1.0 %, respectively) [15]. Nevertheless, in the operable stages, surgery is the most effective treatment modality in these patients, this being similar to those without ILDs. The incidence of postoperative AE of ILDs has been reported as being between 7 and 25 % [16–19]. The Japanese Association for Chest Surgery retrospectively analyzed 1763 patients with NSCLC who had undergone pulmonary resection and presented with a clinical diagnosis of ILD at 61 hospitals in Japan. [20]. Postoperative AE developed within 30 days after the operation in approximately 9.3 % of patients, with a mortality rate of 43.9 %. Multivariate analysis identified the following seven risk factors for AE: male gender, elevated serum levels of KL-6 ($\geq$1000 U/mL), reduced percent predicted vital capacity (<80 %), UIP appearance on CT scan, history of AE, preoperative steroid use, and surgical procedures with an increased risk of AE in proportion to resected lung volume. The positive effect of perioperative prophylactics, such as steroids and sivelestat, was not confirmed in this study. This study has several substantial limitations. Firstly, as the inclusion criteria relied on CT appearance alone, it is not possible to know the underlying ILD type. ILDs include various underlying diseases in which relative AE risk is different, such as IIPs, CVD-IP, sarcoidosis, and occupational pulmonary disease. Secondly, clinical diagnoses of UIP appearance and AE were made by the individual institutions involved. Accordingly, despite using the same criteria based on guidelines, the diagnoses of UIP and AE may not be fully consistent between different institutions.

In unresectable locally advanced lung cancer, chemoradiotherapy is considered as standard of care. However, some reports suggest that ILD is a risk factor for severe radiation pneumonitis. Therefore, the use of thoracic radiotherapy must be carefully considered. In clinical practice, most physicians avoid thoracic radiotherapy for lung cancer patients with ILDs.

Several retrospective studies have reported the cumulative incidence of AE associated with cytotoxic chemotherapy to be from 20 to 28 % [21, 22, 23]. Moreover, our previous report has shown that AE is manifested at a high incidence (30 % rate and death in all cases) in patients receiving best supportive care only [21]. We retrospectively analyzed 396 patients with lung cancer coexisting with IIPs who had received chemotherapy at 19 hospitals in Japan [the 2009 annual report of the

Diffuse Lung Diseases Research Group from the Ministry of Health, Labor and Welfare, Japan]. It was demonstrated that 52 patients (13.1 %) from this cohort developed initial chemotherapy-related AE.

Lung cancer patients receiving chemotherapy alone exhibit minimal evidence of cure. Whether such high-risk chemotherapies are indicated for incurable lung cancer patients with ILDs remains a point of controversy. It is important to determine the most appropriate treatment strategy by comprehensively considering activity and prognosis of ILDs, expected benefit, and adverse events including AE by chemotherapy.

15.2.2 Natural History of Chronic ILDs

AE is also a serious problem in patients with chronic ILD but without lung cancer. In particular with respect to IPF, AE manifests frequently during the clinical course and is a major cause of mortality. Randomized phase II [24] and III [25] studies of pirfenidone in Japan have reported that the rate of AE in the placebo groups was 13.9 % over 9 months and 4.8 % over 52 weeks, respectively. Richeldi and colleagues reported that the incidence of adjudicated AE was 5.4 %, and the incidence of investigator-reported AE in Japanese patients was 12 % in the placebo group during a 52-week observation period in their INPULSIS-1 and INPULSIS-2 studies (pooled data) of nintedanib [26]. In respect to some retrospective studies of ILDs other than IPF, the incidences of AE in nonspecific interstitial pneumonia (NSIP) and collagen-vascular disease-associated interstitial pneumonia (CVD-IP) were 4.2 % and 1.3–3.3 % during the first year, respectively [27, 28]. While manifestation of AE has been confirmed in other ILDs, AE is considered to develop more readily in the case of IPF.

The definitive diagnosis of preexisting ILDs is essential to evaluate the risk of AE associated with chemotherapy. However, in clinical practice, pretreatment diagnosis of ILDs is insufficient in most institutions, as the diagnosis of lung cancer takes priority.

15.3 Chemotherapy for Advanced Lung Cancer Patient with Chronic ILDs

The decision to utilize chemotherapy in advanced lung cancer patient with chronic ILDs is difficult. It is not clear whether chemotherapy is beneficial or harmful for each patient with various clinical backgrounds. It is clear that fatal respiratory failure related to chemotherapy is more frequent in those patients with chronic ILDs compared with those without ILDs. Nevertheless, patients with lung cancer should not be excluded from the application of chemotherapy due solely to preexisting

Table 15.1 The incidence of acute exacerbation of idiopathic interstitial pneumonias related to each first-line chemotherapy regimen

Regimen	N	(%)	AE (N)	AE (%)
Carboplatin + paclitaxel	140	35.4	12	8.6
Carboplatin + etoposide	82	20.7	3	3.7
Cisplatin + etoposide	38	9.6	4	10.5
Vinorelbine	30	7.6	8	26.7
Cisplatin + UFT	17	4.3	5	29.4
Carboplatin + vinorelbine	10	2.5	0	0
Cisplatin + vinorelbine	9	2.3	2	22.2
Docetaxel	7	1.8	1	14.3
Carboplatin + docetaxel	6	1.5	4	66.7
Cisplatin + docetaxel	6	1.5	1	1.7
Gefitinib	6	1.5	5	83.3
Others	51		10	19.6
Total	396		52	13.1

AE acute exacerbation of idiopathic interstitial pneumonias, *UFT* uracil and tegafur

interstitial shadow of the lung, as the outcome of chemotherapy for advanced lung cancer patients without ILDs has been gradually improved over time. Moreover, patients with ILDs who do not receive chemotherapy cannot avoid the risk of AE.

Thus, physicians must be aware that AE of ILDs has a high associated mortality, and there is a necessity of careful surveillance for preexisting interstitial lung shadow.

If coexisting ILD has been suspected, systemic survey, including high-resolution CT scans of the chest, pulmonary function tests, arterial blood gas, and serum KL-6, allowing identification of risk factors, should be warranted to manage risk.

15.3.1 Cytotoxic Chemotherapy Agents

The optimal chemotherapy regimen for lung cancer patients with ILDs remains unclear because the existing evidence is based on a few studies with a comparatively small number of patients at single institution. The most cytotoxic chemotherapy agents applied for lung cancer treatment have a usage restriction in respect to ILDs. In a survey performed by the Diffuse Lung Diseases Research Group, the incidence of AE of ILDs in relation to each chemotherapy regimen is shown in Table 15.1. It was clarified that the combinations of carboplatin with paclitaxel for NSCLC patients with ILDs and platinum agents with etoposide for small cell lung cancer (SCLC) patients with ILDs have spread widely as first-line treatment regimens in Japan; in respect of both regimens, this survey suggested a trend toward a low frequency of AE compared to other chemotherapy regimens.

In our prospective feasibility studies of first-line treatment for lung cancer patients with coexisting IIPs, one (5.6 %) out of the 18 patients who received carboplatin with weekly paclitaxel and one (5.9 %) out of the 17 patients who received carboplatin with etoposide developed chemotherapy-related AE of IIPs [29, 30]. Moreover, the objective response rate (ORR) and progression-free survival (PFS) in these studies were comparable to those observed in previous studies for NSCLC and SCLC patients without ILDs (ORR, 61% and 88%; median PFS, 5.3 months and 5.5 months, respectively). Nevertheless, the extent of overall survival (OS) was poorer for patients without ILDs (median OS, 10.6 months and 8.7 months, respectively). Kenmotsu and colleagues retrospectively analyzed 104 NSCLC patients with ILDs treated by platinum-based chemotherapy. Across all patients, the incidence of first-line chemotherapy-related exacerbation of ILDs was 9 %, while the ORR, PFS, and OS were 38 %, 4.8, and 9.9 months, respectively. Five (8 %) out of the 63 patients treated with carboplatin with paclitaxel developed chemotherapy-related exacerbation of ILDs [31].

In SCLC, Yoshida and colleagues reported that one (1.9 %) out of the 52 SCLC patients who received platinum agents with etoposide developed chemotherapy-related exacerbation of ILDs, and the PFS and OS were 4.5 and 9.4 months, respectively [32]. The combination chemotherapies of carboplatin with paclitaxel for NSCLC and carboplatin with etoposide for SCLC are the most widely used of the established standard regimens for advanced lung cancer without ILDs. Furthermore, both of these regimens are the mostly frequently used for lung cancer patients with chronic ILDs, with comparatively permissible safety. Thus, the combination of carboplatin with paclitaxel and carboplatin with etoposide are currently the most recommended treatment options for lung cancer patients with chronic ILDs. However, other studies have reported high risks with these two regimens, with the incidences of AE in these cases ranging from 16–27 % [33–35].

The knowledge base regarding second-line or subsequent treatment is poorer than that in existence for first-line treatment. Most physicians conclude that survival benefit cannot be commensurate to the risk of AE of ILDs and hesitate in respect to the use of subsequent chemotherapy. Therefore, it is considered that one of the possible reasons for unsatisfactory OS despite comparatively good ORR and PFS is less frequently receive second-line chemotherapy in patients with ILDs compared to those without ILDs. As such, it is also essential for improvement of OS to establish optimal chemotherapy for previously treated lung cancer patients with chronic ILDs.

The Diffuse Lung Diseases Research Group retrospectively analyzed chemotherapy-induced AE and the prognosis for lung cancer patients with IIPs who were treated with second-line chemotherapy [the 2012 and 2013 annual report of the Diffuse Lung Diseases Research Group from the Ministry of Health, Labor and Welfare, Japan]. The incidence of AE of ILDs according to each chemotherapy regimen is shown in Table 15.2. The incidence of second-line chemotherapy-related AE was 45 (16.2 %) out of the 278 patients examined, which is comparable to that seen with first-line chemotherapy. Docetaxel (25.9 %) was the most frequently used agent in the context of subsequent chemotherapy. However, docetaxel, pemetrexed,

Table 15.2 The incidence of acute exacerbation of idiopathic interstitial pneumonias related to each second-line chemotherapy regimen

Regimen	N	(%)	AE (N)	AE (%)
Docetaxel	72	25.9	11	15.3
Carboplatin + paclitaxel	31	11.1	3	9.7
Carboplatin + etoposide	15	5.4	0	0
Vinorelbine	24	8.6	6	25
Pemetrexed	21	7.6	6	28.6
Amrubicin	18	6.5	6	33.3
Topotecan	13	4.9	3	23.1
S-1	14	5.3	0	0
EGFR-TKIs	9	3.2	4	44.4
Paclitaxel	7	2.5	1	14.3
Cisplatin + vinorelbine	6	2.2	0	0
Irinotecan	6	2.2	0	0
Others	42		5	12.5
Total	278		45	16.2

AE acute exacerbation of idiopathic interstitial pneumonias, *EGFR* epidermal growth factor receptor, *TKI* tyrosine kinase inhibitor

topotecan, and amrubicin, when used as standard agents for previously treated NSCLC and SCLC, lead to chemotherapy-induced AE of ILDs (AE rates, 15.3 %, 28.6 %, 23.1 %, and 33.3 %, respectively). On the other hand, the use of carboplatin with paclitaxel (9.7 %), carboplatin with etoposide (0 %), or S-1 monotherapy (0 %) appeared to confer a low risk in respect of AE of ILDs. These treatments have not been established as standard of care for previously treated NSCLC or SCLC. A major limitation to this retrospective analysis was that there were only a small number of patients who received each regimen. In relation to survival analysis in this study, OS from second-line and first-line treatment were 8.6 and 15.7 months for NSCLC patients and 9.0 and 17.3 months for SCLC patients, respectively. These encouraging results are comparable with previous reports of patients without ILDs, suggesting that second-line treatment may contribute to improving the prognosis of advanced lung cancer patients with ILDs.

15.3.2 Molecular-Targeted Agents

Over the past decade, introduction of molecular-targeted agents has dramatically changed the treatment landscape of NSCLC. Specific inhibitors targeting various driver oncogene mutations have demonstrated a higher response rate and longer PFS than platinum doublet-based cytotoxic chemotherapy. Molecular-targeted agents available to lung cancer patients in Japan include the EGFR-TKIs and anaplastic lymphoma kinase (ALK)-TKIs against EGFR-activating mutations and ALK rearrangements, respectively, and bevacizumab, which is an angiogenesis inhibitor

Table 15.3 ILDs related to EGFR-TKI usage in Japanese patients

	Ando et al.[a]	Kudoh et al.[b] [14]	Gemma et al. [39]
Study design	Retrospective	Prospective	Retrospective
Number of patients	1976	1482	9909
ILD (any grade)	70 (3.5 %)	59 (4.0 %)	429 (4.3 %)
ILD (grade 5)	31 (1.6 %)	25 (1.7 %)	153 (1.5 %)
Risk factors for ILD	Smoking history	Smoking history	Smoking history
	Preexisting ILD	Preexisting ILD	Preexisting ILD
	Male gender	Poor PS	Lung infection
		Elderly cardiac disease	Emphysema or COPD

ILD interstitial lung disease, *GFR* epidermal growth factor receptor, *TKI* tyrosine kinase inhibitor
[a]Ando, et al. J Clin Oncol. 2006;24:2549–56
[b]Analyzed risk factors have included both gefitinib and chemotherapy

targeting VEGF activity. Adverse events associated with the use of molecular-targeted agents are generally less severe than that experienced with cytotoxic chemotherapy agents. On the other hand, the rare but potentially fatal adverse event, DILD, has been reported in Japanese patients treated with molecular-targeted agents, such as EGFR-TKIs and ALK-TKIs [36–38]. In addition, the incidence of DILD has been reported to be higher in Japanese patients compared to Caucasians.

With regard to lung cancer patients with chronic ILDs, driver oncogene mutations are rare in this population compared to those without ILDs; other defining characteristics of the former group of patients include male gender, current or former smokers, and low frequency of adenocarcinoma histology. Thus, it is predicted that there are few patients with coexisting ILDs for whom molecular-targeted agents can contribute to positive survival effects.

15.3.2.1 EGFR-TKIs

The use of EGFR-TKI is a standard treatment allowing targeting of NSCLCs harboring *EGFR* active mutations. Currently available EGFR-TKIs in Japan are gefitinib, erlotinib, and afatinib. Three large-scale studies of ILD associated with EGFR-TKI usage have been conducted in patients unselected by EGFR mutations in Japan. These studies suggest that preexisting ILDs and history of smoking were common predictive risk factors (Table 15.3). In a prospective cohort and nested case-control study by Kudoh and colleagues, the cumulative incidence of gefitinib-induced ILD during 12 weeks of treatment was 4.0 %. Gefitinib significantly increased the risk of DILD compared with conventional cytotoxic chemotherapy. These investigators also described that preexisting ILD was an independent risk factor for DILD, regardless of gefitinib or other chemotherapy usage (odds ratio, 4.8–5.6) [17]. With regard to erlotinib, in the post-marketing POLARSTAR surveillance study in Japan, it was reported that 429 (4.3 %) out of the 9907 patients developed ILD (all grade), and the mortality rate due to ILD was 1.5 % [39]. Concurrent/previous ILD and history of smoking were also identified as significant risk factors for developing ILD (odds ratios, 3.19 and 2.23, respectively). Moreover, concomitant honeycombing with interstitial pneumonia was identified as poor prognostic factor for ILD death (odds ratio, 6.67).

Thus, it has been indicated by several large-scale studies that the usage of EGFR-TKIs is particularly problematic for lung cancer patients with chronic ILDs. Pooled analysis of surveillance data by the Diffuse Lung Diseases Research Group indicated that 9 (60 %) out of the 15 patients with IIPs who received EGFR-TKIs developed AE of ILDs. Indeed, this suggests that EGFR-mutated patients coexisting with ILDs may avoid EGFR-TKI use in principal.

15.3.2.2 ALK-TKIs

ALK-TKIs which are clinically available, such as crizotinib and alectinib, have been also identified as inducing severe DILD [36–38]. However, the clinical features of patients developing ILD in this respect has been not investigated in detail due to lack of a relevant large-scale study in Japan. With regard to crizotinib, a multi-target ALK inhibitor, ILD, has been reported as one of the most serious adverse events [40, 41]. Thirty-one (2.5 %) out of 1225 crizotinib-treated patients (across all clinical trials) developed any grade ILD, and 11 (0.5 %) out of the ILD patients experienced fatal events [US Food and Drug Administration]. Watanabe and colleagues reported on ALK-positive NSCLC patients with preexisting ILD who then developed fatal acute ILD induced by crizotinib [42]. With regard to alectinib as a second-generation, highly selective ALK inhibitor, severe acute ILD has also been reported [43]. In Japan, 13 (3.7 %) out of 354 patients who received alectinib manifested DILD (clinical database of Chugai Pharmaceutical CO., Ltd).

Alectinib is a selective ALK-TKI with little or no inhibitory activity against other protein kinases, whereas crizotinib also inhibits MET and ROS1. It is uncertain whether this difference affects the prevalence and severity of ALK-TKI-induced ILD. Therefore, it remains unclear whether the risk factors identified for EGFR-TKI also apply to ALK-TKI-induced ILD. For now, it is advisable to pay as much attention to this drug class in respect to this phenomenon as is done for EGFR-TKIs.

15.3.2.3 Bevacizumab

Bevacizumab is a recombinant humanized monoclonal antibody against VEGF. The addition of bevacizumab significantly extends OS and PFS and improves ORR of patients with advanced non-squamous NSCLC when combined with carboplatin and paclitaxel [44]. However, the safety of this combination chemotherapy with bevacizumab in patients with ILDs remains to be clarified. Some retrospective studies have shown that the addition of bevacizumab to carboplatin and paclitaxel in non-squamous NSCLC patients with ILDs did not increase the incidence of AE of ILDs, while also demonstrating desirable antitumor effects. Enomoto and colleagues and Shimizu and colleagues reported the relevant findings for this regimen to be as follows: ORR, 72 and 40 %; median PFS, 7.2 and 5.3 months; median OS, 8.5 and 16.1 months; and incidence rate of AE, 12 and 10 %, respectively [45, 46]. In a post-marketing investigation, the frequency of bevacizumab-related ILD has been reported to be 0.2 %, which is quite low despite the combination with cytotoxic chemotherapy (clinical database of Chugai Pharmaceutical CO., Ltd). Inhibition of VEGF suppresses revascularization and neovascularization in tumors, which are critical to sustain growth. Although neovascularization is also a fundamental process required for healing after tissue injury, suppression of VEGF activity may be not affected in the context of response to lung injury. Thus, carboplatin and paclitaxel combination with bevacizumab may provide an effective and feasible treatment option even for patients with preexisting ILDs.

15.3.2.4 Pirfenidone and Nintedanib

Recently, two antifibrotic molecules, pirfenidone and nintedanib, have been approved for the treatment of IPF, having shown efficacy to slow functional decline and disease progression, as confirmed by the phase III ASCEND trial [47] and the twin phase III INPULSIS-1 and INPULSIS-2 trials [26]. In respect of preventive effects against AE, the phase II [24] and III [25] trials for pirfenidone in Japanese patients and the INPULSIS trial showed inconsistent results. While all patients who manifested an AE of IPF (14.3 % over 9 months) were exclusively in the placebo group in the Japanese phase II trial for pirfenidone, the previous observation was not confirmed in the following phase III trial. In the pooled analysis of INPULSIS trials, there was a positive trend toward reduced incidence of adjudicated AE of 1.9 % in the nintedanib group compared to 5.6 % in the placebo group, but no significant difference was seen.

Angiogenesis inhibitors also have been developed as potential therapies for solid tumors, including NSCLC. Nintedanib is a multi-target TKI that inhibits pro-angiogenic pathways mediated by VEGF, platelet-derived growth factors (PDGF), and fibroblast growth factor receptor (FGFR). A phase III trial, LUME-Lung 1 [48], showed nintedanib in combination with docetaxel is an effective second-line option for patients with advanced NSCLC previously treated with a platinum-based therapy. In the future, nintedanib may be a key agent for the treatment of advanced NSCLC with ILDs. Further evaluation in a large-scale prospective study is required to confirm the potential preventative effect on AE mediated by these agents.

Regardless of the type of molecular-targeted agents used, studies in this arena are limited by two main issues as follows. Firstly, ILDs that might coexist with lung cancer are a heterogeneous collection of disorders of varying pathogenesis, time phase, severity, and activity, with different inherent risk of treatment-associated AE. In addition, study patients cannot be stratified according to risk of AE, because differential diagnosis of ILDs is often difficult. Secondly, despite the prognosis and incidence of AE in lung cancer patients with ILDs who receive best supportive care only being unclear, placebo-controlled randomized trials are challenging due to the small number of patients divided into ILD type, along with the difficulty of conducting a centrally controlled review of diagnosis of ILDs and AE.

References

1. Natsuizaka M, Chiba H, Kuronuma K et al (2014) Epidemiologic survey of Japanese patients with idiopathic pulmonary fibrosis and investigation of ethnic difference. Am J Respir Crit Care Med 190:773–779
2. Turner-Warwick M, Lebowitz M, Burrows B et al (1980) Cryptogenic fibrosing alveolitis and lung cancer. Thorax 35:496–499
3. Hubbard R, Venn A, Lewis S, Britton J (2000) Lung cancer and cryptogenic fibrosing alveolitis. A population-based cohort study. Am J Respir Crit Care Med 161:5–8
4. Ozawa Y, Suda T, Naito T et al (2009) Cumulative incidence of and predictive factors for lung cancer in IPF. Respirology 14:723–728
5. Matsushita H, Tanaka S, Saiki Y et al (1995) Lung cancer associated with usual interstitial pneumonia. Pathol Int 45:925–932
6. Kawasaki H, Nagai K, Yokose T et al (2001) Clinicopathological characteristics of surgically resected lung cancer associated with idiopathic pulmonary fibrosis. Surg Oncol 76:53–57
7. Samet JM (2000) Does idiopathic pulmonary fibrosis increase lung cancer risk? Am J Respir Crit Care Med 161:1–2
8. Takahashi T, Munakata M, Ohtsuka Y et al (2002) Expression and alteration of ras and p53 proteins in patients with lung carcinoma accompanied by idiopathic pulmonary fibrosis. Cancer 95:624–633
9. Vassilakis DA, Sourvinos G, Spandidos DA et al (2000) Frequent genetic alterations at the microsatellite level in cytologic sputum samples of patients with idiopathic pulmonary fibrosis. Am J RespirCrit Care Med 162:1115–1119
10. Hojo S, Fujita J, Yamadori I et al (1998) Heterogeneous point mutations of the p53 gene in pulmonary fibrosis. Eur Respir J 12:1404–1408
11. Kuwano K, Kunitake R, Kawasaki M et al (1996) P21 Waf1/Cip1/Sdi1 and p53 expression in association with DNA strand breaks in idiopathic pulmonary fibrosis. Am J Respir Crit Care Med 154:477–483
12. Uematsu K, Yoshimura A, Gemma A et al (2001) Aberrations in the fragile histidine triad (FHIT) gene in idiopathic pulmonary fibrosis. Cancer Res 61:8527–8533
13. Kondoh Y, Taniguchi H, Kawabata Y et al (1993) Acute exacerbation in findings in three cases. Chest 103:1808–1812
14. Kudoh S, Kato H, Nishiwaki Y et al (2008) Interstitial lung disease in Japanese patients with lung cancer: a cohort and nested case-control study. Am J Respir Crit Care Med 177:1348–1357
15. Ueda Y, Fujii Y, Kuwano H (2009) thoracic and cardiovascular surgery in Japan during 2007. Gen Thorac Cardiovasc Surg 57:488–513
16. Koizumi K, Hirata T, Hirai K et al (2004) Surgical treatment of lung cancer combined with interstitial pneumonia: the effect of surgical approach on postoperative acute exacerbation. Ann Thorac Cardiovasc Surg 10:340–346
17. Chiyo M, Sekine Y, Iwata T et al (2003) Impact of interstitial lung disease on surgical morbidity and mortality for lung cancer: analyses of short-term and long-term outcomes. J Thorac Cardiovasc Surg 126:1141–1146
18. Watanabe A, Higami T, Ohori S, Koyanagi T, Nakashima S, Mawatari T (2008) Is lung cancer resection indicated in patients with idiopathic pulmonary fibrosis? J Thoracic Surg 136:1357–1363

19. Chida M, Ono S, Hoshikawa Y et al (2008) Subclinical idiopathic pulmonary fibrosis is also a risk factor of postoperative acute respiratory distress syndrome following thoracic surgery. Eur J Cardiothorac Surg 34:878–881
20. Sato T, Teramukai S, Kondo H et al (2014) Impact and predictors of acute exacerbation of interstitial lung diseases after pulmonary resection for lung cancer. J Thorac Cardiovasc Surg 147:1604–1611
21. Minegishi Y, Takenaka K, Mizutani H et al (2009) Exacerbation of idiopathic interstitial pneumonias associated with lung cancer therapy. Intern Med 48:665–672
22. Isobe K, Hata Y, Sakamoto S et al (2010) Clinical characteristics of acute respiratory deterioration in pulmonary fibrosis associated with lung cancer following anti-cancer therapy. Respirology 15:88–92
23. Kenmotsu K, Naito T, Kimura M et al (2011) The risk of cytotoxic chemotherapy-related exacerbation of interstitial lung disease with lung cancer. J Thorac Oncol 6:1242–1246
24. Azuma A, Nukiwa T, Tsuboi E et al (2005) Double-blind, placebo-controlled trial of pirfenidone in patients with idiopathic pulmonary fibrosis. Am J Respir Crit Care Med 171:1040–1047
25. Taniguchi H, Ebina M, Kondoh Y et al (2010) Pirfenidone in idiopathic pulmonary fibrosis. Eur Respir J 35:821–829
26. Richeldi L, du Bois RM, Raghu G et al (2014) Efficacy and safety of nintedanib in idiopathic pulmonary fibrosis. N Engl J Med 370:2071–2082
27. Park IN, Kim DS, Shim TS et al (2007) Acute exacerbation of interstitial pneumonia other than idiopathic pulmonary fibrosis. Chest 132:214–220
28. Suda T, Kaida Y, Nakamura Y, Enomoto N et al (2009) Acute exacerbation of interstitial pneumonia associated with collagen vascular diseases. Respir Med 103:846–853
29. Minegishi Y, Sudoh J, Kuribayashi H et al (2011) The safety and efficacy of weekly paclitaxel in combination with carboplatin for advanced non-small cell lung cancer with idiopathic interstitial pneumonias. Lung Cancer 71:70–74
30. Minegishi Y, Kuribayashi H, Kitamura K et al (2011) The feasibility study of carboplatin plus etoposide for advance small cell lung cancer with idiopathic interstitial pneumonias. J Thorac Oncol 6:801–807
31. Kenmotsu H, Naito T, Mori K et al (2015) Effect of platinum-based chemotherapy for non-small cell lung cancer patients with interstitial lung disease. Cancer Chemother Pharmacol 75:521–526
32. Yoshida T, Yoh K, Goto K et al (2013) Safety and efficacy of platinum agents plus etoposide for patients with small cell lung cancer with interstitial lung disease. Anticancer Res 33:1175–1180
33. Shukuya T, Ishiwata T, Hara K et al (2010) Carboplatin plus weekly paclitaxel treatment in non-small cell lung cancer patients with interstitial lung disease. Anticancer Res 30:4357–4361
34. Kinoshita T, Azuma K, Sasada T et al (2012) Chemotherapy for non-small cell lung cancer complicated by idiopathic interstitial pneumonia. Oncol Lett 4:477–482
35. Watanabe N, Taniguchi H, Kondoh Y et al (2014) Chemotherapy for extensive-stage small-cell lung cancer with idiopathic pulmonary fibrosis. Int J Clin Oncol 19:260–265
36. Tamiya A, Okamoto I, Miyazaki M et al (2013) Severe acute interstitial lung disease after crizotinib therapy in a patient with EML4-ALK-positive non-small cell lung cancer. J Clin Oncol 31:E15–E17
37. Maka VV, Krishnaswamy UM, Anil Kumar N et al (2014) Acute interstitial lung disease in a patient with anaplastic lymphoma kinase-positive non-small cell lung cancer after crizotinib therapy. Oxf Med Case Rep 2014:11–12
38. Shaw AT, Kim D-W, Nakagawa K et al (2013) Crizotinib versus chemotherapy in advanced ALK-positive lung cancer. N Engl J Med 368:2385–2394

39. Gemma A, Kudoh S, Ando M et al (2014) Final safety and efficacy of erlotinib in the phase 4 POLARSTAR surveillance study of 10 708 Japanese patients with non-small-cell lung cancer. Cancer Sci 105:1584–1590
40. Kwak EL, Bang YJ, Camidge DR et al (2010) Anaplastic lymphoma kinase inhibition in non-small cell lung cancer. N Engl J Med 363:1693–1703
41. Camidge DR, Bahg YJ, Kwak EL et al (2012) Activity and safety of crizotinib in patients with ALK-positive non-small cell lung cancer: update results from a phase I study. Lancet Oncol 13:1011–1019
42. Watanabe N, Nakahara Y, Taniguchi H et al (2014) Crizotinib-induced acute interstitial lung disease in a patient with EML4-ALK positive non-small cell lung cancer and chronic interstitial pneumonia. Acta Oncol 53:158–160
43. Yamamoto Y, Okamoto I, Otsubo K et al (2015) Severe acute interstitial lung disease in a patient with anaplastic lymphoma kinase rearrangement-positive non-small cell lung cancer treated with alectinib. Investig New Drugs 33:1148–1150
44. Sandler A, Gray R, Perry MC et al (2006) Paclitaxel-carboplatin alone or with bevacizumab for non-small-cell lung cancer. N Engl J Med 355:2542–2550
45. Enomoto Y, Kenmotsu H, Watanabe N et al (2015) Efficacy and safety of combined carboplatin, paclitaxel, and bevacizumab for patients with advanced non-squamous non-small cell lung cancer with pre-existing interstitial lung disease: a retrospective multi-institutional study. Anticancer Res 35:4259–4263
46. Shimizu R, Fujimoto D, Kato R et al (2014) The safety and efficacy of paclitaxel and carboplatin with or without bevacizumab for treating patients with advanced nonsquamous non-small cell lung cancer with interstitial lung disease. Cancer Chemother Pharmacol 74:1159–1166
47. King TE Jr, Bradford WZ, Castro-Bernardini S et al (2014) A phase 3 trial of pirfenidone in patients with idiopathic pulmonary fibrosis. N Engl J Med 370:2083–2092
48. Reck M, Kaiser R, Mellemgaard A et al (2014) Docetaxel plus nintedanib versus docetaxel plus placebo in patients with previously treated non-small-cell lung cancer (LUME-Lung 1): a phase 3, double-blind, randomised controlled trial. Lancet Oncol 15:143–155

Chapter 16
Management of Adverse Effects by Molecular Targeted Therapy and Immunotherapy

Toshimichi Miya

Abstract Although molecular-targeted therapies have markedly developed as drugs targeting at cancer-specified molecular lesions, there are peculiar adverse effects to be managed including dermatitis, diarrhea, mucositis, and interstitial lung disease. It is important to assess risk-benefit balance maximizing the benefits of patients treated by anticancer drugs. Immune checkpoint inhibitors developed recently have shown clinically significant antitumor response for malignant melanoma and lung cancer. Immune therapies also have particular adverse effect including interstitial lung disease, liver injury, and skin eruption. It is noteworthy that these drugs may produce autoimmune disturbance associated with immune-related adverse events (ir-AEs) such as hypothyroidism, systemic lupus erythematosus, arthritis, and intestinal disorders.

Most of managements of adverse effect by molecular-targeted therapy and immunotherapy have not been established based on evidential data; however, precise medicine for toxicities is required for individual patient with cancer. Managements of adverse effects of molecular-targeted therapy are thought to be beyond one oncologist capacity. Team medicine and proper consult to specialist are essential to appropriate treatment for patients with malignant tumor.

Keywords Adverse effect • Skin toxicity • Interstitial lung disease • Liver injury

16.1 Introduction

Molecular-targeted therapies have been developed as drugs targeting at cancer-specified molecular lesions; therefore, theoretically, normal tissues are not influenced by the therapy. Actually, there are peculiar adverse effects including dermatitis, diarrhea, mucositis, and interstitial lung disease requiring meticulous monitoring and multidisciplinary team medicine for proper care of toxicities. Adverse effect of

T. Miya (✉)
Department of Pulmonary medicine/Medical Oncology, Nippon Medical School, Tamanagayama Hospital, 1-7-1 Nagayama, Tama, Tokyo 206-8512, Japan
e-mail: tmiya@nms.ac.jp

© Springer Science+Business Media Singapore 2017

Y. Takiguchi (ed.), *Molecular Targeted Therapy of Lung Cancer*,
DOI 10.1007/978-981-10-2002-5_16

molecular target drugs such as dermatitis caused by epidermal growth factor receptor tyrosine kinase inhibitor (EGFR-TKI) may be a surrogate marker of clinical response [1]. It is important to assess risk-benefit balance maximizing the benefits of patients treated by anticancer drugs.

The marked development has achieved in the field of immune-oncology. Immune checkpoint blockade using inhibitors of cytotoxic T-lymphocyte-associated antigen 4 (CTLA-4), programmed cell death-1 (PD-1), and programmed cell death-ligand 1 (PD-L1) inhibitors has shown clinically significant antitumor response for malignant melanoma [2] and lung cancer [3]. Immune therapies also have particular adverse effect including interstitial lung disease, liver injury, and skin eruption. It is noteworthy that these drugs may produce autoimmune disturbance associated with immune-related adverse events (ir-AEs) such as hypothyroidism, systemic lupus erythematosus, arthritis, and intestinal disorders.

Most of managements of adverse effect by molecular-targeted therapy and immunotherapy have not been established based on evidential data; however, precise medicine for toxicities is required for individual patient with cancer. Patients should be encouraged to inform their physician of the onset and any worsening of adverse effects following treatments to minimize disadvantage of the therapy.

16.2 Management of Skin Toxicity

Dermatologic adverse events are frequently observed in patients treated by EGRF receptor tyrosine kinase inhibitors including gefitinib [5], erlotinib [6], and afatinib [7]. Multi-kinase inhibitors and immune checkpoint therapy also have similar dermatologic toxicity [4]. The management of the dermatologic toxicities should be considered as serious because they also cause pain and cosmetic discomfort that aggravate compliance of the therapy. Acne-like rash is most common in two thirds of patients receiving EGFR-TKI with median onset around 8 days after initiation of the therapy [8]. Pruritus, xerosis, paronychia, and alopecia are also observed during the treatment period. These toxicities are usually reversible and disappeared smoothly after discontinuing drugs; however, continuous treatments are often required because the skin toxicities reaction may be a predictive marker of the good response of the treatment [9, 10]. It is important to manage the skin toxicity to avoid undesirable dose reduction. The optimal managements of skin reaction are necessary to maximize the effect of molecular-targeted therapy.

First of all, physicians have to inform patients of precautions to protect skin reactions. Patients should be advised to keep the skin and hair clean using less-irritated soap, to cut the nail adequately, and to minimize sunlight when going outside. Skin reactions caused by EGFR-TKIs are usually complex type. Various managements should be adopted depending on type and severity of the therapy-associated dermatologic adverse effects. Pruritis and tender erythematous papules and pustules developed in skin with high density of sebaceous glands (scalp, face, upper chest, and back) [11]. Dry skin or xerosis is often observed, and dermatological agents such as heparinoid, chondroitin sulfate, white petrolatum (Vaseline), and urea are required to keep skin moisture. Although the nail changes or paronychia are usually mild, surgical treatments such as partial resection or removal of the nail are required in the severe cases. Acne-like rash caused by EGFR-TKI consisted of follicular papules considered not infectious, but dermatologic toxicity emerged through inhibition to EGRR expressed on normal cutaneous cells. Thus, steroid ointments are primarily used from the beginning [12]. The grading of dermatologic toxicity according to the National Cancer Institute Common Toxicity Criteria (version 3.0) is listed on Table 16.1 [13], and Table 16.2 displays the recommendations for the management scheme for EGFR-TKI-related dermatologic toxicity of the University of Texas M.D. Anderson Cancer Center [14]. Physicians should decide the optimal treatment according to the patient physical status and grading scheme. Doxycycline and minocycline are recommended to treat the rash of grade 2 or more.

Yamazaki N. et al. demonstrated the algorism of treatments for dermatologic toxicities on the basis of Japanese perspectives (Table 16.3) [12]. They recommend the medium or strong steroid lotion to the acneiform rash on the face and head because of better permeability into the skin compared with those of ointment.

Table 16.1 National Cancer Institute Common Toxicity Criteria (ver 4.0) [13]

Adverse event	Grade 1	Grade 2	Grade 3	Grade 4
Rash acneiform	Papules and/or pustules covering<10 % BSA, which may or may not be associated with symptoms of pruritus or tenderness	Papules and/or pustules coveting 10–30 % BSA, which may or may not be associated with symptoms of pruritus or tenderness; associated with psychosocial impact; limiting instrumental ADL	Papules and /or pustules coveting>30 % BSA, which may or may not be associated with symptoms of pruritus or tenderness; limiting self-care ADL; associated with local superinfection with oral antibiotic indicated	Papules and/or pustules coveting any % BSA, which may or may not be associated with symptoms of pruritus or tenderness and are associated with extensive superinfection with IV antibiotics indicated; life-threatening consequences
Rash macula-papular	Macules/papules covering <10 % BSA with or without symptoms (e.g., pruritus, burning, tightness)	Macules/papules covering 10–30 % BSA with or without symptoms (e.g., pruritus, burning, tightness); limiting instrumental ADL	Macules/papules covering >30 % BSA with or without symptoms; limiting self-care ADL	–

Table 16.2 University of Texas M.D. Anderson Cancer Center Management Scheme for EGFR-TKI-related dermatologic toxicity [14]

Toxicity grade	Macular rash	Pustular rash	Dry skin	Pruritus	Ulcerative lesion
Grade 1	Topical hydrocortisone cream/lotion	Clindamycin gel (for isolated lesions) or lotion (for scattered lesions)	–	–	–
Grade 2	Oral methylprednisolone, if> body lesions or hydrocortisone, if<2 body lesions	Minocycline 100 mg or doxycycline 100 mg orally twice a day for 10–14 days	Apply emollient twice a day	Topical antihistamine or diphenhydramine 25–50 mg orally every 6 hours as needed	–
Grade 3	Oral methylprednisolone	Minocycline 100 mg or doxycycline 100 mg orally twice a day for 10–14 days	Apply emollient twice a day	Diphenhydramine 25–50 mg or hydroxyzine 25–50 mg orally every 6 hours as needed	Silver sulfadiazine ointment or dermatology consult
	Dermatology consult				
Grade 4	Dermatology consult				

16.3 Managements of Interstitial Lung Disease (ILD)

All drugs are potentially able to induce ILD in various degree and severity. The development of ILD subsequent to chemotherapeutic agents has been documented for many years, with the use of standard cytotoxic drugs for treatment of NSCLC being associated with ILD at a prevalence of up to 5 % [15]. ILD is the most serious adverse effect of molecular-targeted drugs with relatively high incidence compared to cytotoxic drugs. The frequency of gefitinib-related ILD is higher in Japanese than Western Caucasians implying that Japanese may have an increased genetic susceptibility to ILD [16]. After introduction of gefitinib licensed in Japan for the inoperable or recurrent non-small cell lung cancer in July 2002, many of life-threatening ILD were reported and became an object of social concern. Although EGFR-TKIs are generally well tolerated with mild or moderate skin reactions, gastrointestinal disturbances, and elevations in liver enzymes, the frequency of ILD was 5.8 % and the mortality rate 38.8 % among the patients with ILD according to Reports on Iressa Tablets 250 prospective study [17]. The prevalence of EGFR-associated ILD is generally estimated 3–5 % and mortality of 1–2 % in all patients [18].

Risk factors for ILD and predictive factors should be initially evaluated. The incidence of ILD increases in patients with a performance status (PS) of 2 or more, a history of smoking, a pre-existing interstitial pneumonia at the time of initial

Table 16.3 Clinical management of dermatological toxicities induced EGFR-TKI: Japanese perspective [12]

Acne-like rash	Seborrheic dermatitis	Dry skin	Nail change	Nail changes (paronychia)
Topical agents	Topical agents	Topical agents	Cleanser	Surgical treatment
Nonsteroid	Nonsteroid	Moisturizing agents	Cold compress	Partial resection/ removal of the nail
Ibuprofen piconol	Ibuprofen piconol	Mucopolysaccharide		
Steroids	Steroids	Polysulfuric acid ester		
Mild to very strong				
Antibiotics	Mild to very	Urea		
	Systemic agents			
Tetracycline hydrochloride	Vitamins	Steroids		Removal of the nail
Nadifloxan	Riboflavin tetrabutylate			
Systemic agents	Pyridoxal 5-phosphate			
Vitamins				
Riboflavin tetrabutylate				
Pyridoxal 5-phosphate				
Antibiotics				
Minocycline hydrochloride				
In case of severe itching	In case of secondary infection			
Antiallergic agents	Antibiotics			
Antihistamine	Minocycline hydrochloride			
Steroids				

administration of the drug, and a history of prior chemotherapy. Regarding the factors predicting a fatal prognosis, a higher mortality rate was suggested for patients with a male gender, a PS of 2 or more, high age, smoking history, pre-existing ILD, reduced lung volume, and extensive areas of adherent to the pleura [17].

Erlotinib is another EGFR-TKI approved in Japan in 2007. The all-case surveillance study conducted after approval revealed that ILD developed in 4.5 % (158/3488 cases) and mortality rate from ILD was 34.8 % (55/158 cases) [20]. ILD developed most often in the first 2 weeks after starting erlotinib administration. Risk factors of ILD were similar those of gefitinib. A multivariate analysis demonstrated that the previous ILD, smoking history, concomitant or previous lung infection, and PS of 2 or more were significant risk factors for development of ILD [20, 21]. Patient harboring EGRF-receptor mutation is considered a good candidate for EGRF-TKI. The poor PS or aged patients who are not indicated for cytotoxic chemotherapy may be able to receive molecular-targeted therapy; however, it should be noted that such patients may be at high risk for development of ILD.

The prevalence of gefitinib-induced ILD differed markedly according to sex and smoking status ranging from 0.4 % in females with no history of smoking to 6.6 % in male smokers [16]. Female sex and the absence of a history of smoking are both predictive factors with a lower risk for ILD, higher response rate, and longer survival. They provide important insight into individual risk-benefit assessment. Patient selection on the basis of this favorable profile will not only increase the clinical benefit of treatment with gefitinib but also reduce the risk of development of this life-threatening toxicity.

It is often difficult to diagnose drug-induced ILD in lung cancer patient. Patients with advanced lung cancer tend to have pre-existing lung disease and respiratory tract infections as well as the progressive malignancy like carcinomatous lymphangitis with high prevalence. There are no specific biomarkers, radiographic findings, or pathological patterns for ILD. Diagnosis of drug-induced ILD relies of rigorous exclusion of all other differential diagnoses. As for the treatment of ILD, the principal management is early detection and discontinuation of the causal drug, and if necessary the administration of corticosteroids is indicated. Exceptionally, mTOR inhibitor is often continuing after ILD appears in patients with asymptomatic ILD [22]. All patients treated with molecular-targeted therapy should be warned to promptly report symptoms such as cough and dyspnea.

The response to corticosteroid therapy depends on the pathological pattern of ILD. Physicians should learn the classification of idiopathic interstitial pneumonia including idiopathic pulmonary fibrosis (IPF), nonspecific interstitial pneumonia (NSIP), cryptogenic organizing pneumonia (COP), desquamative interstitial pneumonia (DIP), and acute interstitial pneumonia (AIP) [23]. IPF-like pattern is relatively rare in drug-induced ILD. AIP is a diffuse alveolar damage (DAD) as pathological findings have rapid aggravate and poor prognosis. Among the classification of interstitial pneumonia, OP, eosinophilic pneumonia, and hypersensitivity pneumonia are expected to show a good response to corticosteroids [24].

Since ILD-associated molecular-targeted drugs usually emerge within 4 weeks after initiation of the therapy, the meticulous monitoring is required for the duration period. When the ground-glass opacity consisted of ILD appears on chest CT of patients treated by a molecular-targeted drug, it is recommended to discontinue drug except mTOR inhibitors. ILDs induced by EGFR-TKIs often develop AIP (DAD) type and rarely recovered by cessation of the causal drug. If the patients developed DAD pattern with hypoxia of PaO_2 less than 80 Torr, corticosteroid is recommended for the treatment. In the case of severe hypoxia of less than 60 Torr or rapidly progressive pattern, steroid pulse therapy consisted with methylprednisolone at dose of 500–1000 mg/day for 3 days followed by prednisolone at a dose of 0.5–1.0 kg/day is recommended. Algorism of managements is schemed on Table 16.4. There is no evidence concerning efficacy of immunosuppressive drugs such as cyclosporine for nonresponder to corticosteroid. Administration of the causal drug after resolve of ILD again is generally not recommended.

Table 16.4 Management of interstitial lung disease associated with molecular target drug other than mTOR inhibitor [25]

Grade	Therapy	Re-administration of the drug
Mild: PaO2>80 Torr asymptomatic	Discontinuation of the drug Meticulous observation and corticosteroid, if necessary	Administration of the drug is not recommended after resolve of ILD other than exceptional case
Moderate: PaO2<60 Torr symptomatic	Discontinuation of the drug Prednisolone 0.5–1.0/kg/day	
Severe: PaO2 <60 Torr DAD pattern	Discontinuation of the drug Methylprednisolone 500 mg–1000 mg/day	Administration of the drug is not recommended after resolve of ILD

16.4 Managements of Liver Injury

Anticancer drugs are associated with risks of drug-induced liver injury and hepatitis B virus (HBV) reactivation which may sometimes be serious and resulting in life-threatening liver failure [26]. It is necessary to make differential diagnosis of liver injury including tumor progression in the liver, obstructive jaundice, viral hepatitis, and other drug-induced liver toxicities. Although severe liver injury is generally rare in molecular target drug compared to cytotoxic agents, asymptomatic increase in liver transaminases is often observed in patients treated by gefitinib and erlotinib [27]. Patients should be monitored regularly for changes in liver functional test (e.g., transaminase, bilirubin, alkaline phosphatase). Liver injury is usually recovered after discontinuation or dose reduction of the drug, and the intervention for liver injury is not often required other than exceptional case [28]. There is a report that corticosteroid is effective for allergy-mediated liver injury caused by a molecular-targeted drug [29].

Dose reduction methods are not established for patients with liver injury. Practically, a reduction of daily dose or administration in every other day is empirically performed.

Reactivation of HBV is also relatively rare in molecular target drugs except for anti-CD20 antibody such as rituximab and imatinib [26]. Physician should evaluate the liver function and screening of HBV status at baseline of the treatment. Patients with HBsAg positive or HBsAg negative while HBcAb and/or HBsAb positive are at a high risk for HBV reactivation so they should be managed according to the guideline for preventing HBV reactivation developed by the Japan Society of Hepatology [30].

16.5 Managements of Nausea and Vomiting

Nausea and vomiting are common adverse effects of anticancer drugs. Regarding with molecular-targeted drugs, nausea and vomiting are also frequent; however, the degree is relatively low. According to the National Comprehensive Cancer Network guideline, most molecular-targeted drugs are classified as low emetogenic except for imatinib and crizotinib which are classified as moderate emetogenic [31]. Antiemetic medicines such as metoclopramide, prochlorperazine, or histamine 2 receptor antagonist or proton pump inhibitor are used for preventing and treating nausea and vomiting.

16.6 Managements of Gastrointestinal Toxicity

Gastrointestinal adverse events are frequently observed in patients receiving molecular-targeted drugs. The grades of toxicity are usually not so high that the majority of patients with NSCLC treated by oral EGFR-TKI do not require treatment interruption. Diarrhea of any grade were experience in 83 % of patients treated with erlotinib, 27 % of patients treated with gefitinib, and 76 % of patients treated with afatinib. Mucositis and stomatitis were experienced in 14.5 %, 6 %, and 17 % of patients treated with erlotinib, gefitinib, and afatinib, respectively [32]. The mechanism of diarrhea caused by EGFR-TKI remains unclear. While it is supposed to be resulted from the damage of gastrointestinal cell on which EGFR are located, multiple factory causes alter gut motility, colonic crypt damage, changes of intestinal microflora, and altered transport in the colon [33]. In the cases of severe gastrointestinal adverse events with fluid and electrolyte losses, dehydration, electrolyte imbalances, and renal insufficiency consequent on diarrhea may ill impact on patient's quality of life.

Management strategies to reduce the severity or eliminate the diarrhea entirely should be performed according to the grade of diarrhea (Table 16.5) to avoid reducing the dose of molecular-targeted drug. Early recognition and management of diarrhea is essential to prevent dose reduction or discontinuation of EGFR-TKI therapy.

Table 16.5 National Cancer Institute Common Toxicity Criteria for diarrhea (ver 3.0) [13]

Adverse event	Grade 1	Grade 2	Grade 3	Grade 4
Patients without colostomy	Increase of <4 stools/day	Increase of 4–6 stools/day or nocturnal stools	Increase of ≥7 stools/day or incontinence; or need for parenteral support for dehydration	Physiologic consequences requiring intensive care; or hemodynamic collapse
Patients with colostomy	Mild increase in loose watery colostomy output compared with pretreatment	Moderate increase in loose watery colostomy output compared with pretreatment, but not interfering with normal activity	Sever increase in loose watery colostomy output compared with pretreatment, interfering with normal activity	Physiologic consequences requiring intensive care; or hemodynamic collapse

Patient education is essential. Physicians should encourage patients to understand the high frequency of diarrhea, the implications of therapy, and the purpose of diarrhea management strategies. It is also advised taking loperamide at the onset of diarrhea. There are no convincing data to support the routine implementation of prophylactic treatment. Loperamide is considered to be the golden standard for the pharmacologic treatment of diarrhea. Loperamide prolongs the transit time of the intestinal contents, reduces the daily fecal volume, increases the viscosity and bulk density, and diminishes the losses of fluid and electrolytes. Recommended guidelines for the treatment of cancer treatment-induced diarrhea published by Bensen et al. are referential to treatment for molecular-targeted drug-induced diarrhea [33]. Initial management should include dietary modifications (e.g., eliminating all lactose-containing product and high-osmolar dietary supplements). Loperamide is recommended to start at an initial dose of 4 mg followed by 2 mg every 4 h or after every unformed stool (not to exceed 16 mg/d). If mild to moderate diarrhea resolves with loperamide, patients are instructed to continue dietary modification and to gradually add solid food to their diet. Patients may discontinue loperamide when they have been diarrhea-free for at least 12 hours. If the diarrhea persists for more than 24 hours, loperamide should be increased to 2 mg every 2 h. If the diarrhea persists for more than 48 hours, physician should make further evaluation including complete stool test and blood work-up. Fluid and electrolytes should be supplemented as needed. Patients should be started on a second-line antidiarrheal agent such as tincture of opium.

16.7 Managements of Toxicities of Antiangiogenic Therapy

Anti-VEGF therapy such as bevacizumab is associated with various toxicities. Hypertension is a frequent toxicity of anti-VEGF agent [34]. The grade 3–4 hypertension has been reported in clinical studies of bevacizumab. The correct evaluation of the levels of hypertension is of critical importance. A basement assessment and follow-up monitoring of blood pressure is considered necessary for all patients. Physicians should educate patients to measure home blood pressure as a monitoring of hypertension.

There is a wide difference in incidence of hypertension among various malignancies and doses of drug, ranging from 2.7 % to 32 % for the low dose of bevacizumab and from 17.6 % to 36 % for the high-dose bevacizumab [35]. The incidence and severity of hypertension in cancer patients are complicated by the type of drugs, dose and schedule used, age of patients, as well as the presence of coexisting cardiovascular disease. Pre-existing hypertension may be a largest risk factor. Hypertension in patients receiving bevacizumab appears 4–6 weeks after the first administration and blood pressure returns to previous values if bevacizumab administration is discontinued [35]. Previous clinical studies described hypertension as easily manageable by common medical treatment; however, up to 15 % of patient experienced severe hypertension requiring multidrug therapy, and a few cases possibly experienced malignant hypertension [36, 37]. There are no evidence-based recommendations for the appropriate antihypertensive agent for anti-VEGF-induced hypertension. Until the evident data is obtained from clinical studies, the guideline for the management of hypertension by the Japanese Society of Hypertension can be used to manage hypertension [38].

There are no evidence-based recommendations regarding which antihypertensive agent is optimal as drug of anti-VEGF-related hypertension. A large number of clinical trials have suggested that the main benefits of antihypertensive treatment are identical among thiazide diuretics, beta-blocker, calcium antagonists, angiotensin-converting enzyme blocker inhibitors (ACEs), and angiotensin receptor blockers (ARBs) [39, 40]. Even specific antihypertensive class for the treatment of anti-VEGF-induced hypertension remains unclear, it was reported that VEGF-mediated vascular hyperpermeability was suppressed in mice given ARBs, implying ARBs may have effect on VEGF-induced proteinuria. It has been also demonstrated that ACEs induced expression of nephrin in diabetic nephropathy and improve endothelia function and microcirculatory density [35, 39]. These data may support the use of ACEs, or ARBs may be suited for the treatment of VEGF inhibitor-associated hypertension. There is also a paper reporting that calcium channel blockers such as amlodipine are useful [40]. Nifedipine possessing the function of VRGF secretion should probably be avoided.

16.8 Managements of Proteinuria and Renal Injury

Proteinuria and renal injury are often observed in patients treated with VEGFR inhibitors [41]. They are usually accompanied with emerge of hypertension. Urine test is required periodically during the VEGF therapy. Proteinuria is usually recovered by discontinuation of the drugs. In the case with persistent proteinuria or nephritic syndrome, physician should consult nephrologists concerning the treatment for renal injury.

16.9 Managements of Perforation of Gastrointestinal Tract

One of the significant complications from bevacizumab is spontaneous bowel perforation which may lead to peritonitis, fistula formation, or intra-abdominal abscess requiring emergency operative intervention resulting in significant morbidity and mortality in cancer patients. The frequency of bowel perforation in patients with lung cancer treated by bevacizumab in clinical trials is less than 1 % and relatively rare compared to colorectal cancer [42]. It is noteworthy to mention that bevacizumab-induced bowel perforation occurs at any place along the GI tract and not only at the tumor site.

The management of bowel perforation is complex and should involve a multidisciplinary approach that includes medical oncologists, surgeons, and interventional radiologists.

16.10 Managements of Pulmonary Hemorrhage

The mechanisms by which anti-VEGF agents induce bleeding are not well understood. It may result from the inhibition of the physiological endothelial repair processes mediated by VEGF [43].

Life-threatening pulmonary hemorrhage occurred in 9.0 % in patients with NSCLC treated with bevacizumab during a phase II trial [44]. Although the phase II data suggested that lung cancer of central location close to major vessels, cavity, or tumor necrosis and squamous cell histology were related with hemorrhage, subsequent studies did not support this conclusion [45]. It is now unclear whether tumor location, histology, size, pre-existing cavitation, cavitation developed after bevacizumab, and vascular involvement are associated with pulmonary hemorrhage. Other reports demonstrated that endobronchial involvement was a significant risk factor [46]. Although the risk factors remain unclear, physicians should discuss with patient about the risk: benefit ratio, and patient should not be excluded from bevacizumab therapy merely because of central location of tumor, age, PS, and anticoagulant use. In present status, patients with squamous histology and/or history of pulmonary hemorrhage should not receive bevacizumab as these groups were excluded from pivotal trials.

As there are no specific recommendations available for the treatment of pulmonary hemorrhage associated with bevacizumab, guidelines of the general management of pulmonary hemorrhage are also used for this purpose. A chest CT scan is needed to identify the bleeding site. Conventional and interventional endobronchoscopic therapies are considered useful including laser coagulation in some cases. Bronchial arterial embolization has an important role as therapy for pulmonary hemorrhage; however, only limited data is available in lung cancer patients. Radiation therapy has been recommended as the management of non-massive pulmonary hemorrhage caused by unresectable lung cancer; however, optimal radiation therapy-associated molecular-targeted drug has not been established.

16.11 Managements of Adverse Effects of Immunotherapy

Immune checkpoint inhibitors such as CTLA-4, PD-1, and PD-L1 inhibitor have promising results with prolonged clinical response making an epoch in the treatment of lung cancer. Nivolumab has been approved in Japan in 2015 for the treatment of advanced NSCLC. Although adverse effects of immune therapy are generally considered mild, there have been reported various adverse events by

Table 16.6 Summary of adverse event of nivolumab in phase II study of Japanese patients with lung cancer [4]

Number of patients		111
Number of patients with adverse effects of any grade		88
Adverse effect	Any grade	Grade 3–4
Total	79.8 %	16.9 %
Anemia	2.7 %	
Arrhythmia	1.8 %	
Vertigo	0.9 %	
Hyper ,hypothyroidism	8.1 %	
Colitis	0.9 %	0.9 %
Constipation	5.4 %	
Nausea	9.9 %	
Vomiting	4.5 %	
Fatigue, asthenia	23.4 %	0.9 %
Infusion reaction	2.7 %	
Liver enzyme	6.3 %	
Lymphpcytopenia	8.1 %	0.9 %
Hyponatremia	3.6 %	1.8 %
Appetite loss	14.4 %	0.9 %
Arthralgia, myalgia	8.1 %	0.9 %
Peripheral neuropathy	1.8 %	
Interstitial lung disease	4.5 %	1.8%
Skin toxicity	32.4 %	0.9 %

clinical studies. Phase II study preformed in Japan demonstrated that grade 3–4 toxicities were observed in 16.2 % of patients [4]. The adverse events to be paid special attention were ILD, colitis, liver injury, neuropathy, adrenal insufficiency, dermopathy, myasthenia, diabetes, thyroid dysfunction, renal dysfunction, encephalopathy, thrombosis, and infusion reaction. Table 16.6 summarized the adverse effect in 111 Japanese patients treated by nivolumab in phase II study.

Immune checkpoint inhibitors have various, sometimes severe adverse effects; thus, meticulous monitoring including vital sign, electrocardiogram, blood sample test, chest x-ray, and performance status is crucial. Immune and endocrinal examinations such as rheumatoid factor, antinuclear antibody, SP-D, KL-6, and thyroid function test (TSH, free T3, free T4) are essential periodically. The adverse effect of the immune checkpoint inhibitor is frequently related with the mechanism of immune activation by the drugs appearing as immune-related adverse events (ir-AEs). Corticosteroid replacement should be considered such as cases of ILD, myasthenia, colitis, and thyroid dysfunction. When the adverse effect appears, discontinuation of the drug and consult to specialists are recommended.

16.12 Other Toxicities

As mentioned above, molecular-targeted therapy is accompanied with various adverse effects. Visual disorder is often observed in patients with lung cancer treated by crizotinib in clinical practice. Managements of adverse effects of molecular-targeted therapy are thought to be beyond one oncologist capacity. Team medicine and proper consult to specialist are essential to appropriate treatment for patients with malignant tumor.

References

1. Wacker B, Nagrani T, Weinberg J et al (2007) Correlation between development of rash and efficacy in patients treated with the epidermal growth factor receptor tyrosine kinase inhibitor erlotinib in two large phase III studies. Clin Cancer Res 13:3913–3921. doi:10.1158/1078-0432.ccr-06-2610
2. Zhu A Liu W, Gotlieb V, et al. (2015) The rapidly evolving therapies for advanced melanoma-towards immunotherapy. Molecular targeted therapy, and beyond. Crit Rev Oncol Hematol. pii: S1040-8428(15)30091-3. doi:016/j.critrevonc.2015.12.002
3. Garon EB (2015) Current perspectives in immunotherapy for non-small cell lung cancer. Semin Oncol 42(Suppl 2):S11-8. doi:10.1053/j.seminoncol.2015.09.019 Epub 2015 Sep 11
4. Ataka S, Nishio M, Hida T, et al (2015) Phase II study of anti-PD-1 antibody nivolumab for Japanese patients with non small cell lung cancer. The 56th annual meeting of lung cancer (abstract) http://www.haigan.gr.jp/journal/am/2015a/15a_ws040WS4-1.html
5. Herbst RS, LoRusso PM, Purdom M et al (2003) Dermatologic side effects associated with gefitinib therapy: clinical experience and management. Clin Lung Cancer 4:366–369
6. Tsimboukis S, Merikas I, Karapanagiotou EM et al (2009) Erlotinib-induced skin rash in patients with non-small-cell lung cancer: pathogenesis, clinical significance, and management. Clin Lung Cancer 10:106–111
7. Lacouture ME, Schadendorg D, Chu CY et al (2013) Dermatologic adverse events associated with afatinib: an oral ErbB family blocker. Expert Rev Anticancer Ther 13(6):721–728. doi:10.1586/era.13.30 Epub 2013 Mar 18
8. Lacouture ME, Anadkat MJ, RJ E et al (2011) Clinical practice guidelines for the prevention and treatment of EGFR inhibitor associated dermatologic toxicities. Support Care Cancer 19:1079–1095
9. Acharyya S, Sau S, Dasgupta P et al (2012) Skin rash as a surrogate marker of clinical response of targeted therapy using gefitinib in advanced or metastatic non-small-cell lung cancer- a retrospective study. J Indian Med Assoc 110:474–493
10. Perez-Soler R, Chachoua A, Hammond LA et al (2004) Determinants of tumor response and survival with erlotinib in patients with non-small-cell lung cancer. J Clin Oncol 22:3238–3247
11. Li T, Perez-Soler R (2009) Skin toxicities associated with epidermal growth factor receptor inhibitor. Target Oncol 4:107–119
12. Yamazaki N, Muro K (2007) Clinical management of EGFRI dermatologic toxicities: the Japanese perspective. Oncology 21(11Suppl 5):27–28
13. Common terminology Criteria for Adverse Events (Version 4.0). National Cancer Institute Web site. https://ctep.cancer,gov/frms/CTCAE-Index.pdf
14. Rhee J, Oishi K, Garey J et al (2001) Management of rash and other toxicities in patients treated with epidermal growth factor receptor-targeted agents. Clin Colorectal Cancer 5(suppl2):S101–S106

15. Cooper JA, White DA, Matthay RA et al (1986) Drug induced pulmonary disease. Part 1: cytotoxic drugs. Am Rev Respir Dis 133:321–340
16. Ando M, Okamoto I, Yamamoto N et al (2006) Predictive factors for interstitial lung disease, antitumor response, and survival in non-small-cell lung cancer patients treated with gefitinib. J Clin Oncol 24:2549–2556
17. Kudo S, Kato H, Nishiwaki Y et al (2008) Interstitial lung disease in Japanese patient with lung cancer. Am J Respir Crit Care Med 177:1348–1357
18. Pharmaceuticals and Meical Devices Agency (2004), Reports on Iressa Tablets 250 retrospective study (special intervention). In: Pharmaceuticals and Medical Devices Safety Information. No.206 October 2004. https://pmda.go.jp./english/service/pdf/precautions/PMDSI-206.pdf
19. Saito Y, Gemma A (2012) Current status of DILD in molecular targeted therapies. Int J Clin Oncol 17:534–541
20. Nakagawa K, Kudo S, Ohe Y et al (2012) Postmarketing surveillance study of erlotinib in Japanese patients with non-small-cell lung cancer(NSCLC): an interim analysis of 3488 patients (POLARSTAR). J Thorac Oncol 7:1296–1303
21. Kubota K, Nishiwaki Y, Tamura T et al (2008) Efficacy and safety of erlotinib monotherapy for Japanese patients with advanced non-small-cell lung cancer: a phase II study. J Thorac Oncol 3:1439–1445
22. Willemsen AE, Grutters JC, Gerritsen WR et al (2015) mTOR inhibitor-induced interstitial lung disease in cancer patients: comprehensive review and a practical management algorithm. Int J Cancer. doi:10.1002/ijc.29887
23. Travis WD, Costabel U, Hansell DM et al (2013) An official American thoracic Society/ European Respiratory Society statement: update of the international multidiscip linary classification on the idiopathic interstitial pneumonia. Am J Repir Crit Care Med 188:733–748
24. Muller NL, White DA, Jiang H et al (2004) Diagnosis and management of drug-associated interstitial lung disease. Br J Cancer Suppl 2:S24–S30
25. Kubo K, Azuma A, Kanazawa M et al (2013) Consensus statement for the diagnosis and treatment of drug-induced lung injury. Respir Investig 51:260–277
26. Perrillo RP, Gish R, Falck-Ytter YT et al (2015) American gastroenterological association institute technical review on prevention and treatment of hepatitis B virus reactivation during immunosuppressive drug therapy. Gastroenterology 148:221–244
27. Shah NT, Kris MG, Pao W et al (2005) Practical management of patients with non-small-cell lung cancer treated with gefitinib. J Clin Oncol 23:165–174
28. Sandler AB (2006) Nondermatologic adverse events associated with anti-EGFR therapy. Oncology Suppl 20:35–40
29. Oonuma H, Kato J, Koyama K et al (2015) Corticosteroid therapy for crizotinib-induced, allergy-mediated liver injury: a case report. JJLC 55:48–52
30. JSH Guidelines for the Management of Hepatitis B Virus Infection (2014) Drafting Committee for Hepatitis Management Guidelines and the Japan Society of Hepatology. https://www.jsh.or.jp/medical/guidelines/jsh_guidlines/hepatitis_b. doi:10.1111/hepr.1226
31. NCCN Clinical Pactice Gidelines in oncology –Antiemesis.-Ver.2 2013. http://www.nccn.org/professionals/physician_gls/f_guidelines.asp
32. Yang JC, Reguart N, Barinoff J et al (2013) Diarrhea associated with afatinib: an oral ErbB family blocker. Expert Rev Anticancer Ther 13:729–736
33. Bensen AB, Ajani JA, Catalano RB et al (2004) Recommended guide lines for the treatment of cancer treatment-induced diarrhea. J Clin Oncol 22:2918–2916
34. Chen HX, Cleck JN (2009) Adverse effects of anticancer agents that target the VEGF pathway. Nat Rev Clin Oncol 6:465–477
35. Syrigos KN, Karapanagiotou E, Boura P et al (2011) Bevacizumab-induced hypertension: pathogenesis and management. BioDrugs 25:159–169
36. Yang JC, Haworth L, Sherry RM et al (2003) A randomized trial of bevacizumab, an anti-vascular endothelial growth factor antibody, for metastatic renal cancer. N Engl J Med 349:427–434

37. Ranpura V, Pulipati B, Chu D et al (2010) Increased risk of high-grade hypertension with bevacizumab in cancer patients: a meta-analysis. Am J Hypertens 23:460–468
38. Guideline for management of hypertension 2014. The Japanese Society of Hypertension. https://www.jpnsh.jp
39. Izzedine H, Ederhy S, Goldwassser F et al (2009) Management of hypertension in angiogenesis inhibitor-related patients. Ann Oncol 20:807–8115
40. Mir O, Coriat R, Ropert S et al (2012) Treatment of bevacizumab-induced hypertension by amlodipine. Investig New Drugs 30:702–707
41. Niho S, Kunitoh H, Nokihara H et al (2012) Randomized phase II study of first-line carboplatin-paclitaxel with or without bevacizumab in Japanese patients with advanced non-squamous non-small-cell lung cancer. Lung Cancer 76:362–367
42. Abu-Hejleh T, Mezhir JJ, Googheart MJ et al (2012) Incidence and management of gastrointestinal perforation from bevacizumab in advanced cancers. Curr Oncol Rep 14:277–284
43. Kamba T, McDonald DM (2007) Mechanisms of adverse effects of anti-VEGF therapy for cancer. Br J Cancer 96:1788–1795
44. Johnson DH, Fehrenbacher L, Novotnt WF et al (2004) Randomized phase II trial comparing bevacizumab plus carboplatin and paclitaxel with carboplatin and paclitaxel alone in previously untreated locally advanced or metastatic non-small -cell lung cancer. J Clin Oncol 22:2184–2191
45. Reck M, Barlesi F et al (2012) Predicting and managing the risk of pulmonary haemorrhage in patients with NSCLC treated with bevacizumab: a consensus report from a panel of experts. Ann Oncol 23:1111–1120
46. Sandler AB, Schiller JH, Gray R et al (2009) Retrospective evaluation of the clinical and radiographic risk factors associated with severe pulmonary hemorrhage in first-line advanced, unresectable non-small-cell lung cancer treated with carboplatin and paclitaxel plus bevacizumab. J Clin Oncol 27:1405–1412

Part III
Evaluation

Chapter 17
Health-Related Quality of Life in Molecular Targeted Therapy

Shinji Nakamichi and Kaoru Kubota

Abstract Quality of life (QOL) in medicine has been evaluated as health-related QOL (HRQOL). HRQOL is a component of patient-reported outcomes (PROs). HRQOL is a true clinical endpoint when validated and reliable QOL instruments are used. Clinical trials often evaluated HRQOL as the secondary endpoint. Improvement of progression-free survival (PFS) with improved HRQOL would be clinically meaningful outcome. Recently, several randomized trials have been conducted with QOL as the primary endpoint. A randomized trial of early palliative care (EPC) integrated with standard oncologic care or standard oncologic care alone in patients with metastatic non-small cell lung cancer (NSCLC) showed that EPC significantly improved QOL and mood. Median overall survival (OS) was longer among patients receiving EPC. The data suggests that QOL is highly related to OS and QOL evaluation should be integrated into oncology practice for patients with advanced lung cancer. To improve patient management, effective communication is necessary. Communication skill training (CST) program based on SHARE protocol is effective for both oncologists and patients with cancer. Because physicians tend to concentrate on cancer-related outcomes and often neglect assessments of QOL, tools to evaluate QOL would be useful to improve quality of care in patients with advanced lung cancer.

Keywords Quality of life (QOL) • Health-related quality of life (HRQOL) • Patient-reported outcomes (PROs) • Early palliative care (EPC) • Communication skill training (CST)

S. Nakamichi, M.D. • K. Kubota, M.D., Ph.D. (✉)
Department of Pulmonary Medicine and Oncology, Graduate School of Medicine, Nippon Medical School, 1-1-5 Sendagi, Bunkyo-ku, Tokyo 113-8603, Japan
e-mail: kkubota@nms.ac.jp

© Springer Science+Business Media Singapore 2017

Y. Takiguchi (ed.), *Molecular Targeted Therapy of Lung Cancer*,
DOI 10.1007/978-981-10-2002-5_17

17.1 Definition of QOL and HRQOL

Definition of quality of life (QOL) is described as the general well-being of a person or society in terms of health and happiness rather than wealth by dictionary. Another textbook describes that QOL is a ubiquitous concept that has different philosophical, political and health-related definitions. Because the concept of QOL includes many areas, evaluation of QOL in patients with lung cancer should be restricted to health, namely, health-related QOL.

Health is defined as a state of complete physical, mental and social well-being. This is not merely the absence of disease or infirmity. WHO definition of health has not been amended since 1948. Therefore, health-related QOL (HRQOL) includes the physical, functional, social, and emotional well-being of an individual. HRQOL is a patient-reported outcome (PRO) usually measured with carefully designed and validated instruments such as questionnaires or semi-structured interview schedules.

Validated QOL instruments should be used when we measure QOL.

17.2 Chemotherapy and QOL in Lung Cancer

Brief history of clinical trials of lung cancer chemotherapy and quality of life would be reviewed. In a prospective randomized study conducted in Thailand [1], 287 patients with advanced non-small cell lung cancer (NSCLC) with ECOG performance status (PS) 0–1 or 2 were randomly assigned to receive either best supportive care (BSC) or BSC plus cisplatin-based combination chemotherapy (IEP regimen; ifosfamide, epirubicin, and cisplatin, or MVP regimen; mitomycin C, cisplatin, and vindesine). Karnofsky Performance Status (KPS), Thai modified Functional Living Index-Cancer (T-FLIC), and Thai modified Quality of Life-Index (T-QLI) were used to estimate the QOL. This study demonstrated that cisplatin-based chemotherapy improves quality of life as well as overall survival (OS) in patients with advanced NSCLC.

TAX JP 301 study was conducted in Japan [2]. Patients with stage IV NSCLC were randomized to 60 mg/m^2 of docetaxel plus cisplatin or vindesine plus cisplatin. OS and objective response (OR) were significantly better for docetaxel plus cisplatin arm. QOL in physical domain also favored in docetaxel plus cisplatin.

Four-arm cooperative study (FACS) was conducted as a randomized trial in Japan to compare the efficacy and toxicity of three platinum-based combination regimens (carboplatin plus paclitaxel, cisplatin plus gemcitabine, and cisplatin plus vinorelbine) against cisplatin plus irinotecan (IP) in patients with untreated advanced NSCLC with a non-inferiority design [3]. The primary endpoint of this study was OS. Although all the four regimens were well tolerated, None of three arms didn't meet the primary outcome. However, using QOL-ACD developed in Japan, all experimental arms were better than the control arm cisplatin plus irinotecan in physical domain. The difference of QOL data would be due to the difference of toxicities.

Phase III study comparing oral S-1 plus carboplatin with paclitaxel plus carboplatin in chemotherapy-naïve patients with advanced NSCLC was conducted in Japan (LETS study) [4]. Carboplatin plus S-1 was non-inferior to carboplatin plus paclitaxel with regard to OS. While there was no difference between the two arms in Functional Assessment of Cancer Therapy-Lung (FACT-L), scores based on the neurotoxicity subscale of the FACT/GOG NTX (11-item FACT/Gynecologic Oncology Group-Neurotoxicity) was favored for carboplatin plus S-1.

A randomized, open-label, phase III, non-inferiority trial that compared S-1 plus cisplatin with docetaxel plus cisplatin (CATS study) was conducted to compare OS in patients with advanced NSCLC as the primary endpoint [5]. QOL was also evaluated as a secondary endpoint. OS for S-1 plus cisplatin was non-inferior to docetaxel plus cisplatin. QOL data were evaluated with the European Organization for Research and Treatment for Cancer Quality of Life Questionnaire-Core 30 (EORTC QLQ-C30) and the 13-item lung cancer-specific questionnaire module (EORTC QLQ-LC13). Global health status/QOL functioning 1 week after the first dose of cisplatin in the EORTC QLQ-C30 favored S-1 plus cisplatin. Thirteen data of EORTC QLQ-LC favored for S-1 plus cisplatin not only at 1 week after the first dose of cisplatin but also at the end of the second course (Fig. 17.1).

Iressa Pan-Asia Study (IPASS) had also evaluated QOL [6]. QOL was assessed with the use of the Functional Assessment of Cancer Therapy-Lung (FACT-L) questionnaire and the Trial Outcome Index (TOI), which is the sum of the physical well-being, functional well-being, and Lung Cancer Subscale (LCS) scores of FACT-L, and symptoms were assessed with the use of the LCS score. QOL was significantly favored gefitinib in patients with EGFR mutation. The results were vice-versa in patients with EGFR mutation negative.

NEJ 002 study confirmed QOL advantage of gefitinib in patients with EGFR mutation positive [7]. QOL data from 148 patients (72 in the gefitinib arm and 76 in the carboplatin plus paclitaxel arm) were analyzed. Time to defined deterioration in physical and life well-being significantly favored gefitinib over chemotherapy.

LUX-Lung 3 study demonstrated PRO data favored afatinib compared to cisplatin plus pemetrexed in patients with EGFR mutation positive NSCLC [8]. Lung cancer symptoms and health-related QOL were assessed using the EORTC QLQ-C30 and EORTC QLQ-LC 13 questionnaires. Analyses of cough, dyspnea, and pain were preplanned, including percentage of patients who improved on therapy, time to deterioration of symptoms, and change in symptoms over time. First-line afatinib was associated with better control of cough and dyspnea compared with chemotherapy, although diarrhea, dysphagia, and sore mouth were worse. Global health status/QOL was also improved over time with afatinib compared with chemotherapy. These data indicate that molecular targeted therapy significantly improves QOL of the patients who have the driver oncogenes.

QOL data has been evaluated as a secondary endpoint so far. However, some recent studies use quality of life as the primary endpoint. ERACLE study was a randomized trial which compared cisplatin plus pemetrexed to carboplatin,

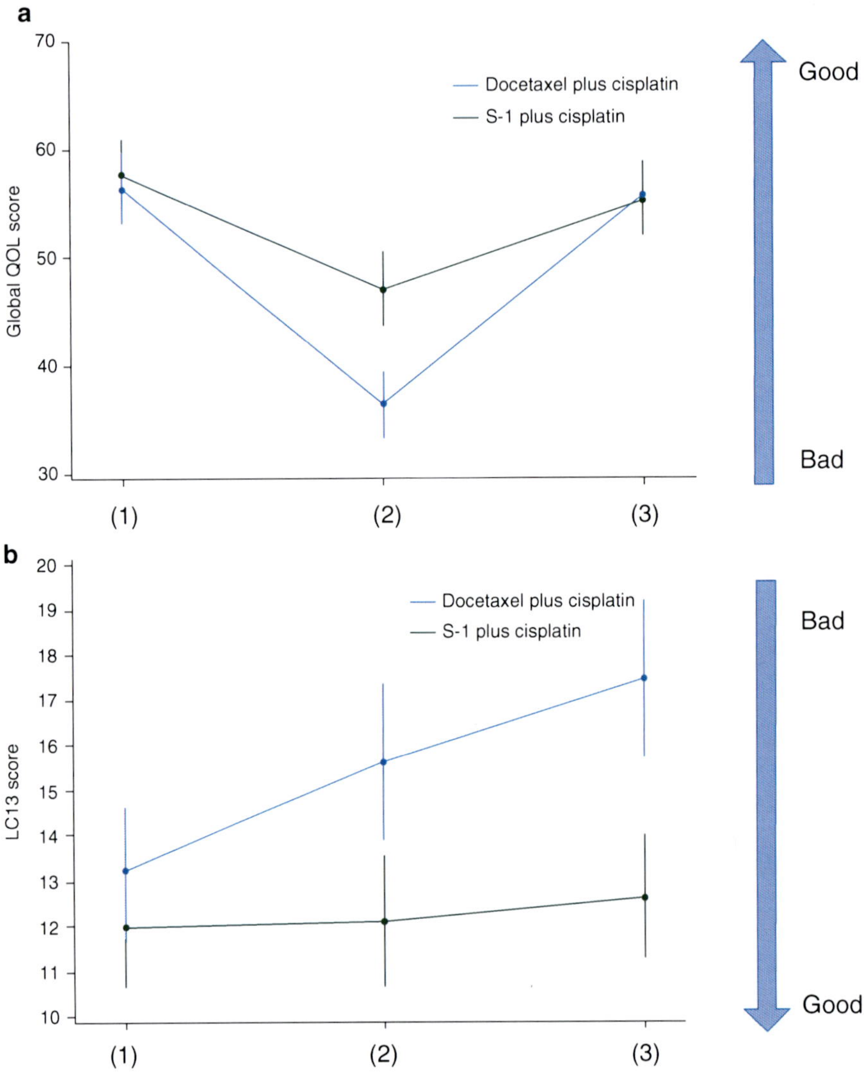

Fig. 17.1 QOL assessments. Patients responded three times: (*1*) before each treatment, (*2*) 1 week after the first dose of cisplatin, and (*3*) at the end of the second course. (**a**) Score changes of global health status/QOL (items 29 and 30) in the EORTC QLQ-C30. (**b**) Score changes in the EORTC QLQ-LC 13 (Modified from Ref. [5])

paclitaxel plus bevacizumab [9]. The primary endpoint was the difference in QOL between the 2 treatment arms after 12 weeks of maintenance, measured using the EuroQoL 5 Dimensions-Index (EQ5D-I) and EQ5D-visual analogue scale (EQ5D-VAS). Although there was no difference between the two arms in EQ5D-visual analogue scale, EQ5D-I favored for cisplatin plus pemetrexed.

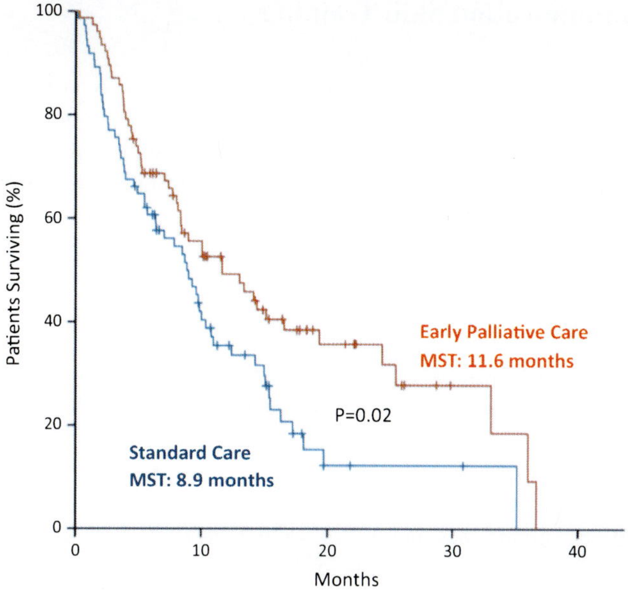

Fig. 17.2 Early palliative care led to improvement of QOL (Modified from Ref. [10])

17.3 Palliative Care in Patients with Advanced Lung Cancer

Temel JS and colleagues conducted a famous randomized trial of early palliative care (EPC) integrated with standard oncology care compared to standard oncologic care in patients with metastatic NSCLC [10]. The primary endpoint of the study was QOL at 12 weeks after randomization. This study revealed that EPC significantly improved QOL. Quality of terminal care was also better in EPC group with lower depression. Median OS was significantly longer in EPC group (11.6 months vs. 8.9 months, $p = 0.02$). Palliative care team conducted physical and psychological symptom control, establishing the goal of care and assisting decision making of patients. Furthermore, patients in EPC group had more accurate perception about prognosis compared to those in standard care group [11]. Patients in EPC group with accurate perception received less cytotoxic chemotherapy within 60 days before death than patients in standard care group. These data indicate that patients in EPC group could make more appropriate decisions (Fig. 17.2).

Another randomized trial results of early integration of palliative care with standard oncologic care versus late (3 months later) was reported [12]. Early-entry participants' PROs and resource use were not statistically different. However, their survival 1 year after enrollment was improved compared with those who began 3 months later.

EPS studies indicate that early symptom management with psychosocial support and early illness understanding and assisting with decision-making of the patients should be considered as evidence-based patient management.

17.4 Communication Skill Training

To improve patient management, effective communication is critical. Based on Japanese patients' preferences regarding breaking bad news, communication skill training program named SHARE protocol has been developed. SHARE consists of four components, including setting up supportive environment for interview, considering how to deliver the bad news, discussing additional information that patient would like to know, and providing reassurance and addressing patient's emotions with empathic responses. SHARE protocol emphasizes reassurance and emotional support based on patients' preference. Using the SHARE protocol, a randomized trial of communication skill training (CST) in oncologists who worked for National Cancer Center in Japan was conducted [13]. Oncologists were randomized to CST or control group. Both groups were evaluated pre- and post-CST by experts and themselves. Their patients were also evaluated regarding depression, anxiety, satisfaction with communication, and trust in oncologists. CST program consists of an hour lecture, 30 minutes of demonstration video, and an hour of role-play with simulated patients, eight times, in total 10 h and 2 days schedule. Thirty oncologists participated in the study. Backgrounds of the oncologists were similar in terms of age, clinical experience, gender, and specialty between the two groups. Performance of physicians was significantly improved by self-evaluation and experts' evaluation. In total, 601 patients were evaluated. Age and gender were similar between the two groups. More patients in CST group were treated at surgical oncology and on current treatment. The HADS is a self-administered and standardized instrument for evaluating patients' distress. HADS depression and trust in oncologist were significantly favored in patients who were treated by oncologists in CST group. Interestingly, time of consultation was not different in doctors after CST group. This is the first study that demonstrated improvement of patients' outcomes by CST. CST program based on patient preference is effective for both oncologists and patients with cancer (Table 17.1).

Table 17.1 SHARE model for communication skill training

Component	Description
S	Setting up supportive environment for interview (e.g., greeting patient cordially, looking at patient's eyes and face)
H	Considering how to deliver bad news (e.g., not beginning bad news without preamble, checking to see whether talk is fast paced)
A	Discussing additional information that patient would like to know (e.g., answering patient's questions fully, explaining second opinion)
RE	Providing reassurance and addressing patient's emotions with empathic responses (e.g., remaining silent out of concern for patient's feelings, accepting patient's expression of emotions)

17.5 Conclusion

The American Society of Clinical Oncology (ASCO) recommended in 2011 that QOL should be an explicit priority throughout the course of advanced cancer care. There are cancer outcome and patient outcome [14]. Cancer outcome includes response, response duration, and relapse. Patient outcome includes survival and quality of life. These are true endpoints.

Unfortunately, physicians tend to concentrate on cancer-related outcomes, often neglecting assessment of QOL. Actually, it is quite difficult to assess multidimensional QOL of patients in busy clinic.

To evaluate effects of QOL measures in daily oncology practice, our group has started a randomized trial with or without QOL measures using care notebook that is currently underway. The primary outcome is content of the patient provider interaction.

It is necessary that the choice of treatment should be determined by the patient's wish; patient's status; social background such as family, job, and life style; and medical condition, because QOL and daily life are significantly changed by the treatment. Future direction would be making tools to understand more about patients and to help physicians in daily practice.

References

1. Thongprasert S, Sanguanmitra P, Juthapan W et al (1999) Relationship between quality of life and clinical outcomes in advanced non-small cell lung cancer: best supportive care (BSC) versus BSC plus chemotherapy. Lung Cancer 24:17–24
2. Kubota K, Watanabe K, Kunitoh H et al (2004) Phase III randomized trial of docetaxel plus cisplatin versus vindesine plus cisplatin in patients with stage IV non-small-cell lung cancer: the Japanese Taxotere Lung Cancer Study Group. J Clin Oncol 22:254–261
3. Ohe Y, Ohashi Y, Kubota K et al (2007) Randomized phase III study of cisplatin plus irinotecan versus carboplatin plus paclitaxel, cisplatin plus gemcitabine, and cisplatin plus vinorelbine for advanced non-small-cell lung cancer: Four-Arm Cooperative Study in Japan. Ann Oncol 18:317–323
4. Okamoto I, Yoshioka H, Morita S et al (2010) Phase III trial comparing oral S-1 plus carboplatin with paclitaxel plus carboplatin in chemotherapy-naïve patients with advanced non-small-cell lung cancer: results of a west Japan oncology group study. J Clin Oncol 28:5240–5246
5. Kubota K, Sakai H, Katakami N et al (2015) A randomized phase III trial of oral S-1 plus cisplatin versus docetaxel plus cisplatin in Japanese patients with advanced non-small-cell lung cancer: TCOG0701 CATS trial. Ann Oncol 26:1401–1408
6. Mok TS, Wu YL, Thongprasert S et al (2009) Gefitinib or carboplatin-paclitaxel in pulmonary adenocarcinoma. N Engl J Med 361:947–957
7. Oizumi S, Kobayashi K, Inoue A et al (2012) Quality of life with gefitinib in patients with EGFR-mutated non-small cell lung cancer: quality of life analysis of North East Japan Study Group 002 Trial. Oncologist 17:863–870
8. Yang JC, Hirsh V, Schuler M et al (2013) Symptom control and quality of life in LUX-Lung 3: a phase III study of afatinib or cisplatin/pemetrexed in patients with advanced lung adenocarcinoma with EGFR mutations. J Clin Oncol 31:3342–3350

9. Galetta D, Cinieri S, Pisconti S et al (2015) Cisplatin/pemetrexed followed by maintenance pemetrexed versus carboplatin/paclitaxel/bevacizumab followed by maintenance bevacizumab in advanced nonsquamous lung cancer: the GOIM (Gruppo Oncologico Italia Meridionale) ERACLE Phase III randomized trial. Clin Lung Cancer 16:262–273
10. Temel JS, Greer JA, Muzikansky A et al (2010) Early palliative care for patients with metastatic non-small-cell lung cancer. N Engl J Med 363:733–742
11. Temel JS, Greer JA, Admane S et al (2011) Longitudinal perceptions of prognosis and goals of therapy in patients with metastatic non-small-cell lung cancer: results of a randomized study of early palliative care. J Clin Oncol 29:2319–2326
12. Bakitas MA, Tosteson TD, Li Z et al (2015) Early versus delayed initiation of concurrent palliative oncology care: patient outcomes in the enable iii randomized controlled trial. J Clin Oncol 33:1438–1445
13. Fujimori M, Shirai Y, Asai M et al (2014) Effect of communication skills training program for oncologists based on patient preferences for communication when receiving bad news: a randomized controlled trial. J Clin Oncol 32:2166–2172
14. DeVita VT Jr, Lawrence TS, Rosenberg SA (2011) DeVita, Hellman, and Rosenberg's cancer: principles and practice of oncology, 9th edn. Wolters Kluwer Health/Lippincott Williams & Wilkins, Philadelphia

Chapter 18
Gene Signature

Hideki Ujiie, Daiyoon Lee, Tatsuya Kato, and Kazuhiro Yasufuku

Abstract In the past decade, there has been seen an increase in the number of cancer therapies that aim to circumscribe the spread and expansion of primary and metastatic tumors. A common characteristic among these therapies is their ability to target cancer progression via different pathways, which is fundamental to preventing successful tumor spreading and dissemination. Recent advancements in gene expression profiling have been fundamental in the identification of new cancer targets, and, consequently, improved targeted therapies have emerged as gene expression arrays, and DNA sequencing have enhanced our understanding of cancer genetics. Modern tumor pathology is now understood and studied at the molecular level ranging from immunohistochemistry (IHC) biomarkers to gene signature classifications and gene mutations, all of which provide significant knowledge about which patients will respond to targeted therapy regimens. We briefly discuss the common types of targeted therapies currently used clinically and provide a brief background on IHC, gene expression, and DNA sequencing technologies. We further provide a discussion on guided therapies and also focus on the appropriate targeted therapies and the pathways they inhibit. A number of prognostic gene expression signatures have been reported to predict survival in non-small cell lung cancer (NSCLC). We focus on the role of gene expression profiling in NSCLC as predictive and prognostic biomarker and its potential use for personalized therapy in the years to come.

Keywords Gene signature • Non-small cell lung cancer (NSCLC) • Next-generation sequencing • Immunohistochemistry (IHC) • Gene expression • Gene profiling

---

H. Ujiie, M.D., Ph.D. (✉)
Division of Thoracic Surgery, Toronto General Hospital, University Health Network, University of Toronto, Toronto, ON, Canada

Latner Thoracic Surgery Research Laboratories, University Health Network, University of Toronto, Toronto, ON, Canada
e-mail: Hideki.Ujiie@uhn.ca; Hideki_Ujiie@hotmail.com

D. Lee, MSc • T. Kato, M.D., Ph.D. • K. Yasufuku, M.D., Ph.D.
Division of Thoracic Surgery, Toronto General Hospital, University Health Network, University of Toronto, Toronto, ON, Canada

© Springer Science+Business Media Singapore 2017

Y. Takiguchi (ed.), *Molecular Targeted Therapy of Lung Cancer*,
DOI 10.1007/978-981-10-2002-5_18

18.1 Introduction

Lung cancer is the leading cause of cancer-related mortality worldwide [1]. With the National Lung Screening Trial results and the recent approval of Centers for Medicare and Medicaid Service coverage for screening CT scans, an increase in our ability to detect and treat early-stage lung cancer is anticipated [2–4]. Despite the curative intent of surgical resection therapy, tumor recurrence and expansion continue to be the primary causes of cancer-related death between patients with early-stage lung cancer [5, 6]. Furthermore, lung cancer is still not detected until it is at an advanced stage, which can make it more challenging to treat due to potential metastasis. In particular, the 5-year survival for patients with regional lymph node metastasis leads to very poor prognosis [7]. The lack of major improvements in the survival rate for lung cancer despite advances in surgery, chemotherapy, and radiotherapy has stimulated a search for alternative strategies to improve lung cancer management. This requires a better understanding of lung cancer pathogenesis and identification of new therapeutic targets.

The hallmarks of cancer as defined by Hanahan and Weinberg (2011) provide a detailed overview of the biological characteristics that tumor cells utilize to successfully colonize primary and metastatic tumor sites [8]. A myriad of intricate signaling pathways underlie these capabilities and, given how fundamental these pathways are to tumor growth and spreading, they represent a successful framework for the development of targeted cancer therapies. This is most obvious when considering the number of therapy regimens that were developed for each cancer hallmark identified [8]. In spite of this considerable development, both researchers and clinicians are still trying to understand when, how, and to whom specific therapies should be given and, particularly, alternative treatment options when patients stop responding to targeted therapies [9].

Prior to the turn of this century, cancer treatment regimens were limited to chemotherapy, radiotherapy, surgery, and endocrine therapy. The latter treatment was the first targeted cancer therapy, originating from seminal studies by George Thomas Beatson in 1896 [10]. Although he did not have an extensive understanding of the mechanism behind estrogen deprivation, Beatson (1896) performed a bilateral oophorectomy in a female subject with recurrent breast cancer who subsequently achieved complete remission and survived for 4 years postsurgery. The past 15–20 years, however, have seen a greatly increased number of targeted treatments, and three main classes of drugs have developed for clinical use: monoclonal antibodies [11], small molecule inhibitors [12], and fusion proteins [13]. These therapies take many forms, but can be further grouped based on their mechanism of action or biological targets, and include hormonal therapies, signal transduction inhibitors, gene expression modulators, angiogenesis inhibitors, immunotherapies, monoclonal antibodies with conjugated therapeutics, and apoptosis-inducing drugs. However, this list is still incomplete, and the classes and forms of targeted therapies are constantly expanding in part due to major contributions from large-scale genomic analyses [14].

Despite the development of large-scale genomic analysis platforms, immunohistochemical (IHC) staining of tumor biomarkers (and their following pathological clarification) still plays a key role in selecting the correct treatment for patients with cancer. After fixation, the tumor is immersed in liquid paraffin, which is allowed to harden for easier sectioning. Thin tissue sections (4–5 μm in diameter) are cut and subjected to incubation with primary and labeled secondary antibodies, after which the resulting tissue staining is evaluated. The intensity and quantity of staining can then be used to make decisions on whether or not a certain therapy is suitable for the patient. Interestingly, it has been shown in the last 10–15 years that, in some cases, the prognostic information provided by protein biomarkers is equivalent or even inferior to that provided by RNA-based gene expression signatures [15]. In 1995 the first study to use cDNA microarrays investigated gene expression patterns [16], and signaled the arrival of a technology that transformed the scientific research community over the course of the next 20 years. There has since been an exponential growth in the use and routine application of expression arrays. Usually, the development of such signatures relies on fresh-frozen tumor samples due to differences in RNA quality of paraffin-embedded material [17], though improvements in isolation techniques and assays have sought to alleviate this problem [18]. Following RNA isolation (and conversion to cDNA), tumor specimens are applied to a gene chip array where the fluorescence intensity of bound cDNA reflects the expression of gene transcripts. This intensity is understood mathematically by a microarray scanner, and, following data standardization, bioinformaticians use one of the three main strategies to define a gene signature: top-down, bottom-up, and candidate gene approaches [19].

Similar to gene expression arrays, the use of DNA sequencing technologies has increased dramatically in the past decade. From initial sequencing of bacteriophage DNA in 1977 [20] to the release of its next-generation counterpart for whole human genome annotation in 2005, DNA sequencing has had a profound impact on our understanding of tumor biology [21]. For the sequencing of cancer specimens, DNA is extracted from both the tumor and germ line to determine which mutations is tumor specific. Once obtained, DNA is typically amplified, and the identity of individual DNA bases is determined by a sequencer [22]. Following arrangement to a reference genome, mutations specific to the tumor DNA are used to understand which genes or pathways are changed, and their relationship to patient outcome is interpreted through analyses of progression-free survival (PFS) and overall survival (OS). The fundamental aim of both microarray and sequencing technologies has been to identify patient subgroups that (I) could be spared unnecessary exposure to toxic chemotherapeutic or targeted drug regimens, (II) are in need of more aggressive or targeted treatment strategies, and, most importantly, (III) are sensitive or resistant to specific drugs based on the genetic characteristics of their tumor. Clinical problems will always happen if the specific targets of a drug are not known or are poorly characterized. A better understanding of the mechanism of action of the drug and its associated biological effects is required. This demonstrates the importance of evaluating the effect of the clinical trial drugs being managed, which should be the objective for modern analysis strategies including genomics, proteomics, and functional image analysis. The worldwide cancer problem is expected to increase from 14.1 million (in 2012) to over 20 million new cases per year by 2025 [23, 24]. It has been estimated from current mortality statistics that cancer was the cause of 8.2 million deaths in 2012. Furthermore, lung, prostate and colorectal cancers are the most common cancers in men, while cancers in the breast, colorectal, and cervix uteri are the most common in women.

Recently, there are several articles that mentioned the gene signature of lung cancer [25].

Here, we present an overview of NSCLC, for which the use and development of targeted therapies is rapidly developing. We describe how treatment decisions are currently made, the relevant targeted therapies, and, where possible, how recent detections from large-scale genomic profiling may be incorporated into current and future treatment decisions.

18.2 Non-small Cell Lung Cancer

18.2.1 Background

Lung cancer is the most common cause of cancer in the world. It was the most common cancer in men and the third most common in women. In terms of mortality, lung cancer is the number one cause of cancer-related deaths worldwide [24]. More than 50 % of patients with lung cancer already have metastatic disease at diagnosis, and the tumor is only localized and suitable for surgery in 25 % of patients. Cisplatin-based adjuvant chemotherapy is managed in patients with node-positive disease (stage II and III) and provides a 5-year OS advantage of 4 %. Platinum is the central supporting part of lung cancer treatment even in metastatic disease. In this clinical setting, median and 5-year OS are 10 months and <5 %, respectively. From a histological perspective, approximately 85 % of patients with lung cancer have a non-small cell phenotype such as adenocarcinoma (ADC), which accounts for more than 50 % of cases, squamous cell carcinoma (SqCC), or non-small cell lung cancer (NSCLC) not otherwise quantified.

While the major cause of lung cancer is smoking, approximately 15 % of patients with lung cancer have never been or have only intermittently been exposed to tobacco smoke. It is within this specific group of topics that it is more general to detect the targetable mutations, that is, mutations in the epidermal growth factor receptor (*EGFR*) gene and rearrangements in the anaplastic lymphoma kinase (*ALK*) gene (Fig. 18.1). As these mutations infrequently (<1 %) occur in pure SqCC, targeted molecular testing is recommended in the routine diagnostic work-up of non-squamous NSCLC [26]. In addition, evaluating all molecular targeted agents is useful for metastatic lung cancer. Even in the presence of targetable genetic differences, recent evidence does not support the use of specific inhibitors in the adjuvant setting, primarily due to the absence of trials designed with adequate power to detect

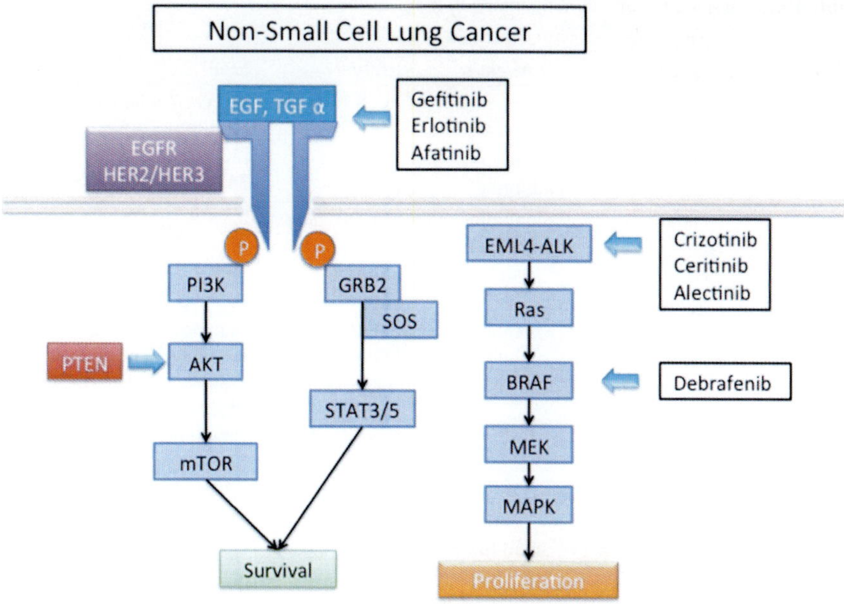

Fig. 18.1 Therapy-targeted pathways in lung cancer
Targeted therapy in lung cancer and the drugs that influence them. *EGF*, epidermal growth factor; *TGF-α*, transforming growth factor-α; *EGFR*, epidermal growth factor receptor; *HER2/3*, human epidermal growth factor receptor 2/3; *PI3K*, phosphatidylinositol-4,5-bisphosphate 3-kinase; *SOS*, son of sevenless guanine nucleotide exchange factor; *GRB2*, growth factor receptor-bound protein 2; *PTEN*, phosphatase and tensin homologue; *mTOR*, mammalian target of rapamycin; *STAT 3/5*, signal transducer and activator of transcription 3/5; *EML4-ALK*, echinoderm microtubule-associated protein-like 4-anaplastic lymphoma kinase; *BRAF*, B-Raf proto-oncogene, serine/threonine kinase; *MEK*, MAPK/Erk kinase; *MAPK*, mitogen-activated protein kinase

significant survival differences. A summary of the targeted therapies used to treat patients with NSCLC is presented in Table 18.1.

18.3 Targeted Therapies

18.3.1 Anti-EGFR

Activation of the EGFR pathway influences several oncogenic processes, including cell proliferation, resistance to apoptosis, migration, invasion, and angiogenesis. The EGFR protein is expressed in approximately 85 % of NSCLC, and its gene has become an interesting target for lung cancer therapy as a result of the development of the small tyrosine kinase inhibitors (TKIs) gefitinib, erlotinib, and afatinib. Gefitinib and erlotinib are reversible inhibitors that specifically target the EGFR protein, while afatinib is an irreversible inhibitor that binds covalently to EGFR and

Table 18.1 Targeted therapy in non-small cell lung cancer (NSCLC)

Therapy type	Therapy agent	Type/class	Target	Evidence
Anti-EGFR therapy	Gefitinib	Small molecule inhibitor	EGFR protein	In *EGFR*-mutated tumors, erlotinib [28, 52], gefitinib [30–32], and afatinib [33, 34] all improve PFS but not OS when compared to chemotherapy alone
	Erlotinib	Small molecule inhibitor	EGFR protein	
	Afatinib	Small molecule inhibitor	EGFR and other ERBB family members	
ALK inhibitors	Crizotinib	Small molecule inhibitor	*EML4-ALK* kinase activity	In tumors positive for *ALK* rearrangements, crizotinib provides a better RR and PFS compared with chemotherapy in both the first- and second-line setting, but no difference in OS [38, 39]
Multikinase inhibitors	Dabrafenib	Small molecule inhibitor	*BRAF* V600E mutations	Dabrafenib improves RR and PFS in lung cancer patients with *BRAF* V600E mutations [44]
	Vandetanib	Small molecule inhibitor	*RET* kinase activity	Vandetanib has been associated with significant antitumoral activity [46]
Checkpoint inhibitors	Nivolumab	Monoclonal antibody	PD1	Nivolumab improves OS compared to docetaxel in the second-line setting in patients with squamous cell lung cancer [47]

NSCLC non-small cell lung cancer, *RR* response rate, *PFS* progression-free survival, *OS* overall survival, *EGFR* epidermal growth factor receptor, *PD1* programmed cell death 1, *EML4-ALK* echinoderm microtubule-associated protein-like 4-anaplastic lymphoma kinase, *RET* rearranged during transfection.

the other members of the ERBB family, including HER2, ERBB3, and ERBB4. Investigation has shown that the presence of a mutation in the exons coding for the tyrosine kinase domain provides with a protein that functions as a cancer driver that is sensitive to TKI activity. More than 40 different mutation sites have been identified in the *EGFR* gene, the most common being a deletion in exon 19 (Del19) and the point mutation in exon 21 (L585R); both of these account for more than 85 % of all detected mutations [27]. *EGFR* mutations can be found in approximately 17 % of Caucasian and 40 % of East-Asian patients with lung ADC and are more common in nonsmokers. PCR-based *EGFR* mutation testing is routinely performed in the diagnostic develop of non-squamous NSCLC [26]. The clinical activity of EGFR-TKI in patients harboring an *EGFR* mutation has been determined in a number of clinical trials in which more than 1800 patients with advanced *EGFR* mutation-positive lung cancer have been randomly assigned to receive either EGFR-TKI (erlotinib [28, 29], gefitinib [30–32] or afatinib [33]) or conventional platinum-based chemotherapy. All studies have presented considerable advantages for TKIs in terms of PFS compared with chemotherapy alone. The median PFS was 9.2–13.6 versus 4.6–6.9 months for TKIs compared to chemotherapy alone, respectively. Conclusively, in terms of OS, all trials have presented similar results for TKIs and chemotherapy alone, probably due to the large number of patient intersecting to TKIs after progression on chemotherapy. Nevertheless, it is notable that a combined analysis of the two trials comparing afatinib to chemotherapy has presented little but significant difference of 3 months in median OS favoring afatinib in the sub-group of patients (89 %) harboring the most common *EGFR* mutations, Del19 and L858R [34]. To improve the activity of TKIs in the first-line chemotherapy, combination strategies have also been investigated. In a recent study, erlotinib was managed alone or combined with the anti-vascular endothelial growth factor (VEGF) antibody bevacizumab as first-line chemotherapy in patients with *EGFR*-mutated lung ADC [35]. Median PFS was almost doubled by the addition of bevacizumab to erlotinib (16.0 vs. 9.7 months with erlotinib alone). OS data were not established at the time of publication. Further trials investigating the potential of this regimen are ongoing. Despite the high level of activity demonstrated by EGFR-TKI, patients' tumor eventually progressed after a median period of nearly 10 months. Considerable research is currently focused on understanding the mechanism underlying the development of resistance to EGFR-TKI. The major mechanism is the acquirement of another mutation in exon 20 (T790M), which moderates the ATP-binding domain and significantly reduces the inhibition capabilities of TKIs. A T790M mutation is detected in more than 50 % of patients progressing on TKI. Other potential molecular changes with TKI resistance are *MET* amplification, *HER2* amplification, *PIK3CA* mutations, and histological transformation into small-cell lung cancer.

Recently, third-generation EGFR inhibitors AZD9291, rociletinib, and HM61713 have produced extremely suggesting results. These drugs are specifically designed to target the T790M mutation but are also effective toward the other more common essential *EGFR* mutations. Expansion programs as well as randomized trials for this chemical substance are currently ongoing [36].

18.3.2 Anti-ALK

The discovery of *ALK* rearrangements represents another major breakthrough in the area of targeted therapies for NSCLC. Although this molecular change is detected in only 3–5 % of patients with lung ADC, considerable research has been dedicated to the clinical development of potent and specific ALK inhibitors, resulting in a time distance of only 4 years between the first account of *ALK* in lung cancer (in 2007) and approval by the Food and Drug Administration (FDA) of the first inhibitor, crizotinib (in 2011). The *ALK* gene is expressed as a result of fusion with another gene, the most common being *EML4* in lung cancer. This fusion generates the expression of a kinase with high oncogenic potential that is mainly involved in cell proliferation (Fig. 18.1). The gold standard and FDA-approved method to detect *ALK* alterations in lung cancer is a break-apart fluorescence in situ hybridization (FISH) assay. However, FISH is relatively expensive and requires highly trained pathologists, since it may be difficult to identify and properly interpret the presence of split signals within the same chromosome, as often occurring with the *ALK* gene. Recently, an IHC assay targeting the effector protein has been developed and validated on a large scale and is likely to replace FISH as the standard diagnostic method for ALK expression in lung cancer [37]. In terms of clinical activity, crizotinib provides a better RR and PFS compared to chemotherapy in both the first-line (10.9 vs. 7.0 months with chemotherapy alone) and second-line setting (7.7 vs. 3.0 months with chemotherapy alone) [38, 39]. Again no differences in OS were observed between the crizotinib and chemotherapy arms, probably due to the frequent crossover to crizotinib at progression for patients assigned to receive chemotherapy.

As described with *EGFR* mutations, resistance to first-generation inhibitors in patients harboring *ALK* rearrangements is a matter of concern. The most common resistance mechanisms are the development of additional *ALK* mutations, *ALK* amplification, activation of *EGFR* signaling, and *KIT* amplification [40]. Furthermore, because penetration of crizotinib to the central nervous system (CNS) is poor, it is common to observe isolated CNS progression while other tumor localizations are still in reduction. Second-generation inhibitors, such as ceritinib, alectinib, and AP26113, have shown substantial activity against multiple *ALK* mutations and clinical effect in patients progressing on crizotinib. Thus far, ceritinib is the only approved ALK inhibitor besides crizotinib [41]. Several clinical trials comparing second-generation inhibitors to chemotherapy as well as adjacent association between the ALK inhibitors are ongoing to assess the best sequence of systemic therapy for *ALK*-positive lung cancer patients.

Approximately 1 % of lung ADC cases harbor a rearrangement in another fusion gene, *ROS1*. *ALK* and *ROS1* share 70 % homology to crizotinib with similar response profiles [42]. It is very probable that the development of the other ALK inhibitors will lead to considerable benefit for the management of *ROS1*-positive lung cancer patients.

18.4 Other Targets

Using highly sensitive methods, *BRAF* mutations are detectable in up to 5 % of patients with lung ADC [43]. Treatment with specific inhibitors may be a valid option in this subgroup of patients, as confirmed by early experience of dabrafenib in lung cancer patients with *BRAF* V600E mutation (32 % RR and PFS of 5.5 months) [44].

RET rearrangements have also been described in lung cancer [45]. These are very infrequent genetic alterations (<2 %), but interesting nevertheless due to the availability of the specific inhibitor vandetanib, which has been associated with significant antitumor activity [46].

Finally, it is worth highlighting the recent breakthroughs that immunotherapy-based checkpoint inhibitors have had in lung cancer. Several monoclonal antibodies targeting the programmed cell death 1 (PD1) protein and its receptor PDL1 are currently under clinical development. Although no robust biomarkers have been identified yet as companion diagnostics for optimal patient selection, these drugs have shown a unique response and survival outcomes in lung cancer patients. Nivolumab, an anti-PD1 antibody, has been approved recently by the FDA for the treatment of SqCC. Approval was based on the preliminary results of a phase III trial in which nivolumab showed improved OS compared to docetaxel in the second-line setting [47].

18.5 Development of Anticancer Drugs and Strategies of the Success of Targeted Therapies

While targeted therapies have certainly improved survival rates among patients with cancer, the prognosis for those with metastatic disease remains extremely poor. This has focused interest on the best way to develop anticancer drugs while also maintaining a positive benefit risk ratio. One such approach is to ensure that tumor biopsies are obligatory, particularly in exploratory clinical studies [48]. This finding is closely linked to the multifaceted mutational processes operating in tumors [49, 50], whereby one small clone of a heterogeneous primary tumor may acquire metastatic abilities that are not seen in the surrounding tumor majority [51], thus potentially making translation IHC analysis insufficient.

Additional approaches ensure that patient baseline samples from clinical trials are carefully stored and catalogued, thus permitting initially unplanned retrospective analysis of newly characterized tumor biomarkers that may have appeared over the course of a long 4–5 year trial period and for cross validation purposes [48]. This would have the additional benefit of allowing subgroup analysis based on newly discovered and useful biomarkers/mutations and, as was the case with panitumumab and gefitinib, could result in approval of a drug that otherwise would have been declined.

Final approach to producing a successful targeted therapy is the addition of a translational research stage before the clinical trial starts. This is not as simple as taking samples and passing them to researchers, but must involve multidisciplinary team discussions to highlight firstly the potential mechanisms of primary or secondary drug resistance and secondly how these can be analyzed at a basic research aspect. Furthermore, this type of interaction can take the form of trying to produce predictive classifiers that can be used to identify patient subgroups that may not respond to treatment using some of the techniques. For example, a gene expression signature of responders versus nonresponders or DNA sequencing analysis to determine whether patients with specific gene mutations do not respond to the treatment can be performed.

Taken together, these fundamental ideas highlight the importance of patient selection and treatment-predictive biomarkers in confirming the success of targeted therapies [52] and signal the new age of personalized therapy. As part of this, clinical studies may in future include patients based on common molecular signatures and mutational patterns or functional properties before subgrouping based on histopathological diagnosis.

18.6 Conclusions

In this chapter, we have highlighted firstly how molecular markers can be examined and secondly how the data can be applied in clinical studies and routine clinical management of patients with NSCLC. Molecular markers are highly relevant in the adjuvant setting for NSCLC. These markers are also used to guide treatment in advanced lung cancer patients. Targeted therapies are not currently recommended in the adjuvant setting for lung cancers; however, a number of drugs are used in metastatic cases. Testing of the mutational status of the *KRAS* gene is recommended before commencing treatment with anti-EGFR therapies, as patients with mutations in this gene will not respond to therapy. The angiogenic inhibitor bevacizumab has also been approved for use in patients. However, the lack of a formal predictive biomarker for the drug means that it is not possible to select patients who are more likely to respond to treatment. Bevacizumab does not improve OS for several malignancies when combined with the best chemotherapy mainstays, and its added value is therefore questionable. EGFR and ALK inhibitors are commonly used to treat patients with NSCLC, meaning that diagnostic tests (such as qPCR and FISH/immunohistochemistry) to determine the presence of mutations and genetic deviations in these genes have become clinically useful before starting targeted treatment.

Generally, this evidence strongly suggests that in the immediate future, clinical diagnostics will use IHC and RNA and DNA-based methods to select patients who will benefit most from targeted treatment regimens.

Conflict of Interest Statement No conflict of interest to declare.

References

1. Siegel R, Ma J, Zou Z, Jemal A (2014) Cancer statistics, 2014. CA Cancer J Clin 64(1):9–29. doi:10.3322/caac.21208
2. National Lung Screening Trial Research, Aberle DR et al (2011) Reduced lung-cancer mortality with low-dose computed tomographic screening. N Engl J Med 365(5):395–409. doi:10.1056/NEJMoa1102873

3. Kovalchik SA, Tammemagi M, Berg CD, Caporaso NE, Riley TL, Korch M, Silvestri GA, Chaturvedi AK, Katki HA (2013) Targeting of low-dose CT screening according to the risk of lung-cancer death. N Engl J Med 369(3):245–254. doi:10.1056/NEJMoa1301851
4. Black WC, Gareen IF, Soneji SS, Sicks JD, Keeler EB, Aberle DR, Naeim A, Church TR, Silvestri GA, Gorelick J, Gatsonis C, National Lung Screening Trial Research Team. (2014) Cost-effectiveness of CT screening in the National Lung Screening Trial. N Engl J Med 371(19):1793–1802. doi:10.1056/NEJMoa1312547
5. Martini N, Bains MS, Burt ME, Zakowski MF, McCormack P, Rusch VW, Ginsberg RJ (1995) Incidence of local recurrence and second primary tumors in resected stage I lung cancer. J Thorac Cardiovasc Surg 109(1):120–129
6. Ujiie H, Kadota K, Chaft JE, Buitrago D, Sima CS, Lee MC, Huang J, Travis WD, Rizk NP, Rudin CM, Jones DR, Adusumilli PS (2015) Solid predominant histologic subtype in resected stage I lung adenocarcinoma is an independent predictor of early, extrathoracic, multisite recurrence and of poor postrecurrence survival. J Clin Oncol 33(26):2877–2884. doi:10.1200/jco.2015.60.9818
7. Surveillance, Epidemiology, and End Results Program http://seer.cancer.gov/statfacts/html/lungb.html
8. Hanahan D, Weinberg RA (2011) Hallmarks of cancer: the next generation. Cell 144(5):646–674. doi:http://dx.doi.org/10.1016/j.cell.2011.02.013
9. Holohan C, Van Schaeybroeck S, Longley DB, Johnston PG (2013) Cancer drug resistance: an evolving paradigm. Nat Rev Cancer 13(10):714–726. doi:10.1038/nrc3599
10. Beatson G (1896) On the treatment of inoperable cases of carcinoma of the mamma: suggestions for a new method of treatment, with illustrative cases. Lancet 148(3803):162-165. doi: http://dx.doi.org/10.1016/S0140-6736(01)72384-7
11. Weiner LM, Surana R, Wang S (2010) Monoclonal antibodies: versatile platforms for cancer immunotherapy. Nat Rev Immunol 10(5):317–327. doi:10.1038/nri2744
12. Zhang J, Yang PL, Gray NS (2009) Targeting cancer with small molecule kinase inhibitors. Nat Rev Cancer 9(1):28–39. doi:10.1038/nrc2559
13. Czajkowsky DM, Hu J, Shao ZF, Pleass RJ (2012) Fc-fusion proteins: new developments and future perspectives. EMBO Mol Med 4(10):1015–1028. doi:10.1002/emmm.201201379
14. Hoadley KA, Yau C, Wolf DM, Cherniack AD, Tamborero D, Ng S, Leiserson MD, Niu B, McLellan MD, Uzunangelov V, Zhang J, Kandoth C, Akbani R, Shen H, Omberg L, Chu A, Margolin AA, Van't Veer LJ, Lopez-Bigas N, Laird PW, Raphael BJ, Ding L, Robertson AG, Byers LA, Mills GB, Weinstein JN, Van Waes C, Chen Z, Collisson EA, Benz CC, Perou CM, Stuart JM (2014) Multiplatform analysis of 12 cancer types reveals molecular classification within and across tissues of origin. Cell 158(4):929–944. doi:10.1016/j.cell.2014.06.049
15. Dowsett M, Sestak I, Lopez-Knowles E, Sidhu K, Dunbier AK, Cowens JW, Ferree S, Storhoff J, Schaper C, Cuzick J (2013) Comparison of PAM50 risk of recurrence score with oncotype DX and IHC4 for predicting risk of distant recurrence after endocrine therapy. J Clin Oncol 31(22):2783–2790. doi:10.1200/jco.2012.46.1558
16. Schena M, Shalon D, Davis RW, Brown PO (1995) Quantitative monitoring of gene expression patterns with a complementary DNA microarray. Science 270(5235):467–470
17. Oberli A, Popovici V, Delorenzi M, Baltzer A, Antonov J, Matthey S, Aebi S, Altermatt HJ, Jaggi R (2008) Expression profiling with RNA from formalin-fixed, paraffin-embedded material. BMC Med Genomics 1:9. doi:10.1186/1755-8794-1-9
18. Mittempergher L, de Ronde JJ, Nieuwland M, Kerkhoven RM, Simon I, Rutgers EJ, Wessels LF, Van't Veer LJ (2011) Gene expression profiles from formalin fixed paraffin embedded breast cancer tissue are largely comparable to fresh frozen matched tissue. PLoS One 6(2):e17163. doi:10.1371/journal.pone.0017163
19. Chibon F (2013) Cancer gene expression signatures – the rise and fall? Eur J Cancer 49(8):2000–2009. doi:10.1016/j.ejca.2013.02.021
20. Sanger F, Nicklen S, Coulson AR (1977) DNA sequencing with chain-terminating inhibitors. Proc Natl Acad Sci U S A 74(12):5463–5467

21. Chmielecki J, Meyerson M (2014) DNA sequencing of cancer: what have we learned? Annu Rev Med 65:63–79. doi:10.1146/annurev-med-060712-200152
22. Shendure J, Ji H (2008) Next-generation DNA sequencing. Nat Biotechnol 26(10):1135–1145. doi:10.1038/nbt1486
23. Bray F, Jemal A, Grey N, Ferlay J, Forman D (2012) Global cancer transitions according to the Human Development Index (2008–2030): a population-based study. Lancet Oncol 13(8):790–801. doi:10.1016/s1470-2045(12)70211-5
24. Ferlay J, Soerjomataram I, Dikshit R, Eser S, Mathers C, Rebelo M, Parkin DM, Forman D, Bray F (2015) Cancer incidence and mortality worldwide: sources, methods and major patterns in GLOBOCAN 2012. Int J Cancer 136(5):E359–E386. doi:10.1002/ijc.29210
25. Tobin NP, Foukakis T, De Petris L, Bergh J (2015) The importance of molecular markers for diagnosis and selection of targeted treatments in patients with cancer. J Intern Med. doi:10.1111/joim.12429
26. Reck M, Popat S, Reinmuth N, De Ruysscher D, Kerr KM, Peters S (2014) Metastatic non-small-cell lung cancer (NSCLC): ESMO clinical practice guidelines for diagnosis, treatment and follow-up. Ann Oncol 25(Suppl 3):iii27–iii39. doi:10.1093/annonc/mdu199
27. Lynch TJ, Bell DW, Sordella R, Gurubhagavatula S, Okimoto RA, Brannigan BW, Harris PL, Haserlat SM, Supko JG, Haluska FG, Louis DN, Christiani DC, Settleman J, Haber DA (2004) Activating mutations in the epidermal growth factor receptor underlying responsiveness of non-small-cell lung cancer to gefitinib. N Engl J Med 350(21):2129–2139. doi:10.1056/NEJMoa040938 NEJMoa040938 [pii]
28. Zhou C, Wu YL, Chen G, Feng J, Liu XQ, Wang C, Zhang S, Wang J, Zhou S, Ren S, Lu S, Zhang L, Hu C, Hu C, Luo Y, Chen L, Ye M, Huang J, Zhi X, Zhang Y, Xiu Q, Ma J, Zhang L, You C (2011) Erlotinib versus chemotherapy as first-line treatment for patients with advanced EGFR mutation-positive non-small-cell lung cancer (OPTIMAL, CTONG-0802): a multicentre, open-label, randomised, phase 3 study. Lancet Oncol 12(8):735–742. doi:10.1016/S1470-2045(11)70184-X
29. Rosell R, Carcereny E, Gervais R, Vergnenegre A, Massuti B, Felip E, Palmero R, Garcia-Gomez R, Pallares C, Sanchez JM, Porta R, Cobo M, Garrido P, Longo F, Moran T, Insa A, De Marinis F, Corre R, Bover I, Illiano A, Dansin E, de Castro J, Milella M, Reguart N, Altavilla G, Jimenez U, Provencio M, Moreno MA, Terrasa J, Munoz-Langa J, Valdivia J, Isla D, Domine M, Molinier O, Mazieres J, Baize N, Garcia-Campelo R, Robinet G, Rodriguez-Abreu D, Lopez-Vivanco G, Gebbia V, Ferrera-Delgado L, Bombaron P, Bernabe R, Bearz A, Artal A, Cortesi E, Rolfo C, Sanchez-Ronco M, Drozdowskyj A, Queralt C, de Aguirre I, Ramirez JL, Sanchez JJ, Molina MA, Taron M, Paz-Ares L (2012) Erlotinib versus standard chemotherapy as first-line treatment for European patients with advanced EGFR mutation-positive non-small-cell lung cancer (EURTAC): a multicentre, open-label, randomised phase 3 trial. Lancet Oncol 13(3):239–246. doi:10.1016/s1470-2045(11)70393-x
30. Mok TS, Wu YL, Thongprasert S, Yang CH, Chu DT, Saijo N, Sunpaweravong P, Han B, Margono B, Ichinose Y, Nishiwaki Y, Ohe Y, Yang JJ, Chewaskulyong B, Jiang H, Duffield EL, Watkins CL, Armour AA, Fukuoka M (2009) Gefitinib or carboplatin-paclitaxel in pulmonary adenocarcinoma. N Engl J Med 361(10):947–957. doi:10.1056/NEJMoa0810699
31. Maemondo M, Inoue A, Kobayashi K, Sugawara S, Oizumi S, Isobe H, Gemma A, Harada M, Yoshizawa H, Kinoshita I, Fujita Y, Okinaga S, Hirano H, Yoshimori K, Harada T, Ogura T, Ando M, Miyazawa H, Tanaka T, Saijo Y, Hagiwara K, Morita S, Nukiwa T, North-East Japan Study Group (2010) Gefitinib or chemotherapy for non-small-cell lung cancer with mutated EGFR. N Engl J Med 362(25):2380–2388. doi:10.1056/NEJMoa0909530
32. Mitsudomi T, Morita S, Yatabe Y, Negoro S, Okamoto I, Tsurutani J, Seto T, Satouchi M, Tada H, Hirashima T, Asami K, Katakami N, Takada M, Yoshioka H, Shibata K, Kudoh S, Shimizu E, Saito H, Toyooka S, Nakagawa K, Fukuoka M (2010) Gefitinib versus cisplatin plus docetaxel in patients with non-small-cell lung cancer harbouring mutations of the epidermal growth factor receptor (WJTOG3405): an open label, randomised phase 3 trial. Lancet Oncol 11(2):121–128. doi:10.1016/s1470-2045(09)70364-x

33. Sequist LV, Yang JC, Yamamoto N, O'Byrne K, Hirsh V, Mok T, Geater SL, Orlov S, Tsai CM, Boyer M, Su WC, Bennouna J, Kato T, Gorbunova V, Lee KH, Shah R, Massey D, Zazulina V, Shahidi M, Schuler M (2013) Phase III study of afatinib or cisplatin plus pemetrexed in patients with metastatic lung adenocarcinoma with EGFR mutations. J Clin Oncol 31(27):3327–3334. doi:10.1200/jco.2012.44.2806
34. Yang JC, Wu YL, Schuler M, Sebastian M, Popat S, Yamamoto N, Zhou C, Hu CP, O'Byrne K, Feng J, Lu S, Huang Y, Geater SL, Lee KY, Tsai CM, Gorbunova V, Hirsh V, Bennouna J, Orlov S, Mok T, Boyer M, Su WC, Lee KH, Kato T, Massey D, Shahidi M, Zazulina V, Sequist LV (2015) Afatinib versus cisplatin-based chemotherapy for EGFR mutation-positive lung adenocarcinoma (LUX-Lung 3 and LUX-Lung 6): analysis of overall survival data from two randomised, phase 3 trials. Lancet Oncol 16(2):141–151. doi:10.1016/s1470-2045(14)71173-8
35. Seto T, Kato T, Nishio M, Goto K, Atagi S, Hosomi Y, Yamamoto N, Hida T, Maemondo M, Nakagawa K, Nagase S, Okamoto I, Yamanaka T, Tajima K, Harada R, Fukuoka M, Yamamoto N (2014) Erlotinib alone or with bevacizumab as first-line therapy in patients with advanced non-squamous non-small-cell lung cancer harbouring EGFR mutations (JO25567): an open-label, randomised, multicentre, phase 2 study. Lancet Oncol 15(11):1236–1244. doi:10.1016/s1470-2045(14)70381-x
36. Steuer CE, Khuri FR, Ramalingam SS (2015) The next generation of epidermal growth factor receptor tyrosine kinase inhibitors in the treatment of lung cancer. Cancer 121(8):E1–E6. doi:10.1002/cncr.29139
37. Thunnissen E, Bubendorf L, Dietel M, Elmberger G, Kerr K, Lopez-Rios F, Moch H, Olszewski W, Pauwels P, Penault-Llorca F, Rossi G (2012) EML4-ALK testing in non-small cell carcinomas of the lung: a review with recommendations. Virchows Archiv 461(3):245–257. doi:10.1007/s00428-012-1281-4
38. Shaw AT, Kim DW, Nakagawa K, Seto T, Crino L, Ahn MJ, De Pas T, Besse B, Solomon BJ, Blackhall F, Wu YL, Thomas M, O'Byrne KJ, Moro-Sibilot D, Camidge DR, Mok T, Hirsh V, Riely GJ, Iyer S, Tassell V, Polli A, Wilner KD, Janne PA (2013) Crizotinib versus chemotherapy in advanced ALK-positive lung cancer. N Engl J Med 368(25):2385–2394. doi:10.1056/NEJMoa1214886
39. Solomon BJ, Mok T, Kim DW, Wu YL, Nakagawa K, Mekhail T, Felip E, Cappuzzo F, Paolini J, Usari T, Iyer S, Reisman A, Wilner KD, Tursi J, Blackhall F (2014) First-line crizotinib versus chemotherapy in ALK-positive lung cancer. N Engl J Med 371(23):2167–2177. doi:10.1056/NEJMoa1408440
40. Camidge DR, Pao W, Sequist LV (2014) Acquired resistance to TKIs in solid tumours: learning from lung cancer. Nat Rev Clin Oncol 11(8):473–481. doi:10.1038/nrclinonc.2014.104
41. Shaw AT, Kim DW, Mehra R, Tan DS, Felip E, Chow LQ, Camidge DR, Vansteenkiste J, Sharma S, De Pas T, Riely GJ, Solomon BJ, Wolf J, Thomas M, Schuler M, Liu G, Santoro A, Lau YY, Goldwasser M, Boral AL, Engelman JA (2014) Ceritinib in ALK-rearranged non-small-cell lung cancer. N Engl J Med 370(13):1189–1197. doi:10.1056/NEJMoa1311107
42. Shaw AT, Ou SH, Bang YJ, Camidge DR, Solomon BJ, Salgia R, Riely GJ, Varella-Garcia M, Shapiro GI, Costa DB, Doebele RC, Le LP, Zheng Z, Tan W, Stephenson P, Shreeve SM, Tye LM, Christensen JG, Wilner KD, Clark JW, Iafrate AJ (2014) Crizotinib in ROS1-rearranged non-small-cell lung cancer. N Engl J Med 371(21):1963–1971. doi:10.1056/NEJMoa1406766
43. Marchetti A, Felicioni L, Malatesta S, Grazia Sciarrotta M, Guetti L, Chella A, Viola P, Pullara C, Mucilli F, Buttitta F (2011) Clinical features and outcome of patients with non-small-cell lung cancer harboring BRAF mutations. J Clin Oncol 29(26):3574–3579. doi:10.1200/jco.2011.35.9638
44. Planchard D, Kim TM, Mazieres J, Quoix E, Riely GJ, Barlesi F, Souquet P, Smit EF, Groen HJM, Kelly RJ, Cho B, Socinski MA, Tucker C, Ma B, Mookerjee B, Curtis CM, Johnson BE (2014) LBA38_PR DABRAFENIB in patients with braf V600E-mutant advanced non-small cell lung cancer (NSCLC): a multicenter, open-label, phase II trial (BRF113928). Ann Oncol 25(suppl 4). doi:10.1093/annonc/mdu438.46

45. Wang R, Hu H, Pan Y, Li Y, Ye T, Li C, Luo X, Wang L, Li H, Zhang Y, Li F, Lu Y, Lu Q, Xu J, Garfield D, Shen L, Ji H, Pao W, Sun Y, Chen H (2012) RET fusions define a unique molecular and clinicopathologic subtype of non-small-cell lung cancer. J Clin Oncol 30(35):4352–4359. doi:10.1200/jco.2012.44.1477
46. Gautschi O, Zander T, Keller FA, Strobel K, Hirschmann A, Aebi S, Diebold J (2013) A patient with lung adenocarcinoma and RET fusion treated with vandetanib. J Thorac Oncol 8(5):e43–e44. doi:10.1097/JTO.0b013e31828a4d07
47. (2015) Nivolumab approved for lung cancer. Cancer Discov 5(5):OF1. doi:10.1158/2159-8290.cd-nb2015-042
48. Jonsson B, Bergh J (2012) Hurdles in anticancer drug development from a regulatory perspective. Nat Rev Clin Oncol 9(4):236–243. doi:10.1038/nrclinonc.2012.14
49. Alexandrov LB, Nik-Zainal S, Wedge DC, Campbell PJ, Stratton MR (2013) Deciphering signatures of mutational processes operative in human cancer. Cell Rep 3(1):246–259. doi:10.1016/j.celrep.2012.12.008
50. Nik-Zainal S, Alexandrov LB, Wedge DC, Van Loo P, Greenman CD, Raine K, Jones D, Hinton J, Marshall J, Stebbings LA, Menzies A, Martin S, Leung K, Chen L, Leroy C, Ramakrishna M, Rance R, Lau KW, Mudie LJ, Varela I, McBride DJ, Bignell GR, Cooke SL, Shlien A, Gamble J, Whitmore I, Maddison M, Tarpey PS, Davies HR, Papaemmanuil E, Stephens PJ, McLaren S, Butler AP, Teague JW, Jonsson G, Garber JE, Silver D, Miron P, Fatima A, Boyault S, Langerod A, Tutt A, Martens JW, Aparicio SA, Borg A, Salomon AV, Thomas G, Borresen-Dale AL, Richardson AL, Neuberger MS, Futreal PA, Campbell PJ, Stratton MR (2012) Mutational processes molding the genomes of 21 breast cancers. Cell 149(5):979–993. doi:10.1016/j.cell.2012.04.024
51. Gerlinger M, Rowan AJ, Horswell S, Larkin J, Endesfelder D, Gronroos E, Martinez P, Matthews N, Stewart A, Tarpey P, Varela I, Phillimore B, Begum S, McDonald NQ, Butler A, Jones D, Raine K, Latimer C, Santos CR, Nohadani M, Eklund AC, Spencer-Dene B, Clark G, Pickering L, Stamp G, Gore M, Szallasi Z, Downward J, Futreal PA, Swanton C (2012) Intratumor heterogeneity and branched evolution revealed by multiregion sequencing. N Engl J Med 366(10):883–892. doi:10.1056/NEJMoa1113205
52. Bergh J (2009) Quo vadis with targeted drugs in the 21st century? J Clin Oncol 27(1):2–5. doi:10.1200/jco.2008.18.8342

Part IV
Novel Approach

Chapter 19
Targeting Epithelial-Mesenchymal Transition and Cancer Stem Cell

Ryota Kurimoto and Yuichi Takiguchi

Abstract Despite standard chemotherapy, resistant cells were appeared in many lung cancer cases. Although the mechanisms to acquire chemoresistance were well studied, it is actually difficult to get over the resistance. Epithelial-to-mesenchymal transition (EMT) is a well-known phenomenon to promote cancer cell to invasiveness, metastasis, and chemoresistance in lung cancer. Underlying mechanisms to induce and restore EMT have been studied, and several drugs could reverse EMT and its drug resistance. Cancer stem cell (CSC) model is recently described as the mechanism to initiate tumor and form the intratumoral heterogeneity in several cancers. In lung cancer, several studies indicated the presence of CSC population. It was also reported that CSCs have the resistance to chemotherapy. The treatment of anti-EMT/CSC has been considered as the key strategy to overcome the resistance through target for residual cancer cells after standard chemotherapy.

Keywords EMT • Cancer stem cell • Lung cancer • Drug resistance

19.1 Introduction

Lung cancer is the leading cause of death worldwide [1]. Many of patients with this cancer are diagnosed at advanced stage and treated with chemotherapy. Recent chemotherapy has significantly longer survival time especially for non-small cell lung cancer (NSCLC) [2, 3]. However, the resistance to these drugs usually occurred in many cases. Although tumor cells were treated by the effective chemotherapy, some tumor cells had originally tolerance or acquired resistance for treated chemotherapy

R. Kurimoto, M.D. (✉) • Y. Takiguchi, M.D., Ph.D.
Department of Medical Oncology, Graduate School of Medicine,
Chiba University, Chiba, Japan
e-mail: rkurimoto.syst@tmd.ac.jp

© Springer Science+Business Media Singapore 2017

Y. Takiguchi (ed.), *Molecular Targeted Therapy of Lung Cancer*,
DOI 10.1007/978-981-10-2002-5_19

in short terms without genetic mutations [4]. These non-mutational mechanisms to drug resistance remain still fully unclear. It has been thought that epithelial-to-mesenchymal transition (EMT) and cancer stem cell (CSC) are mainly causes of these resistance processes [5].

EMT is key pathogenesis on the invasion and metastasis of malignant cancer. In several cancers, EMT induces drug resistance including cytotoxic drugs and molecular targeting drugs [6, 7]. In NSCLC, EMT has also resistance for chemotherapy [8]. Therefore, reversing EMT may be an effective way to improve the response to chemotherapy. Indeed, it has been reported that several drugs could reverse EMT and its drug resistance.

The CSC population was initially proposed in hematopoietic malignancies, which have the self-renewing potentials and maintain the tumor initiation and tumor growth [9]. It has been also investigated that some solid cancers, such as breast cancer [10] and brain tumor [11], had the population of CSC. However, it has been not fully understood in lung cancer. Recently, some populations have expressed specific CSC markers and drug resistances in non-small cell lung cancer cells [12]. Similar to EMT, targeting CSC might overcome the non-mutational drug resistance through target for residual cancer cells after standard chemotherapy.

In this review, we focused on EMT and CSC in lung cancer and its possibility for targeted therapies.

19.2 EMT

19.2.1 Characteristics of EMT

EMT was firstly described as a process that occurred at several scenes of embryonic development [7]. Epithelial cells have a phenotype of tightly adhesive to around cells and a potential to separate the internal organs by formation of "the epithelial barrier." On the other hand, mesenchymal cells have a phenotype of less adhesive and a spindle-like shape. These phenomena of mesenchymal cells enable to make cells highly motile and invasive. In EMT, originally epithelial cells change over to the cells with mesenchymal characteristics [7]. EMT cells have less expression of epithelial markers (E-cadherin and claudin-1) and high expression of mesenchymal markers (vimentin, fibronectin, and N-cadherin). Because of these alterations of cells, EMT plays an important role in the development of organogenesis.

19.2.2 EMT in Cancer

EMT also participates in cancer invasion and metastasis [13]. Many evidences to support the important parts of EMT in carcinogenesis had been reported in vitro, in vivo, and in clinical studies [7, 13, 14]. For induction of EMT, several inducing

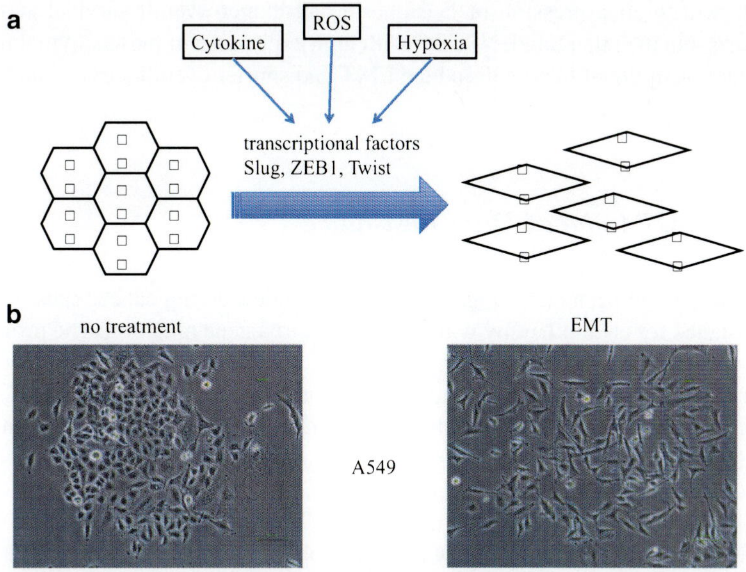

Fig. 19.1 EMT induction. (**a**) Several cytokines, ROS, and hypoxic environment induce EMT through transcriptional factors, slug, ZEB1, twist, and others. (**b**) Lung adenocarcinoma cell line (A549) is induced EMT by TGF-β treatment (10 ng/mL) for 48 h

factors were reported. TGF-ß is the main inducing factor of the EMT in many cancers [15]. TGF-ß signaling activates several cascades including the Smad3, PI3K/Akt/mTOR, and MEK/Erk through different patterns, depending on the cancer cells [16–19]. These activated signaling increases the expression of transcriptional factors such as zinc finger proteins (ZEB1 and ZEB2), basic helix-loop-helix protein (twist), and the snail family (snail, slug). FGF [20], HGF [21], and IL-6 [22] had been also suggested to induce EMT in several cancers. Moreover, the situations of high reactive oxygen species (ROS) [23] and hypoxic environment [24] also induce EMT. Recently, microRNAs, which are noncoding RNAs with the functions of regulating gene expression, also play crucial roles in EMT. MicroRNA-200 family regulates the transcriptional factors ZEB1 and ZEB2, which repress E-cadherin expression [25] (Fig. 19.1).

19.2.3 EMT in Lung Cancer

As similarly to other cancers, it had been also indicated the EMT in lung cancer in vitro and in vivo. In vitro, lung cancer cells altered to EMT cells in some lung cancer cells by means of TGF-β [26], IL-6 [22], and HGF [27]. These EMT cells had high motile and invasive potential. Moreover, poorly prognostic clinicopathological features had been described in NSCLC with EMT features. For example,

NSCLC with high expression of E-cadherin had longer overall survival and less metastasis. On the other hand, NSCLC with high expression of mesenchymal markers and transcriptional factors to induce EMT had shorter overall survival and poor differentiation [8].

19.2.4 EMT-Induced Drug Resistance

In addition, it had been also suggested that EMT-induced lung cancer cells acquire the resistance for chemotherapy without any acquired gene mutation. The profile of less expression of E-cadherin and high expression of N-cadherin was indicated in patients with resistance to chemoradiotherapy including cisplatin in NSCLC [28]. The knockdown of snail inhibited the induction of EMT and its drug resistance to cisplatin in NSCLC cells [29]. These findings strongly support the ability of EMT to induce resistance to cytotoxic drugs in NSCLC. In addition, resistance to EGFR-TKIs is also demonstrated by accumulating evidences [5]. Transcriptional factors including ZEB1 and slug might play a crucial role in these resistances. TGF-β- and EGF-induced EMT acquired resistance to EGFR-TKIs through mTOR and MEK/Erk pathways in vitro [27, 30]; HGF-induced EMT cells acquired drug resistance for gefitinib in NSCLC cells and etoposide in small cell lung cancer (SCLC) cells [21, 31]. IL-6-induced EMT cells acquired drug resistance for gefitinib in NSCLC cells. Shien et al. reported that chronic exposure of gefitinib enables to induce EMT and resistance to EGFR-TKIs without any known mutations [32]. Clinical observations also indicated that several NSCLC with resistance to EGFR-TKIs demonstrated the EMT features in patients with NSCLC harboring EGFR mutation [33] (Fig. 19.2).

19.2.5 Restoration of EMT-Induced Drug Resistance

The restorations of EMT-induced drug resistance were also reported. A MEK/Erk inhibitor and mTOR inhibitor suppressed EMT and improved drug sensitivity in lung cancer cells [27, 34–37]. It was reported that metformin, a widely used drug for diabetes, might decrease the frequency of several cancers [38, 39] and be associated with improved survival among patients with diabetes with stage IV NSCLC [38]. In vitro, it has been reported that metformin suppressed the proliferation of several cancer cells and reverse TGF-ß induced EMT in breast cancer cells [40–43]. In lung cancer, metformin significantly suppressed the regrowth of the tumors after withdrawing gefitinib treatment in xenograft mouse model of lung cancer cell in vivo [44]. Recently, Li L. et al. showed that metformin restored IL-6-induced EMT and drug resistance through inhibition of JAK/STAT3 pathways [22]. In addition, TTF-1 [45], crizotinib [31], which inhibits ALK and MEK in NSCLC, and the eukaryotic initiation factor inhibitor GC7 (N1-guanyl-1,7-diaminoheptane), [46] also decrease

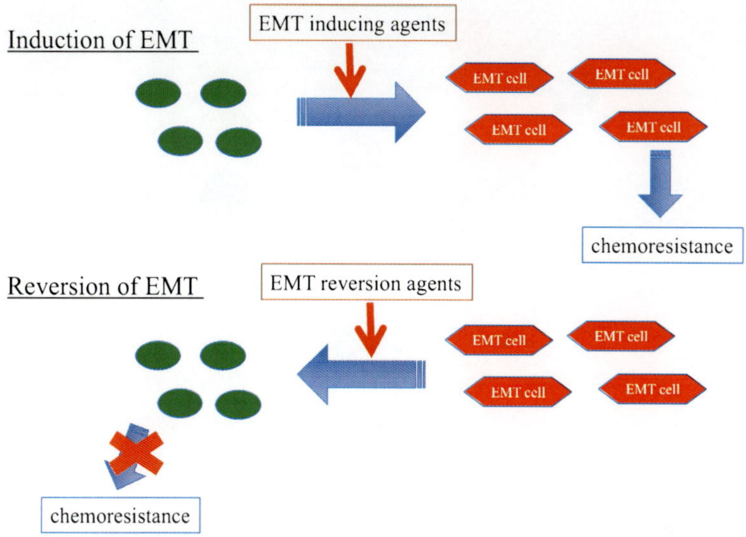

Fig. 19.2 Drug resistance in EMT. Induction of EMT promotes chemoresistance in lung cancer. Restoration of EMT also restores chemoresistance. Targeting EMT is a promising strategy to overcome drug resistance

EMT-induced drug resistance in lung cancer cells. In other cancer, other drugs such as tranilast [47], resveratrol [48], propolis [49], and eribulin [50] have been also reported to affect EMT-induced drug resistance (Fig. 19.2).

19.3 Cancer Stem Cell

19.3.1 Characteristics of Cancer Stem Cell

The concept of CSC has been recently described as one of the mechanisms to establish the intratumoral heterogeneity [51]. Based on the hypothesis, the population of CSCs has the potential to self-renewal and to maintain cancer. Firstly, leukemia-initiating cells in acute myeloid leukemia were identified and these cells enabled to form tumor in SCID mice [9]. These cells had the profile of cellular surface antigen with positive for CD34 and negative for CD38 [52]. Afterward, solid tumor-initiating cells had been also isolated in succession. Cell surface markers vary from types of cancer. Breast cancer-initiating cells describe $CD44^{+}$ and $CD24^{-/low}$ [10]. Brain tumor-initiating cells and colon cancer-initiating cells describe $CD133^{+}$ [11, 53]. Moreover, other markers were also reported in several solid cancers. Any of these tumor-initiating cells enable to originate tumor, guide cells differentiation, and create the heterogeneity of tumor [51] (Fig. 19.3).

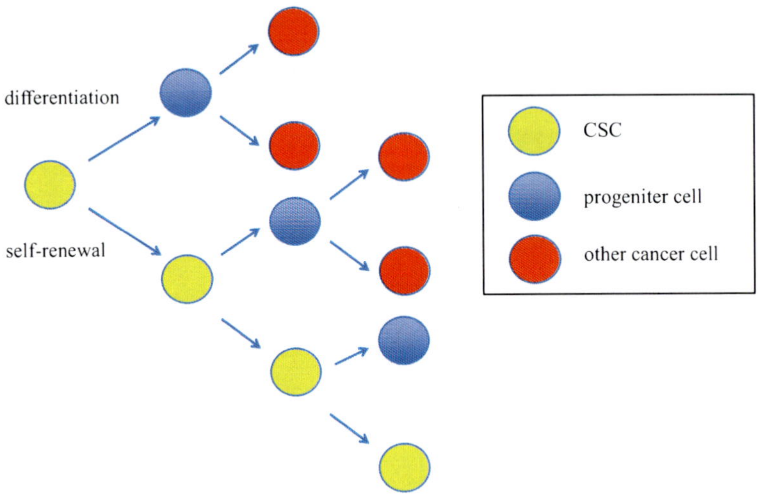

Fig. 19.3 Cancer stem cell model. Cancer stem cell is capable to self-renewal and differentiation. This characteristic causes initiating tumor and intratumoral heterogeneity

19.3.2 Association of EMT and CSC

It has been also described the association between EMT and CSC. In breast cancer cells, TGF-β-induced EMT cells included the cells with the stemness to create mammosphere forming and expression of the specific cell surface markers of $CD44^{+}$ and $CD24^{-/low}$ [54]. CD44 involves in Wnt pathway, which plays an important role in EMT induction in breast cancer cells [55]. Cells with expression of CD133 and CXCR4, a chemokine receptor, had the high invasive potential in mesenchymal pancreas cancer cells [56]. Moreover, it had been also indicated that CSCs had drug resistance in several cancers, which was similar to EMT. These findings might suggest the potent association between EMT and CSCs.

19.3.3 CSC in Lung Cancer

In lung cancer cells, several reports suggested the presence of CSCs with cell surface markers of CD133, CD44, aldehyde dehydrogenase (ALDH), ABCB1, and CXCR4 [12, 57–59]. The profile of these markers still varies from reports. Moreover, it has been also suggested that lung cancer-initiating cells linked to EMT phenotype. Ectopic expressions of Oct4 and Nanog, which were some of transcriptional factors to make cells stemness in several cancers, induced high frequency of CD133-positive cells, sphere formation, drug resistance, and EMT with slug expression in lung cancer [60]. TGF-β treatment induced not only EMT but also sphere formation and expression of Oct4, Nanog, and CD133 in lung cancer cells [61].

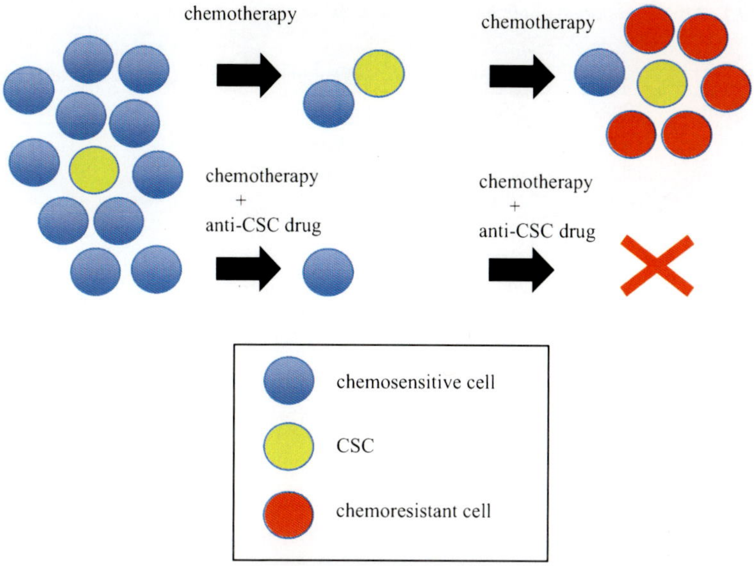

Fig. 19.4 Targeting cancer stem cell strategy. It has been thought that cancer stem cells (CSCs) have originally resistance to standard chemotherapies. Residual cancer cells after chemotherapy might induce acquired resistance to standard chemotherapies. Targeting CSC might overcome this process

19.3.4 Overcome CSC-Induced Drug Resistance

Recently, some treatment strategies for CSCs were indicated. Inhibition of TGF-β type I receptor with paclitaxel treatment improved the antitumor efficacy in CSC of breast cancer cells [62]. Inhibition of nodal/Activin receptor Alk4/Alk7 suppressed CSCs and increased drug sensitivity to gemcitabine in pancreas cancer [63]. BMP4 was one of the differentiation factors of colorectal cancer, and it suppressed the CSCs and improved drug sensitivity to 5-fluorouracil and oxaliplatin in colorectal cancer cells [64]. Histone deacetylates (HDAC) was also reported to differentiate cells to mesenchymal and induce CSCs [65, 66]. HDAC inhibitor was reported to suppress the CSCs in chronic myelogenous leukemia and some solid tumors in preclinical studies [67]. Notch signaling and cyclin-D1 pathway, which regulated cell cycle, induced EMT and are related to CSCs in breast cancer cells [68, 69]. Notch inhibitor was also studied. In NSCLC cells, it was reported that inhibition of Notch signaling [70, 71]; checkpoint protein kinase (Chk1) [72]; Bcl-XL, one of the anti-apoptosis factors [73]; all-trans retinoic acid [74]; and trifluoperazine, one of an antipsychotic agents [75], inhibited CSC growth and suppressed the acquisition of drug resistance in preclinical studies (Fig. 19.4).

In clinical, some drugs have been studied in a phase 1 or 2 clinical trial in patients with lung cancer (Table 19.1). The combination with romidepsin, a HDAC inhibitor, and erlotinib was reported in phase 1 clinical trial [76]. The combination of

Table 19.1 Already reported clinical trials of anti-CSC drug for lung cancer

Drug	Target	Histology	Combination	Clinical trial	References
Romidepsin	HDAC	NSCLC	Erlotinib	1	[76]
Decitabine	HDAC	NSCLC	Valproic acid	1	[77]
Demcizumab	DLL4 (notch)	non-Sq NSCLC	CBDCA+PEM	1	[78]
Selumetinib	MEK	NSCLC with KRAS mutation	DOC	2	[79, 80]
Tarextumab	Notch	SCLC	CDDP+ETP	1	[81]

CSC cancer stem cell, *HDAC* histone deacetylates, *DLL4* delta-like ligand 4, *CBDCA* carboplatin, *PEM* pemetrexed, *DOC* docetaxel, *CDDP* cisplatin, *ETP* etoposide

decitabine, a HDAC inhibitor, and valproic acid was also studied in phase 1 clinical trial for NSCLC [77]. Phase 1 clinical trial of demcizumab, which inhibits delta-like ligand 4 (DLL4) in the Notch signaling pathway, with carboplatin and pemetrexed for non-squamous NSCLC has been reported at ASCO 2015 Annual Meeting [78]. Phase 2 clinical trial of demcizumab has been initiated. Selumetinib, a MEK inhibitor, with docetaxel was studied for patients with NSCLC harboring KRAS mutation in phase 2 clinical trial [79, 80]. Selumetinib was also reported for maintenance therapy after platinum doublet therapy for patients with NSCLC at ASCO 2015 Annual Meeting [81]. Phase 1 clinical trial of tarextumab, a Notch inhibitor, was presented in patients with SCLC at ASCO 2015 Annual Meeting [81]. Phase 2 clinical trial of tarextumab has been initiated. Moreover, several other drugs were also studied in phase 1 trial for solid tumors.

19.4 Immuno-protective in EMT and CSCs

The immune-check point therapy is noteworthy treatment in several cancers. Programmed death -1 (PD-1) receptor is one of the targets of these treatments [82]. PD-1 receptor and PD-L1 expression play the important role for immune escape mechanism [82]. However, the characteristics of tumor cells with sensitivity for immune-check point therapy are still unknown. PD-L1 expression is now being watched with interest. Recently, a link between the induction of EMT and the overexpression of PD-L1 was reported. Chen L. et al. reported that ZEB1 promoted metastasis and increased expression of PD-L1 through inhibition of microRNA-200 in lung cancer cells [83]. Alsuliman A. et al. indicated that PD-L1 expression increased in TGF-ß-driven EMT cell and decreased by inhibition of PI3K or Erk in breast cancer cells [84]. In lung cancer, Ota K. et al. indicated that PI3K/Akt and MEK/Erk pathways mediate PD-L1 expression in NSCLC cells with either ALK-translocation or EGFR mutation [85]. Recently, we also demonstrated a close relationship between PD-L1 expression and EMT induction/reversion by several drugs [86]. These findings might suggest the possibility of EMT-related immune suppression and modification of immune-check point therapy. Further investing is needed.

19.5 Conclusion

EMT and CSCs were indicated as the key phenomena of acquisition or potentially resistance to chemotherapies. Therefore, the new strategies of combination with anti-EMT/CSCs treatment with chemotherapy might reduce the residual cells after chemotherapy and overcome the acquisition of resistance. Driving cancer cells into extinction is expected by further investigations.

References

1. Ferlay J, Shin HR, Bray F et al (2010) Estimates of worldwide burden of cancer in 2008: GLOBOCAN 2008. Int J Cancer 127:2893–2917
2. Rosell R, Carcereny E, Gervais R et al (2012) Erlotinib versus standard chemotherapy as first-line treatment for European patients with advanced EGFR mutation-positive non-small-cell lung cancer (EURTAC): a multicentre, open-label, randomised phase 3 trial. Lancet Oncol 13:239–246
3. Maemondo M, Inoue A, Kobayashi K et al (2010) Gefitinib or chemotherapy for non-small-cell lung cancer with mutated EGFR. N Engl J Med 362:2380–2388
4. Sharma SV, Lee DY, Li B et al (2010) A chromatin-mediated reversible drug-tolerant state in cancer cell subpopulations. Cell 141:69–80
5. Thomson S, Buck E, Petti F et al (2005) Epithelial to mesenchymal transition is a determinant of sensitivity of non-small-cell lung carcinoma cell lines and xenografts to epidermal growth factor receptor inhibition. Cancer Res 65:9455–9462
6. Kajiyama H, Shibata K, Terauchi M et al (2007) Chemoresistance to paclitaxel induces epithelial-mesenchymal transition and enhances metastatic potential for epithelial ovarian carcinoma cells. Int J Oncol 31:277–283
7. Thiery JP, Acloque H, Huang RY et al (2009) Epithelial-mesenchymal transitions in development and disease. Cell 139:871–890
8. Sato M, Shames DS, Hasegawa Y (2012) Emerging evidence of epithelial-to-mesenchymal transition in lung carcinogenesis. Respirology 17:1048–1059
9. Bonnet D, Dick JE (1997) Human acute myeloid leukemia is organized as a hierarchy that originates from a primitive hematopoietic cell. Nat Med 3:730–737
10. Al-Hajj M, Wicha MS, Benito-Hernandez A et al (2003) Prospective identification of tumorigenic breast cancer cells. Proc Natl Acad Sci U S A 100:3983–3988
11. Singh SK, Hawkins C, Clarke ID et al (2004) Identification of human brain tumour initiating cells. Nature 432:396–401
12. Liu J, Xiao Z, Wong SK et al (2013) Lung cancer tumorigenicity and drug resistance are maintained through ALDH(hi)CD44(hi) tumor initiating cells. Oncotarget 4:1698–1711
13. Thiery JP (2002) Epithelial-mesenchymal transitions in tumour progression. Nat Rev Cancer 2:442–454
14. Prall F (2007) Tumour budding in colorectal carcinoma. Histopathology 50:151–162

15. Gavert N, Ben-Ze'ev A (2008) Epithelial-mesenchymal transition and the invasive potential of tumors. Trends Mol Med 14:199–209
16. Grande M, Franzen A, Karlsson JO et al (2002) Transforming growth factor-beta and epidermal growth factor synergistically stimulate epithelial to mesenchymal transition (EMT) through a MEK-dependent mechanism in primary cultured pig thyrocytes. J Cell Sci 115:4227–4236
17. Chen XF, Zhang HJ, Wang HB et al (2012) Transforming growth factor-beta1 induces epithelial-to-mesenchymal transition in human lung cancer cells via PI3K/Akt and MEK/Erk1/2 signaling pathways. Mol Biol Rep 39:3549–3556
18. Shi Y, Massague J (2003) Mechanisms of TGF-beta signaling from cell membrane to the nucleus. Cell 113:685–700
19. Derynck R, Zhang YE (2003) Smad-dependent and Smad-independent pathways in TGF-beta family signalling. Nature 425:577–584
20. Shirakihara T, Horiguchi K, Miyazawa K et al (2011) TGF-beta regulates isoform switching of FGF receptors and epithelial-mesenchymal transition. EMBO J 30:783–795
21. Ishikawa D, Takeuchi S, Nakagawa T et al (2013) mTOR inhibitors control the growth of EGFR mutant lung cancer even after acquiring resistance by HGF. PLoS One 8:e62104
22. Li L, Han R, Xiao H et al (2014) Metformin sensitizes EGFR-TKI-resistant human lung cancer cells in vitro and in vivo through inhibition of IL-6 signaling and EMT reversal. Clin Cancer Res 20:2714–2726
23. Zhang KH, Tian HY, Gao X et al (2009) Ferritin heavy chain-mediated iron homeostasis and subsequent increased reactive oxygen species production are essential for epithelial-mesenchymal transition. Cancer Res 69:5340–5348
24. Mimeault M, Batra SK (2013) Hypoxia-inducing factors as master regulators of stemness properties and altered metabolism of cancer- and metastasis-initiating cells. J Cell Mol Med 17:30–54
25. Gibbons DL, Lin W, Creighton CJ et al (2009) Expression signatures of metastatic capacity in a genetic mouse model of lung adenocarcinoma. PLoS One 4:e5401
26. Zhang HJ, Wang HY, Zhang HT et al (2011) Transforming growth factor-beta1 promotes lung adenocarcinoma invasion and metastasis by epithelial-to-mesenchymal transition. Mol Cell Biochem 355:309–314
27. Buonato JM, Lazzara MJ (2014) ERK1/2 blockade prevents epithelial-mesenchymal transition in lung cancer cells and promotes their sensitivity to EGFR inhibition. Cancer Res 74:309–319
28. Shintani Y, Okimura A, Sato K et al (2011) Epithelial to mesenchymal transition is a determinant of sensitivity to chemoradiotherapy in non-small cell lung cancer. Ann Thorac Surg 92:1794–1804; discussion 1804
29. Zhuo W, Wang Y, Zhuo X et al (2008) Knockdown of Snail, a novel zinc finger transcription factor, via RNA interference increases A549 cell sensitivity to cisplatin via JNK/mitochondrial pathway. Lung Cancer 62:8–14
30. Yu HG, Wei W, Xia LH et al (2013) FBW7 upregulation enhances cisplatin cytotoxicity in non- small cell lung cancer cells. Asian Pac J Cancer Prev 14:6321–6326
31. Canadas I, Rojo F, Taus A et al (2014) Targeting epithelial-to-mesenchymal transition with Met inhibitors reverts chemoresistance in small cell lung cancer. Clin Cancer Res 20:938–950
32. Shien K, Toyooka S, Yamamoto H et al (2013) Acquired resistance to EGFR inhibitors is associated with a manifestation of stem cell-like properties in cancer cells. Cancer Res 73:3051–3061
33. Sequist LV, Waltman BA, Dias-Santagata D et al (2011) Genotypic and histological evolution of lung cancers acquiring resistance to EGFR inhibitors. Sci Transl Med 3:75ra26
34. Lamouille S, Derynck R (2007) Cell size and invasion in TGF-beta-induced epithelial to mesenchymal transition is regulated by activation of the mTOR pathway. J Cell Biol 178:437–451

35. Lamouille S, Connolly E, Smyth JW et al (2012) TGF-beta-induced activation of mTOR complex 2 drives epithelial-mesenchymal transition and cell invasion. J Cell Sci 125:1259–1273
36. Maru S, Ishigaki Y, Shinohara N et al (2013) Inhibition of mTORC2 but not mTORC1 up-regulates E-cadherin expression and inhibits cell motility by blocking HIF-2alpha expression in human renal cell carcinoma. J Urol 189:1921–1929
37. Kim EY, Kim A, Kim SK et al (2014) Inhibition of mTORC1 induces loss of E-cadherin through AKT/GSK-3beta signaling-mediated upregulation of E-cadherin repressor complexes in non-small cell lung cancer cells. Respir Res 15:26
38. Lin JJ, Gallagher EJ, Sigel K et al (2015) Survival of patients with stage IV lung cancer with diabetes treated with metformin. Am J Respir Crit Care Med 191:448–454
39. Evans JM, Donnelly LA, Emslie-Smith AM et al (2005) Metformin and reduced risk of cancer in diabetic patients. BMJ 330:1304–1305
40. Gotlieb WH, Saumet J, Beauchamp MC et al (2008) In vitro metformin anti-neoplastic activity in epithelial ovarian cancer. Gynecol Oncol 110:246–250
41. Rocha GZ, Dias MM, Ropelle ER et al (2011) Metformin amplifies chemotherapy-induced AMPK activation and antitumoral growth. Clin Cancer Res 17:3993–4005
42. Ben Sahra I, Laurent K, Loubat A et al (2008) The antidiabetic drug metformin exerts an antitumoral effect in vitro and in vivo through a decrease of cyclin D1 level. Oncogene 27:3576–3586
43. Cufi S, Vazquez-Martin A, Oliveras-Ferraros C et al (2010) Metformin against TGFbeta-induced epithelial-to-mesenchymal transition (EMT): from cancer stem cells to aging-associated fibrosis. Cell Cycle 9:4461–4468
44. Kitazono S, Takiguchi Y, Ashinuma H et al (2013) Effect of metformin on residual cells after chemotherapy in a human lung adenocarcinoma cell line. Int J Oncol 43:1846–1854
45. Saito RA, Watabe T, Horiguchi K et al (2009) Thyroid transcription factor-1 inhibits transforming growth factor-beta-mediated epithelial-to-mesenchymal transition in lung adenocarcinoma cells. Cancer Res 69:2783–2791
46. Xu G, Yu H, Shi X et al (2014) Cisplatin sensitivity is enhanced in non-small cell lung cancer cells by regulating epithelial-mesenchymal transition through inhibition of eukaryotic translation initiation factor 5A2. BMC Pulm Med 14:174
47. Darakhshan S, Ghanbari A (2013) Tranilast enhances the anti-tumor effects of tamoxifen on human breast cancer cells in vitro. J Biomed Sci 20:76
48. Shi XP, Miao S, Wu Y et al (2013) Resveratrol sensitizes tamoxifen in antiestrogen-resistant breast cancer cells with epithelial-mesenchymal transition features. Int J Mol Sci 14:15655–15668
49. Kao HF, Chang-Chien PW, Chang WT et al (2013) Propolis inhibits TGF-beta1-induced epithelial-mesenchymal transition in human alveolar epithelial cells via PPARgamma activation. Int Immunopharmacol 15:565–574
50. Yoshida T, Ozawa Y, Kimura T et al (2014) Eribulin mesilate suppresses experimental metastasis of breast cancer cells by reversing phenotype from epithelial-mesenchymal transition (EMT) to mesenchymal-epithelial transition (MET) states. Br J Cancer 110:1497–1505
51. Nguyen LV, Vanner R, Dirks P et al (2012) Cancer stem cells: an evolving concept. Nat Rev Cancer 12:133–143
52. Lapidot T, Sirard C, Vormoor J et al (1994) A cell initiating human acute myeloid leukaemia after transplantation into SCID mice. Nature 367:645–648
53. Todaro M, Alea MP, Di Stefano AB et al (2007) Colon cancer stem cells dictate tumor growth and resist cell death by production of interleukin-4. Cell Stem Cell 1:389–402
54. Mani SA, Guo W, Liao MJ et al (2008) The epithelial-mesenchymal transition generates cells with properties of stem cells. Cell 133:704–715
55. Wielenga VJ, Smits R, Korinek V et al (1999) Expression of CD44 in Apc and Tcf mutant mice implies regulation by the WNT pathway. Am J Pathol 154:515–523

56. Hermann PC, Huber SL, Herrler T et al (2007) Distinct populations of cancer stem cells determine tumor growth and metastatic activity in human pancreatic cancer. Cell Stem Cell 1:313–323
57. Freitas DP, Teixeira CA, Santos-Silva F et al (2014) Therapy-induced enrichment of putative lung cancer stem-like cells. Int J Cancer 134:1270–1278
58. Sugano T, Seike M, Noro R et al (2015) Inhibition of ABCB1 overcomes cancer stem cell-like properties and acquired resistance to MET inhibitors in non-small cell lung cancer. Mol Cancer Ther 14:2433–2440
59. Koren A, Motaln H, Cufer T (2013) Lung cancer stem cells: a biological and clinical perspective. Cell Oncol (Dordr) 36:265–275
60. Chiou SH, Wang ML, Chou YT et al (2010) Coexpression of Oct4 and Nanog enhances malignancy in lung adenocarcinoma by inducing cancer stem cell-like properties and epithelial-mesenchymal transdifferentiation. Cancer Res 70:10433–10444
61. Pirozzi G, Tirino V, Camerlingo R et al (2011) Epithelial to mesenchymal transition by TGFbeta-1 induction increases stemness characteristics in primary non small cell lung cancer cell line. PLoS One 6:e21548
62. Bhola NE, Balko JM, Dugger TC et al (2013) TGF-beta inhibition enhances chemotherapy action against triple-negative breast cancer. J Clin Invest 123:1348–1358
63. Lonardo E, Hermann PC, Mueller MT et al (2011) Nodal/Activin signaling drives self-renewal and tumorigenicity of pancreatic cancer stem cells and provides a target for combined drug therapy. Cell Stem Cell 9:433–446
64. Lombardo Y, Scopelliti A, Cammareri P et al (2011) Bone morphogenetic protein 4 induces differentiation of colorectal cancer stem cells and increases their response to chemotherapy in mice. Gastroenterology 140:297–309
65. Bolden JE, Peart MJ, Johnstone RW (2006) Anticancer activities of histone deacetylase inhibitors. Nat Rev Drug Discov 5:769–784
66. Zhang B, Strauss AC, Chu S et al (2010) Effective targeting of quiescent chronic myelogenous leukemia stem cells by histone deacetylase inhibitors in combination with imatinib mesylate. Cancer Cell 17:427–442
67. Lane AA, Chabner BA (2009) Histone deacetylase inhibitors in cancer therapy. J Clin Oncol 27:5459–5468
68. Schott AF, Landis MD, Dontu G et al (2013) Preclinical and clinical studies of gamma secretase inhibitors with docetaxel on human breast tumors. Clin Cancer Res 19:1512–1524
69. Qiu M, Peng Q, Jiang I et al (2013) Specific inhibition of Notch1 signaling enhances the antitumor efficacy of chemotherapy in triple negative breast cancer through reduction of cancer stem cells. Cancer Lett 328:261–270
70. Yuan X, Wu H, Han N et al (2014) Notch signaling and EMT in non-small cell lung cancer: biological significance and therapeutic application. J Hematol Oncol 7:87
71. Maraver A, Fernandez-Marcos PJ, Herranz D et al (2012) Therapeutic effect of gamma-secretase inhibition in KrasG12V-driven non-small cell lung carcinoma by derepression of DUSP1 and inhibition of ERK. Cancer Cell 22:222–234
72. Bartucci M, Svensson S, Romania P et al (2012) Therapeutic targeting of Chk1 in NSCLC stem cells during chemotherapy. Cell Death Differ 19:768–778
73. Zeuner A, Francescangeli F, Contavalli P et al (2014) Elimination of quiescent/slow-proliferating cancer stem cells by Bcl-XL inhibition in non-small cell lung cancer. Cell Death Differ 21:1877–1888
74. Moro M, Bertolini G, Pastorino U et al (2015) Combination treatment with all-trans retinoic acid prevents cisplatin-induced enrichment of CD133+ tumor-initiating cells and reveals heterogeneity of cancer stem cell compartment in lung cancer. J Thorac Oncol 10:1027–1036
75. Yeh CT, Wu AT, Chang PM et al (2012) Trifluoperazine, an antipsychotic agent, inhibits cancer stem cell growth and overcomes drug resistance of lung cancer. Am J Respir Crit Care Med 186:1180–1188

76. Gerber DE, Boothman DA, Fattah FJ et al (2015) Phase 1 study of romidepsin plus erlotinib in advanced non-small cell lung cancer. Lung Cancer 90:534–541
77. Chu BF, Karpenko MJ, Liu Z et al (2013) Phase I study of 5-aza-2′-deoxycytidine in combination with valproic acid in non-small-cell lung cancer. Cancer Chemother Pharmacol 71:115–121
78. ASCO, Hidalgo M, Cooray P, Jameson MB et al (2015) A phase Ib study of the anti-cancer stem cell agent demcizumab (DEM) & gemcitabine (GEM) +/– paclitaxel protein bound particles (nab-paclitaxel) in pts with pancreatic cancer. J Clin Oncol 33 (suppl; abstr 4118)
79. Janne PA, Smith I, McWalter G et al (2015) Impact of KRAS codon subtypes from a randomised phase II trial of selumetinib plus docetaxel in KRAS mutant advanced non-small-cell lung cancer. Br J Cancer 113:199–203
80. Janne PA, Shaw AT, Pereira JR et al (2013) Selumetinib plus docetaxel for KRAS-mutant advanced non-small-cell lung cancer: a randomised, multicentre, placebo-controlled, phase 2 study. Lancet Oncol 14:38–47
81. ASCO, Pietanza MC, Spira AI, Jotte RM et al (2015) Final results of phase Ib of tarextumab (TRXT, OMP-59R5, anti-Notch2/3) in combination with etoposide and platinum (EP) in patients (pts) with untreated extensive-stage small-cell lung cancer (ED-SCLC). Clin Oncol 33 (suppl; abstr 7508)
82. Ribas A (2012) Tumor immunotherapy directed at PD-1. N Engl J Med 366:2517–2519
83. Chen L, Gibbons DL, Goswami S et al (2014) Metastasis is regulated via microRNA-200/ZEB1 axis control of tumour cell PD-L1 expression and intratumoral immunosuppression. Nat Commun 5:5241
84. Alsuliman A, Colak D, Al-Harazi O et al (2015) Bidirectional crosstalk between PD-L1 expression and epithelial to mesenchymal transition: significance in claudin-low breast cancer cells. Mol Cancer 14:149
85. Ota K, Azuma K, Kawahara A et al (2015) Induction of PD-L1 Expression by the EML4-ALK oncoprotein and downstream signaling pathways in non-small cell lung cancer. Clin Cancer Res 21:4014–4021
86. Ryota Kurimoto, Shunichiro Iwasawa, Takahiro Ebata, et al (2016) Drug resistance originating from a TGF-ß/FGF-2-driven epithelial-tomesenchymal transition and its reversion in human lung adenocarcinoma cell lines harboring an EGFR mutation. Int J Oncol 48:1825–1836

Chapter 20
Targeting the Lung Cancer Microenvironment: Harnessing Host Responses

Mark M. Fuster

Abstract Understanding the host response to lung cancer is critical in the development of long-term therapeutic responses and cures for advanced-stage disease. While state-of-the-art treatments that target the tumor cell directly are effective as initial antitumor approaches, strategies that augment antitumor host responses are highly appealing, and may overcome resistance through novel discoveries. These involve (1) discovery of basic mechanisms by which the tumor "hijacks" host immune regulation and vascular homeostasis (thus promoting tumor growth), and (2) discovery of tumor-resistance pathways that counter immune- and/or vascular-targeting therapies. Major mechanisms by which lung carcinoma is able to usurp host mechanisms include both the tumor's manipulation of immune checkpoint regulatory pathways (with a cytokine and dendritic cell balance that maintains a high suppressor/effector T-cell ratio) and the remodeling of blood and lymphatic vasculature by multiple endothelial mitogens, thereby promoting tumor growth and dissemination. Lymphatic dissemination in particular involves not only tumor cells but also immunosuppressive dendritic cell trafficking to tumor-draining lymph nodes. Novel approaches to overcome these challenges include immune checkpoint-blocking strategies (e.g., PD-1/PD-L1 or CTLA4 blockade which inhibit T-effector suppression) or agonists to T-stimulatory pathways, such as OX40 or 4-1BB. They also include vaccine development and/or approaches to manipulate dendritic cells or engineer T cells (e.g., CAR-T cells) against antigens that are (preferably) clonally expressed by the entire tumor. Major limitations to these approaches include poor tumor-antigen recognition or presentation by dendritic cells or hyporesponsive T cells in the immunosuppressive tumor microenvironment. Moreover, autoimmune-type side effects of immune checkpoint T-cell targeting or T-cell engineering present therapeutic challenges. Finally, the discovery of tumor neo-antigens, which are known to be more abundantly expressed in tumors initiated by environmental stimuli (e.g., melanoma or squamous lung carcinoma), as well as their ability to

M.M. Fuster, M.D. (✉)
Division of Pulmonary & Critical Care, Department of Medicine, VA San Diego Healthcare System, University of California, San Diego,
3350 La Jolla Village Dr., San Diego, CA 92161, USA
e-mail: mfuster@ucsd.edu

© Springer Science+Business Media Singapore 2017

Y. Takiguchi (ed.), *Molecular Targeted Therapy of Lung Cancer*,
DOI 10.1007/978-981-10-2002-5_20

predict T cell responsiveness, is another important development in the quest to augment host immune responses to lung cancer. These discoveries will be valuable in promoting a set of strategies that markedly improve the chances for durable remissions or cures in the setting of advanced-stage lung cancer or even recurrent disease following definitive treatments.

Keywords Lung cancer • Immunity • Lymphatic • Neo-antigen • Host

20.1 Introduction: Lung Cancer Aggression and Mortality – The Rise of Host-Modifying Therapies

20.1.1 Lung Cancer Mortality and Metastatic Mechanisms

Lung cancer is the leading cause of cancer death in the USA and in the world [1], with 5-year survival rates that have not improved appreciably above 17 % and remaining <20 % for several decades. While tobacco cessation, improvements in industrial agent exposure controls, and lung cancer screening provide some of the most obvious modifiable factors for society, their advancement has historically remained slow and frustrating. In the USA, the overwhelming majority of lung cancer presentations are cases of advanced-stage disease, with metastatic disease (>50 %) or regional spread (~25 %), with 5-year survivals <10 % and markedly under 50 %, respectively [2]. To some extent, while environmental insults remain challenging to curb, low-dose CT (LDCT) screening has an increasing role in preventing presentation with advanced-stage disease. This technology has not only demonstrated category-1 evidence for a lowering of lung cancer mortality but also the ability to shift the stage of lung cancer detection so as to intervene curatively at an earlier stage [3]. Despite this, the metastatic aggression of lung cancer remains so high that detection at an early stage results in 5-year survivals that are modest at best. For these reasons, novel therapies that revolutionize beyond direct cancer cell-targeted metabolic or toxic inhibitors remain a great promise. A key concept we among others envision is achieving a marked increase in cancer control through harnessing the host response to cancer. The latter involves a variety of mechanisms from vascular to immunologic [4, 5]. We focus herein upon an overview of these potential approaches, along with current best practices in augmenting the host response as well as novel considerations in strategies to promote exogenous and endogenous immune augmentation through new molecular driving systems.

We know that two major host responses in lung cancer are the growth of vasculature into the tumor as well as infiltration of immunologic cells derived from the host. The latter are not always "good," and as we shall discuss, the tumor drives both vascular conduit and infiltration by a mostly subversive collection of dendritic immune cells as well as domination by macrophages that promote angiogenesis (M2 type) [6]. The subversive dendritic cells that infiltrate tumors drive apoptosis of effector T cells (through co-inhibitory pathway activation via the immune checkpoint pathway) as well as infiltration by suppressor-type T-regulatory (Treg) cells that effectively inhibit tumor immunity [4].

20.1.2 Modifying Vasculature: The Conduit for Metastasis, Therapy, and Immunity

VEGF-blocking approaches and challenges in targeting vascular tyrosine kinase signaling: In the last decade, therapies targeting VEGF-A have shown real, albeit modest efficacy in the inhibition of tumor endothelial remodeling [5]. This began with the humanized anti-VEGF antibody, bevacizumab, with clinical efficacy noted originally in colon cancer, followed by modest gains that it provided in non-squamous advanced-stage non-small cell lung cancer (NSCLC). Further developments led to receptor-blocking antibodies, such as ramucirumab, and most recently clinical trials have examined the efficacy of multi-target tyrosine kinase inhibitors such as nintedanib [7]. This triple angiokinase inhibitor of VEGF, FGF-2, and PDGF also has some lesser signaling inhibition of RET, Src, and Flt-3 and has been used in combination with systemic chemotherapy regimens for added antitumor efficacy in lung cancer [8]. The current use of angiogenesis-blocking therapies in lung cancer includes applications in advanced-stage non-squamous NSCLC as part of multidrug cytotoxic therapy [5]. An attractive area of future development also examines possible biomarkers that predict responders to standard VEGF–VEGFR2-blocking therapies [9]. There has also been some early development of lymphatic endothelial growth receptor (VEGFR-3) blocking agents that might be used in future adjuvant treatment platforms to interfere with lymphatic remodeling during tumor growth and progression [10].

20.2 Harnessing the Host: What Is Available and What Efforts Are Under Way?

20.2.1 First, A Focus on the Cancer-Cell Revolution: Promise and Challenge of Targeted Therapies

The emergence of targeted therapies that antagonize overexpression of specific mutant kinase pathways in the cancer cell has revolutionized therapeutics for advanced-stage NSCLC. This is particularly applicable to non-squamous NSCLC, where the incidence of two of the most prevalent and treatable "driving" mutations (i.e., EGFR and ALK mutations) are found at incidence rates of anywhere from 7 % in American smokers to 39 % in American never-smokers, to as high as 73 % in Asian never-smokers [11]. On the other hand, tumors with diffuse histologic positivity for squamous cell carcinoma uniformly lack sensitivity to these agents and thus are generally omitted from testing in clinical practice. Newer molecular approaches are under development to target unique KRAS mutations which are also expressed by squamous cell carcinomas; and novel panels of miRNA prognostic markers are emerging as promising predictors of prognosis in squamous NSCLC [12]. These discoveries bring a new wave of opportunity in the challenge of treating advanced-stage squamous NSCLC.

The IPASS study by Mok, et al. (2009) demonstrated that so long as we are careful to test for EGFR mutations that confer unique sensitivity to kinase inhibitors such as erlotinib and gefitinib, first-line treatment with these agents in mutant-positive metastatic NSCLC offers equal or better outcomes as conventional chemotherapy with lesser overall toxicity [13]. This was particularly true in a high-incidence Asian population, where initial phase III randomized trials were carried out. Since outcomes for first-line therapy with these agents were highly favorable in patients harboring "sensitizing" mutations, while outcomes with the same therapy in mutation-negative patients were very poor (wherein conventional chemotherapy was more favorable), the availability of testing was critical for subsequent clinical decision-making [13]. A concept that we must keep in mind is that the introduction of these agents has allowed for marked improvements in progression-free survival (PFS). In randomized trials, while there have been challenges in extending overall survival (OS) with a single agent in a respective cohort of sensitive mutants (comparing to conventional-chemotherapy controls), we highlight that significant improvements in the quality of life during the PFS period by such agents are important. Moreover, innovations in the efficacy of newer agents along with the ability to switch/add new targeted therapies as the tumor evolves (and develops resistance to a given agent) may now contribute to gains in OS as a result of these novel approaches, when compared to conventional chemotherapy [14]. Further studies will be important in establishing and improving upon these new trends.

20.2.2 Enter Host Immunity: Beyond Targeted Therapy

In the "ideal" immune response against cancer, host dendritic cells (DCs) would detect tumor antigens at the primary tumor site, and traffic to the draining lymph node (DLN) where an "effector-type" "education" of T cells against tumor antigens would take place. At the DLN, the DCs arriving in the lymphatic conduit from the tumor would present antigen in the context of MHC-I to T-cell receptors on CD8+ T cells, resulting in antitumor CD8+ T cells. This would also promote memory T-cell responses in systemic secondary lymph node organs, allowing for cytolytic antitumor responses in not only the DLN and primary tumor but even at remote metastatic sites accessible to primed blood-borne CD8+ T cells. Moreover, CD4+ T cells primed by tumor antigen in the context of MHC-II would promote helper responses to expand B cells with humoral antitumor responses [15, 16]. In all of this, the promotion of antitumor responses by NK cells and NK-T cells as well as other T-cell subtypes that have been found to play antitumor roles (including some effects of Th17 cells) would serve to augment the antitumor response [17]:

1. General immune considerations: The above ideal situation is a sort of "Disneyland-type" story for antitumor immunity: Put another way, it would be wonderful if the cellular immunity system could be rapidly primed to destroy any early nest of neoplastic cells through efficient T-cell-mediated responses while avoiding the actions (or even exaggerated responses) of homeostatic responses that put a "brake" on the antitumor immune response. Unfortunately, human cancer is characterized by a microenvironment that promotes immune subversion and tolerogenic responses by both DCs and T cells that essentially suppress immunity [15, 18]. This occurs in both the primary tumor and, with less evidence, the DLN as well [19]. In lung cancer, among other tumors, tolerance of tumor presence by the immune system is characterized by increased levels of immunosuppressive cytokines, such as TGF-b and IL-10 in the tumor, along with influx of T-regulatory (Treg) cells and altered function of DCs [18–20]. The latter includes not only immaturity of DCs but also reduced antigen presentation, along with tumor cells which dominate the expression of immune checkpoint co-inhibitory signals that promote tolerance through inhibition in effector T-cell responses. Perhaps the most major pathway promoting the latter is expression of the "programmed death" ligand PD-L1 by suppressive DCs (maintained immature and suppressive by the milieu of cytokines in the tumor) as well as tumor cells [15]. This ligand, including PD-L2 that has been discovered more recently, induces T-effector repression, including T-cell apoptotic signals that thereby inhibit the antitumor response through T-effector cell loss. In addition, there appear to be important roles for myeloid-derived suppressor cells (MDSCs) and even other host myeloid-derived cells such as macrophages serving in a tumor- and angiogenesis-promoting role (M2 phenotype; as opposed to the M1 subtype of tumor-associated macrophages/TAMs also found in cancer), and even neutrophils through other novel mechanisms [18, 21].
2. Cancer immuno-editing and "keeping up" with anticancer immunity: In any carcinoma, the concept of cancer "immuno-editing" arises and serves as a fundamental paradigm through which we might understand the general timeline of cancer versus anticancer immune "pressures" that develop during cancer growth and progression [15, 22]. The process can essentially be summed up as the "3-E's": elimination, equilibrium, and escape [22]. During elimination, cancer cells initially destroyed by competent immunosurveillance mechanisms are able to sensitize immunity, resulting in the inhibition of cancer growth through recognition and responses against two forms of tumor antigens: (i) overexpressed (self-antigens) or (ii) entirely new antigens that are essentially foreign epitopes, so-called neo-antigens. [20] The pace is maintained into an equilibrium phase, whereby the control of tumor outgrowth by adaptive immunity (as a sort of tumor "dormancy") is balanced against the growth of tumor cells that begin to survive against immunosurveillance. The mechanisms for the latter resistance to immunity may involve the key components of the third phase of this process,

which poses the greatest challenge to our therapeutic repertoire: Escape. This component is characterized by the elaboration of immunosuppressive cytokines (such as TGF-b and IL-10 in lung cancer, serving as major inhibitors) as well as the actions of tumor- and possibly vascular growth factors that elaborate in parallel [15, 18, 20]. Some of the latter may serve in redundant manners to stimulate endothelium and angiogenesis to overcome targeted blockade of growth factors [9]. In addition, immune escape is characterized by recruitment of Treg cells that suppress cellular immunity in the tumor and possibly in the DLN. Thus, a goal to address this progression might be to force a "cycle" of immuno-editing back toward elimination through novel strategies. These might include approaches to improve the quantity of response through inhibition of T-effector loss/apoptosis (immune checkpoint directed) or possibly cell-based "reprogramming" methods that recruit immune responses to newly expressed neo-epitopes or new waves of self-epitope overexpression during tumor progression [15]. Regardless, it is essential to recognize the steps in cancer immuno-editing in order to understand the evolution of a poor antitumor response in the host during cancer progression and to assist in the immunotherapeutic approach.

3. Harnessing immune checkpoint pathways: Given their current and growing clinical importance in approved therapies, it is central to consider immune checkpoint pathways in some detail as part of this discussion. In a broad sense, there are parallel co-stimulatory and co-inhibitory pathways that may engage DCs and T cells during DC-mediated presentation of antigen to the T-cell receptor (TcR) in the context of MHC [18]. A variety of antagonists of co-inhibitory pathways have been under development to thereby stimulate effector T-cell engagement with cancer antigens presented in context of DC MHC. On the other hand, a number of agonists for co-stimulatory pathways are also under study and offer promising therapeutic strategies to mobilize T cells against lung carcinomas via the immune checkpoint pathway. We introduce these separately:

   (a) Antagonist approaches

   One of the most important "targets" for immunotherapy is the PD-1/PD-L1 pathway involved as a co-inhibitory immune checkpoint mechanism. In that sense, it can be considered to exist as a homeostatic mechanism to dampen T-effector proliferation when immunity to a foreign antigen is activated, thus serving to but a "brake" on T cells during the engagement of DC MHC-antigen with the TcR on effector T cells [23, 24]. This occurs through engagement of PD-L1 with the PD-1 receptor expressed on T cells. Another similar ligand-receptor co-inhibitory pair is CD80/86 on DCs that interacts with CTLA-4 on T cells: This tends to occur throughout cellular immune compartments, including in the central lymphoid centers, while PD-1 engagement takes place in peripheral compartments where T cells might engage with peripheral DCs (e.g., Langerhans cells in the skin or infiltrating tumor DCs in any peripheral tumor) during antigen presentation at those peripheral sites. We may thus consider antibody approaches that block PD-1 (or PD-L1) or CTLA4 [23], for example, "antagonist" approaches to thereby achieve immune checkpoint blockade. In cancer, such blockade has now been well demonstrated to result in stimulation of antitumor T-cell responses and is currently the most active area of immunotherapy in lung cancer (which includes FDA approval of the anti-PD1 antibody nivolumab) [25]. Greater levels of PD-L1 expression by the tumor correlated with improved responsiveness to antibody blockade in trials of either pembrolizumab [26] or atezolizumab [27] randomized against docetaxel in second-line therapy against NSCLC. The CTLA4 blockade approach (with the antibody ipilimumab) preceded the PD-1 blockade approach; and while both are effective in achieving durable remissions in a subset of NSCLC patients (typically under 25%) treated after failure of second-line therapy for metastatic lung cancer, there appears to be lesser autoimmune side effects with PD-1 blockade [28, 29]. This is likely due to the fact that CTLA4 blockade may inhibit immune checkpoint T-cell repression in both central lymphoid immune compartments as well as the periphery, while PD-1 blockade is likely to exert its actions in peripheral-tissue T cells, thus restricted to the site of "action." Beyond these antagonist approaches, other co-inhibitory pathways may become targets for blocking antibodies in lung cancer. One example is the adenosine receptor, A2AR, which is expressed on T cells, whereby relatively high levels of adenosine expressed in the tumor microenvironment may activate A2AR on cytotoxic T-lymphocyte (CTL) effector cells and DCs as well [30]. This effectively inhibits the antitumor efficacy of these immune cells, particularly under high concentrations of adenosine in the tumor; and thus blocking approaches to this pathway are attractive as novel immune checkpoint inhibitors.

(b) Agonist approaches

Several antibodies that function in an "agonist" or stimulatory manner to stimulate co-stimulatory receptors on T cells at the immune checkpoint level are under development. These include approaches that stimulate a variety of co-stimulatory molecules that belong to the TNF receptor family. Examples that are under current investigation include OX40, 4-1BB, and CD27 which are expressed on activated T cells [15, 31]. In a sense, these serve as a form of "positive-feedback" system that promotes or amplifies effector responses during an immunologic stimulus. The latter may be any cancer antigen (self- or neo-antigen) for which an effective antitumor therapeutic approach might involve boosting the expression or signaling by this family of co-stimulatory molecules on effector T cells. Some antibody-based agonists for these molecules, which include MEDI6469 for OX40, urelumab for 4-1BB, and varlilumab for CD27, are currently under testing as hopeful agents that may be used either alone or in combination with other adjuvant therapies in advanced-stage lung cancer [31].

4. Combining immunotherapy with existing antitumor therapies in advanced NSCLC: It is possible that some of the gains that are rapidly growing in both progression-free survival and even overall survival through the use of targeted

therapies in advanced NSCLC might be augmented by introducing immune checkpoint-blocking approaches at the right times [32]. Indeed, in cases of metastatic progression in NSCLC, where immunotherapy is appropriate, there may be temporary gains in progression-free survival on targeted therapy that may improve quality of life for extended periods [33]. With the eventual development of resistance, the addition of immunotherapy may further extend survival with extended gains in quality of life. Moreover, in a subset of such patients that have failed multiple standard or targeted therapies, it is possible that they would otherwise show long-term responses to immunotherapy, within the range of typical responses (15–25 % from most series) [23, 24]. Thus, this "later" addition may offer gains that otherwise would not have been realized in this group of patients in recent years; however, less is known about how multiple rounds of targeted therapy might modify the response to immunotherapy: Improved biomarkers are needed in this arena. Outside of independent testing for changes in targetable mutations and/or PD-L1 status over time during a patient's long-term treatment course (and treating on their own merit), evidence is lacking on any positive correlation between expression of sensitizing EGFR or ALK mutations and positive responses to PD-1/PD-L1 blockade.

20.2.3 Biomarkers and Identifying Immune Checkpoint Targeting "Responders"

- The use of PD-1/PD-L1 expression as a biomarker: Since immune checkpoint blockade of the PD-1/PD-L1 pathway is now becoming a standard form of immunotherapy, it is of high interest to determine how assessment of the expression of these molecules might facilitate the utility and choice of therapy when it becomes indicated. Initial studies in both melanoma and lung cancer examining expression of PD-L1 in tumor-biopsy material showed that expression of PD-L1 indeed correlated with improved responsiveness to PD-1 as well as PD-L1-blocking antibodies [24, 34]. An important concept that emerged during these studies came from the realization that both of these forms of cancer have a high incidence of initiation as a result of environmental mutagenesis (i.e., solar radiation and cigarette smoke); and for this reason, the focus on responses in the setting of squamous NSCLC evolved. Indeed, the impact of immunotherapy on outcomes in a significant subset of metastatic squamous NSCLC patients that progressed beyond standard therapies (when randomized to nivolumab versus docetaxel) showed improved overall survival with durable remissions in 15–20 % of nivolumab-treated patients. Accordingly, the use of nivolumab in this setting had been approved regardless of tumor PD-1 status [25, 31]. Nevertheless, since other trials examining PD-1-blocking therapies had shown poorer responses to treatment with PD-1 blocking therapy in the absence of tumor PD-L1 expression, testing of PD-L1 as a biomarker prior to use has become commonplace.

This is particularly true in the case of non-squamous NSCLC, where the immune response to tumor antigens expressed as a result of environmental mutagenesis (as a driver of neoplasia) might not contribute as greatly to robust T-cell responses (discussed further in the following section). It may thus be concluded that assessment of tumor PD-L1 status serves as a useful and important biomarker to guide the use of therapies that block this important immune checkpoint axis.

- Environmental mutagenesis and tumor neo-antigen expression: A key biomarker for antitumor effector T-cell responses that has emerged is the level of expression of non-synonymous mutations as well as a variety of neo-antigens by carcinomas [35]. This is noted with higher frequency in tumors that are associated with transformation resulting from environmental stimuli such as melanoma or tobacco-associated squamous NSCLC [36, 37]. A validation of immune reactivity to such antigens from the same patient cohorts demonstrated that candidate neo-antigens could be detected in such tumors; and the proliferation of peripheral blood-derived T cells from the same patients responded to respective MHC-antigen probes (without response to control wildtype peptides corresponding to the respective neo-antigens) [35]. Importantly, neo-antigen load correlated with PD-1 blockade responsiveness in such patients, independent of PD-L1 tumor status, suggesting that priming of T cells to neo-antigens occurs with parallel increases in immune checkpoint sensitivity. The identification of putative neo-antigens in such tumors raises the question of how these novel peptides could be incorporated into possible vaccine approaches (below) or other methods to stimulate immunity in the tumor microenvironment. Finally, whether one is considering how to augment the adaptive immune response to NSCLC or how to optimally design engineered T cells to attack specific NSCLC neo-antigens, an especially interesting recent observation is that clonal expression of a neo-antigen (i.e., expression throughout all cells of a tumor, as opposed to a small subset) predicts responsiveness to PD-1 blockade [38]. This may be associated with recognition of the antigen more broadly throughout a tumor, once T cells are sufficiently stimulated through effector-promoting approaches (i.e., PD-1 blockade). Moreover, as an effective biomarker, one would also suspect that this pattern would predict responsiveness to other immune checkpoint-targeting methods (as well as vaccines, following identification of unique antigens).

20.3 Turning a "Tide" of Tolerant Traffic?

Outside of a rapidly growing wave of immune checkpoint-blocking strategies, it is important to focus on other mechanisms whereby we have attempted to "turn" the tolerant immune-milieu that dominates the primary lung carcinoma and its draining lymph nodes into a microenvironment with better cellular and possibly humoral antitumor mechanisms. The approaches include vaccines, exogenous/adoptive dendritic and T-cell engineering strategies, and possibly novel endogenous mechanisms to improve dendritic cell function, and appropriate antitumor-antigen presentation and associated immunologic activation.

20.3.1 Vaccines

1. Cancer-overexpressed antigens – self- and nonself antigens: To date a variety of vaccine approaches for NSCLC had been examined. More generally, trials have ranged from employing whole cell-based approaches (i.e., harnessing an "empiric" mixture of self/nonself antigens introduced by the tumor cell; using irradiated tumor cell lines or autologous cells) or antigen-based strategies [23]. In terms of the latter, approaches have targeted an overexpressed self-epitope such as a 25aa fragment of MUC-1 conjugated to a lipopeptide (including immunoadjuvant delivery) [39], or neo-antigens that are under investigation for their unique expression in some NSCLC tumors [40]. Ongoing studies include combination approaches, such as the MUC-1 epitope together with cytokine IL-2 (as an immunostimulant) delivered in a poxvirus platform [41]. Most of this had been demonstrated in mouse studies. In a current study of racotumomab, where the vaccine antigen is a sialic acid containing ganglioside [23, 42], randomization to vaccine versus placebo would take place in stage IIIB/IV NSCLC patients that show response or stabilization of disease following standard induction therapy.
2. Effects of radiation and vaccines: With either cell-based or antigen-based approaches at this time, overall results have been marginal; however, the responses of certain subgroups remain highly interesting and attractive. For example, while whole-cell approaches to date had not demonstrated sufficient efficacy in phase III clinical trials to approve them for clinical use, there had been efficacy noted in post hoc analyses of the MUC-1-based vaccine under certain important circumstances. For example, some previous phase III trial data involving novel MUC-1 vaccine (liposomal BLP-25; tecemotide) with unique adjuvant approaches showed unique responses in irradiated patients. In one trial, analysis of the subset of patients with stage IIIB disease that received prior radiotherapy showed that vaccine had a significant effect in improving outcomes (with median survivals >30 months in the irradiated group; L-BLP 25 trial) [39]. This is particularly important as a biological principle, since tumor antigens might be released during radiotherapy treatment, with the opportunity to activate effective antitumor CD8+ T-cell responses. Moreover, the "abscopal" effect of radiation is one in which other immunologic effects of irradiation may improve immunotherapy responsiveness through the ability of radiation to stimulate chemokine-driven T-cell migration to tumors while increasing cytotoxic T-cell (CTL) activation in parallel [43].
3. Novel-combined effects and considerations: The promise of efficacy in the above approaches is mixed, with some important biological lessons that should be harnessed in future studies, including the pairing/timing with radiotherapy. Some possibilities for limited efficacy include insufficient action by the immunogen (i.e., dose or proper adjuvant choice) or late timing in the phase of disease, where vaccine application in the setting of an overly suppressed microenvironment might pose overwhelming challenge. To overcome the latter, some approaches

employed antisense technology (such as antisense TGF-β) in-line with the vaccine delivery, including trials wherein the immunogen was an inactivated whole tumor-cell mixture and where efficacy was also improved following radiation [44]. Further refinements in timing and vaccine development, however, are still very much needed. Consideration of pairing PD-1 or PD-L1 blockade with vaccines is a novel concept [23] since ionizing radiation can upregulate PD-L1 expression, and enhancement of radiation efficacy is seen with PD-L1 blockade [45]. Moreover, increases in tumor-specific T cells as well as optimized tumor-antigen cross presentation in draining lymph nodes appear to occur in response to combined radiation with anti-PD-1/anti-CTLA-4 therapy [46]. Given these observations, a new concept may thus be the emergence of radiation-induced immune-mediated personalized therapy [23]. Finally, with the introduction of sequencing platforms that can rapidly identify neo-antigens, the use of neo-antigens as the peptide immunogen paired with the use of appropriate adjuvants might emerge as novel efficacious vaccine approaches.

20.3.2 Key Advances in T-Cell Engineering

A variety of approaches involve direct harnessing of T cells from the host in order to improve antitumor immunity through unique strategies wherein antigen-dependent priming (including DC-dependent antigen presentation) may be bypassed, and where induction of antitumor T cells may occur through various methods [23]. The following are some of the most active approaches wherein lung cancer has been studied or is under active consideration:

1. Adoptive cell therapies (ACT) involve the expansion of CD8+ and CD4+ tumor-infiltrating lymphocytes (TILs) derived from tumors, with cell expansion and conditioning (with immunostimulatory cytokines), prior to reinfusion back into the host [47]. Typically, a lymphodepleting conditioning regimen is used to improve the duration of a response and appears to lessen tumor infiltration by suppressive cells such as Tregs and myeloid-derived suppressor cells (MDSCs). Objective response rates were noted for melanoma with this form of cell therapy, with durable responses in over 20 % of patients in small series; however, there have not been such responses documented in lung cancer. Nevertheless, modifications to TIL infusion for lung cancer may be introduced into in future regimens, with greater knowledge of differences in the immune-microenvironment of melanoma versus that of NSCLC or even SCLC. It should be cautioned that the infusion of TIL cells combined with a conditioning regimen is not without life-threatening toxicity risks in some cases, and patient selection needs greater work at this time [48].
2. TcR T-cell engineering and CAR-T cells: While our knowledge of highly tumor specific neo-antigens increases, we may consider two specific types of T-cell engineering: TcR T-cell engineering as well as CAR-T-cell production. These unique T cells are engineered to recognize specific antigens. In this case,

relevant antigens might not only be unique to specific carcinoma types as a whole (e.g., MAGE-A1 expressed by a large number of NSCLCs) but for an individual patient's tumor, where sequencing and RNAseq technology might reveal tumor-specific (nonself) antigens presented in the context of MHC [35]. These would be ideal "targets" for TcR T-cell engineering, wherein identification of a unique tumor peptide can be used to generate a TcR (with expression of unique alpha and beta chains) that recognizes peptide in the context of human leukocyte antigen, so long as the latter is expressed by tumor cells [49]. This carries some risk of cross-reactivity with normal tissues and "autoimmune" toxicity [31]. An approach that does not depend on target neo-peptide presentation in the context of MHC is the use of CAR-T cells: In this technology, an Ig variable extracellular domain specific to the tumor peptide in question is fused to a TcR constant domain; and the engineered T cell may thus bind and kill tumor cells expressing the Ig-targeted neo-antigen without the need for tumor cell display antigen in any HLA-dependent context. The "kill" then results from the activation of internal co-stimulatory signaling intermediates that have been engineered (as part of the "chimeric antigen receptor") downstream to the specific cell surface Ig molecule [50, 51]. Again, the need for tumor-unique antigens is important since "off-target" fatal complications have occurred during such treatments. While CAR-T-cell therapies have been especially successful in patients with B-cell malignancies (e.g., dramatic responses to anti-CD19 CAR-T) as well as melanoma, the ability to expand this technology for thoracic malignancies while considering relatively unique/overexpressed antigens (such as mesothelin or perhaps entirely neo/novel antigens) will be a challenge. Thus, safety issues center on the selection of tumor-unique targets and on the paucity of such targets, while the duration of responses remain among other challenges. Harnessing this technology appropriately for solid tumors nevertheless remains promising.

20.3.3 A Collective Consideration of Antigen-Dependent and T-Cell "Harnessing" Mechanisms in Tumor Therapy

Through all that has been considered, in the tumor microenvironment, if one considers the functional importance of dendritic cells during the employment of vaccines along with the immunosuppressive "pressure" by tumor cells, one may ultimately enhance antitumor T-cell functions through new antigen-dependent mechanisms or through direct "harnessing" of T cells through novel engineering approaches. One must, however, realize a variety of limitations associated with each approach. The efficacy and limitations associated with these various approaches are illustrated and listed in Fig. 20.1. It should be mentioned that these limitations also include the induction of autoimmune-type side effects (including life-threatening reactions)

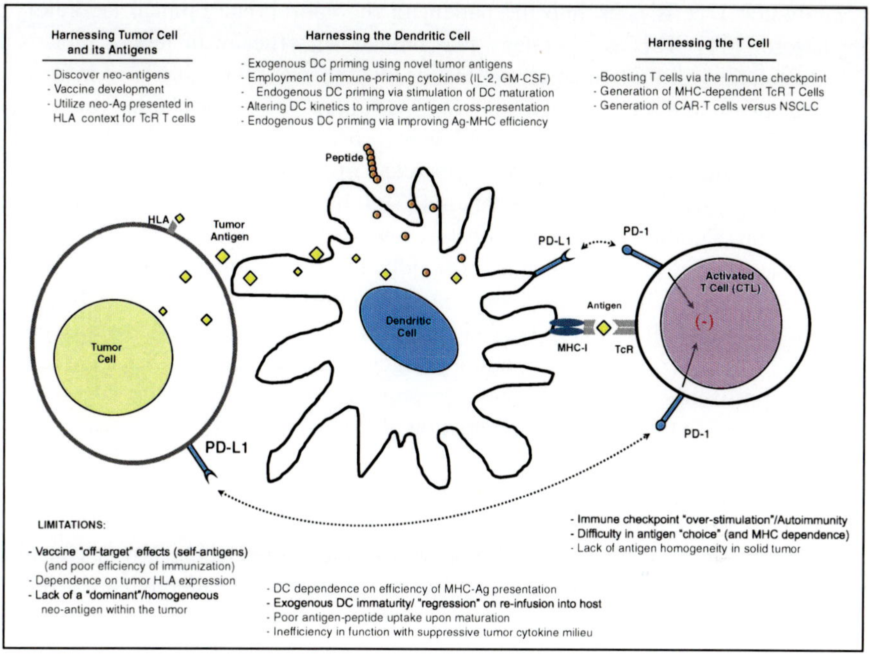

Fig. 20.1 Harnessing dendritic cells and T cells in the tumor microenvironment: efficacy and limitations

that are typically more prevalent when T-cell effector mechanisms are augmented in the central lymph node priming phase (e.g., CTLA-4 blockade during immune checkpoint targeting) as opposed to the effector phase in peripheral tissue at the location of the primary tumor (e.g., PD-1 blockade). These side effects are highly important considerations in clinical medicine, and while their detailed coverage lies beyond the scope of this review, several reports have outlined their prevalence, distribution, and association with various immunologic approaches in cancer, as exemplified in [52].

20.3.4 Exogenous and Endogenous Dendritic Cell Engineering

1. Exogenous approaches: One approach to enhancing cytotoxic T-cell activity against tumors is the harnessing of dendritic cells through exogenous approaches. In this manner, purifying DCs from the patient and using them to boost sensitization of adoptive T cells to tumor antigens (thus presented optimally through MHC on the DCs) ex vivo may allow reinfusion of a highly efficacious population of

antitumor T cells back into the patient in the appropriate clinical laboratory platform [31]. This is a strategy that has shown efficacy in mouse models, although thus far for human carcinoma, its most successful use has been in metastatic prostate carcinoma, employing peripheral blood mononuclear cells (PBMCs) along with delivery of antigen (prostatic acid phosphatase) and an adjuvant cytokine (GM-CSF) to facilitate DC maturation and effector functions prior to reinfusion back into the patient [53]. It has been difficult to demonstrate consistent durable antitumor effects, however, with exogenous DC-based approaches. One major challenge (even if the DC is conditioned very efficiently, with successful achievement of a strong antitumor phenotype ex vivo) is the possibility of rapid alteration in DC behavior and DC expression upon reentry into a host. Moreover, when ideal "tumor antigens" are paired with DCs ex vivo in such approaches, not all antigens are presented efficiently in the context of MHC-I; and so the downstream induction of cytotoxic T-cell responses may still be very limited [54].

2. Endogenous DC engineering: This is a concept that might be considered broadly as one that modulates DC behavior through strategies that do not involve isolation of DCs outside of the host. Of course, vaccines would technically be considered as one form of "endogenous" DC modulation [23], although we would reserve the word "engineering" for non-antigen approaches to specifically modulate DC behavior in a way that improves antitumor responses (where the DC employs "natural" endogenous tumor antigens to induce T-cell responses). Our own laboratory has interest in this approach, from a glycobiology standpoint: An example of such engineering in this context might involve small-molecule inhibition of a key glycan co-receptor involved in driving immature DC traffic and regulating or inhibiting DC maturation responses. Indeed, we have found that DCs induced by certain cytokines such as TNF-α strongly upregulate the heparan sulfate proteoglycan syndecan-4. Silencing of this molecule or the sulfation of its glycan chains (which facilitate the actions of chemokines that drive DC traffic) appears to slow DC traffic [55], which may allow for greater cross presentation of tumor antigens to T cells in tumor-draining lymphoid organs (e.g., tumor DLNs). Glycan mutations in such DCs also appear to strikingly increase DC maturation [55]. Figure 20.2 illustrates how targeting a major proteoglycan on DCs might increase maturation while inhibiting the rate of trafficking of immature DCs in the tumor and draining lymph nodes, ultimately resulting in improvements in T-cell immunity that result in tumor growth inhibition. While this might also impact activated DCs outside of the tumor microenvironment (not necessarily in a detrimental manner, however), it would serve as a promising way to endogenously alter DCs that uniquely express this proteoglycan in the tumor and DLN microenvironments.

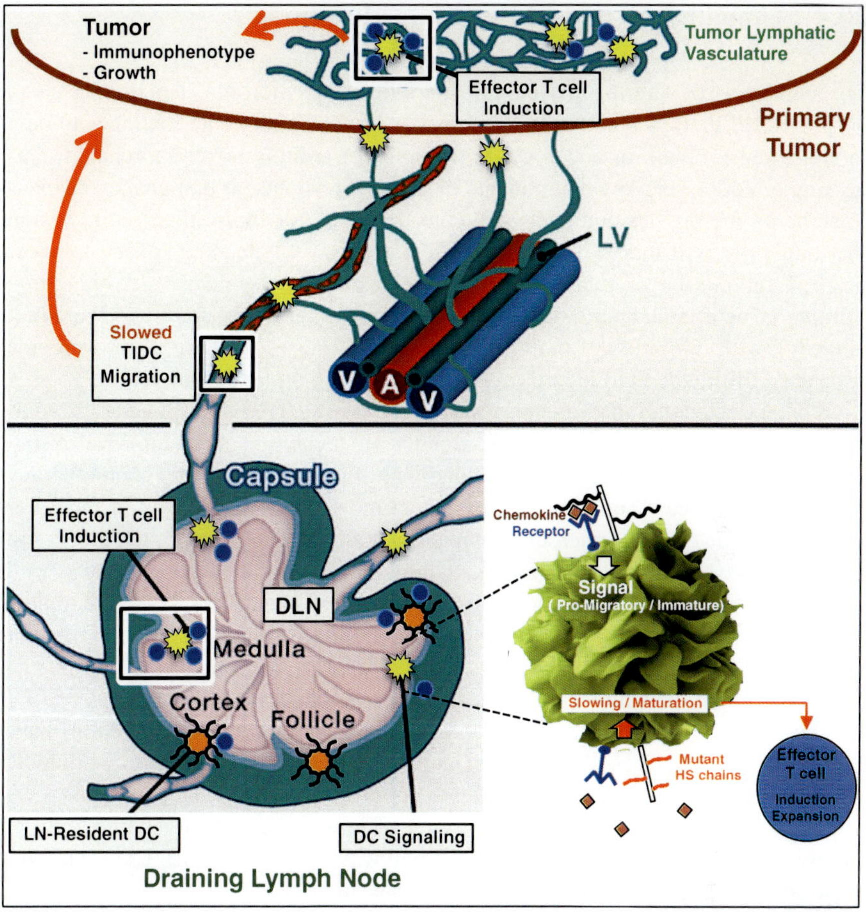

Fig. 20.2 Example of a novel glycan-targeting approach to augment antitumor immunity through endogenous dendritic cell functions: genetic reduction in the sulfation of glycans expressed on the surface of dendritic cells (shown in *yellow*) that traffic within the tumor microenvironment may alter their kinetics through slowing of glycan-mediated chemokine-dependent trafficking of immature dendritic cells from tumor to draining lymph node, with mechanisms illustrated in "magnified" dendritic cell shown within the inset at *lower right*. In particular, the altered glycan heparan sulfate (HS), targeted through mutation in this case, may be associated with a phenotype of DC slowing and increased maturation (*lower right*). This targeting may also increase maturation of tumor-associated dendritic cells in a manner that promotes effector T-cell functions following efficient tumor antigen presentation. Further work is needed to elucidate mechanisms for the latter; however, the ultimate result may contribute to an antitumor immune phenotype [55] and inhibited tumor growth (*red arrows, implying inhibition*). This illustrates one of possibly several novel ways to endogenously augment dendritic cell functions in the lung carcinoma and its draining lymph nodes (*bottom*); illustration and zoom of DC modified from Refs. [56, 57]

20.4 Conclusions and Future Considerations

The challenge of altering the dismal mortality (and morbidity) of lung cancer, despite novel therapies aimed to "personalize" tumor treatment, continues to be a monumental problem in oncology. In the spirit of entirely novel approaches, harnessing or addressing key components of the host response to malignancy takes us to some promising developments and considerations for future therapeutic design that hopefully will markedly improve our impact on this leading cause of cancer death. Targeting the vascular response to tumor growth now involves co-inhibition of other growth mediators beyond VEGF-A (including PDGF and FGF-2 simultaneously), with inhibition of multiple downstream kinases in attempts to overcome resistance. Biomarkers to identify responders are also under consideration.

While targeted therapies against tumor cells have evolved to respond to the common problem of acquired resistance, along with developments to improve overall survival and quality of life further beyond that offered by combined chemotherapy in metastatic disease, immunotherapy has come online. Challenges in the latter arena in the lung cancer microenvironment include immunosuppression as a result of PD-L1/PD-1 immune checkpoint signaling by tumor–T-cell interactions as well as a suppressive cytokine milieu (TGF-β and IL-10 as examples) within the tumors that promote tolerance by dendritic cells. This inhibits efficient priming of effector CD8+ T cells by tumor antigens. This is in addition to macrophages, MDSCs, and other cells that induce tumor tolerance. Antibodies that target either PD-1 or PD-L1 (or the CTLA-4 molecule that serves a similar T-cell suppressive function) have boosted T-cell responses, with trials in metastatic NSCLC that demonstrate durable curative responses as second- or third-line approaches in 20–25 % of patients. This is exciting and naturally begs the question of how we could target immunity to improve T-cell responses further. Agonist pathways (OX40, 4-1BB, CD27) to promote co-stimulatory T-effector responses (alongside PD-1/PD-L1 axis antagonism) as well as combination approaches are being considered. Novel attempts to target as well as use tumor neo- (unique) antigens are under consideration; and both neoantigens and self-antigens overexpressed by tumors (e.g., MUC-1 peptide) in vaccines have shown greater efficacy when paired with chemo- or radiation therapy. Finally, a variety of cell-based therapies have been under development to improve specific antitumor T-cell responses. These may include novel exogenous (ex vivo/in vitro) as well as endogenous ways to improve antitumor dendritic cell functions and more direct T-cell therapies using adoptive transfer of cytokine-modified T cells or even engineering of chimeric antigen receptor T (CAR-T) cells against novel lung carcinoma targets. The latter do not depend on presentation of antigen in the context of MHC, which may overcome a significant limitation in NSCLC immunity; however they are limited to single-molecule targets that stand the risk of tumor evolution/heterogeneity and immunologic escape.

References

1. Siegel RL, Miller KD, Jemal A (2016) Cancer statistics, 2016. CA Cancer J Clin 66(1):7–30
2. Detterbeck FC, Postmus PE, Tanoue LT (2013) The stage classification of lung cancer: diagnosis and management of lung cancer, 3rd ed: American College of chest physicians evidence-based clinical practice guidelines. Chest 143(5 Suppl):e191S–e210S
3. National Lung Screening Trial Research Team, Aberle DR, Adams AM, Berg CD, Black WC, Clapp JD et al (2011) Reduced lung-cancer mortality with low-dose computed tomographic screening. N Engl J Med 365(5):395–409
4. Gajewski TF, Schreiber H, Fu YX (2013) Innate and adaptive immune cells in the tumor microenvironment. Nat Immunol 14(10):1014–1022
5. Hall RD, Le TM, Haggstrom DE, Gentzler RD (2015) Angiogenesis inhibition as a therapeutic strategy in non-small cell lung cancer (NSCLC). Transl Lung Cancer Res 4(5):515–523
6. Yuan A, Hsiao YJ, Chen HY, Chen HW, Ho CC, Chen YY et al (2015) Opposite effects of M1 and M2 macrophage subtypes on lung cancer progression. Sci Rep 5:14273
7. Das M, Wakelee H (2014) Angiogenesis and lung cancer: ramucirumab prolongs survival in 2(nd)-line metastatic NSCLC. Transl Lung Cancer Res 3(6):397–399
8. Reck M, Mellemgaard A (2015) Emerging treatments and combinations in the management of NSCLC: clinical potential of nintedanib. Biologics 9:47–56
9. Chu BF, GA O (2016) Incorporation of antiangiogenic therapy into the non-small-cell lung cancer paradigm. Clin Lung Cancer 17:493–506
10. Chang YW, Su CM, Su YH, Ho YS, Lai HH, Chen HA et al (2014) Novel peptides suppress VEGFR-3 activity and antagonize VEGFR-3-mediated oncogenic effects. Oncotarget 5(11):3823–3835
11. Couraud S, Zalcman G, Milleron B, Morin F, Souquet PJ (2012) Lung cancer in never smokers–a review. Eur J Cancer 48(9):1299–1311
12. Skrzypski M, Czapiewski P, Goryca K, Jassem E, Wyrwicz L, Pawlowski R et al (2014) Prognostic value of microRNA expression in operable non-small cell lung cancer patients. Br J Cancer 110(4):991–1000
13. Mok TS, Wu YL, Thongprasert S, Yang CH, Chu DT, Saijo N et al (2009) Gefitinib or carboplatin-paclitaxel in pulmonary adenocarcinoma. N Engl J Med 361(10):947–957
14. Sellmann L, Fenchel K, Dempke WC (2015) Improved overall survival following tyrosine kinase inhibitor treatment in advanced or metastatic non-small-cell lung cancer-the Holy Grail in cancer treatment? Transl Lung Cancer Res 4(3):223–227
15. Chen DS, Mellman I (2013) Oncology meets immunology: the cancer-immunity cycle. Immunity 39(1):1–10
16. Melero I, Gaudernack G, Gerritsen W, Huber C, Parmiani G, Scholl S et al (2014) Therapeutic vaccines for cancer: an overview of clinical trials. Nat Rev Clin Oncol 11(9):509–524
17. Zamarron BF, Chen W (2011) Dual roles of immune cells and their factors in cancer development and progression. Int J Biol Sci 7(5):651–658
18. Thomas A, Giaccone G (2015) Why has active immunotherapy not worked in lung cancer? Ann Oncol 26(11):2213–2220
19. Schneider T, Hoffmann H, Dienemann H, Schnabel PA, Enk AH, Ring S et al (2011) Non-small cell lung cancer induces an immunosuppressive phenotype of dendritic cells in tumor microenvironment by upregulating B7-H3. J Thorac Oncol 6(7):1162–1168
20. Finn OJ (2008) Cancer immunology. N Engl J Med 358(25):2704–2715
21. Noy R, Pollard JW (2014) Tumor-associated macrophages: from mechanisms to therapy. Immunity 41(1):49–61
22. Mittal D, Gubin MM, Schreiber RD, Smyth MJ (2014) New insights into cancer immunoediting and its three component phases–elimination, equilibrium and escape. Curr Opin Immunol 27:16–25

23. Anagnostou VK, Brahmer JR (2015) Cancer immunotherapy: a future paradigm shift in the treatment of non-small cell lung cancer. Clin Cancer Res 21(5):976–984
24. Garon EB, Rizvi NA, Hui R, Leighl N, Balmanoukian AS, Eder JP et al (2015) Pembrolizumab for the treatment of non-small-cell lung cancer. N Engl J Med 372(21):2018–2028
25. Brahmer J, Reckamp KL, Baas P, Crino L, Eberhardt WE, Poddubskaya E et al (2015) Nivolumab versus docetaxel in advanced squamous-cell non-small-cell lung cancer. N Engl J Med 373(2):123–135
26. Herbst RS, Baas P, Kim DW, Felip E, Perez-Gracia JL, Han JY et al (2016) Pembrolizumab versus docetaxel for previously treated, PD-L1-positive, advanced non-small-cell lung cancer (KEYNOTE-010): a randomised controlled trial. Lancet 387(10027):1540–1550
27. Fehrenbacher L, Spira A, Ballinger M, Kowanetz M, Vansteenkiste J, Mazieres J et al (2016) Atezolizumab versus docetaxel for patients with previously treated non-small-cell lung cancer (POPLAR): a multicentre, open-label, phase 2 randomised controlled trial. Lancet 387(10030):1837–1846
28. Rizvi NA, Mazieres J, Planchard D, Stinchcombe TE, Dy GK, Antonia SJ et al (2015) Activity and safety of nivolumab, an anti-PD-1 immune checkpoint inhibitor, for patients with advanced, refractory squamous non-small-cell lung cancer (CheckMate 063): a phase 2, single-arm trial. Lancet Oncol 16(3):257–265
29. Topalian SL, Drake CG, Pardoll DM (2015) Immune checkpoint blockade: a common denominator approach to cancer therapy. Cancer Cell 27(4):450–461
30. Leone RD, Lo YC, Powell JD (2015) A2aR antagonists: next generation checkpoint blockade for cancer immunotherapy. Comput Struct Biotechnol J 13:265–272
31. Farkona S, Diamandis EP, Blasutig IM (2016) Cancer immunotherapy: the beginning of the end of cancer? BMC Med 14:73
32. Sharma P, Allison JP (2015) Immune checkpoint targeting in cancer therapy: toward combination strategies with curative potential. Cell 161(2):205–214
33. Rooney C, Sethi T (2015) Advances in molecular biology of lung disease: aiming for precision therapy in non-small cell lung cancer. Chest 148(4):1063–1072
34. Topalian SL, Hodi FS, Brahmer JR, Gettinger SN, Smith DC, McDermott DF et al (2012) Safety, activity, and immune correlates of anti-PD-1 antibody in cancer. N Engl J Med 366(26):2443–2454
35. Rizvi NA, Hellmann MD, Snyder A, Kvistborg P, Makarov V, Havel JJ et al (2015) Cancer immunology. Mutational landscape determines sensitivity to PD-1 blockade in non-small cell lung cancer. Science 348(6230):124–128
36. Alexandrov LB, Nik-Zainal S, Wedge DC, Aparicio SA, Behjati S, Biankin AV et al (2013) Signatures of mutational processes in human cancer. Nature 500(7463):415–421
37. Snyder A, Makarov V, Merghoub T, Yuan J, Zaretsky JM, Desrichard A et al (2014) Genetic basis for clinical response to CTLA-4 blockade in melanoma. N Engl J Med 371(23):2189–2199
38. McGranahan N, Furness AJ, Rosenthal R, Ramskov S, Lyngaa R, Saini SK et al (2016) Clonal neoantigens elicit T cell immunoreactivity and sensitivity to immune checkpoint blockade. Science 351(6280):1463–1469
39. Butts C, Socinski MA, Mitchell PL, Thatcher N, Havel L, Krzakowski M et al (2014) Tecemotide (L-BLP25) versus placebo after chemoradiotherapy for stage III non-small-cell lung cancer (START): a randomised, double-blind, phase 3 trial. Lancet Oncol 15(1):59–68
40. Gubin MM, Artyomov MN, Mardis ER, Schreiber RD (2015) Tumor neoantigens: building a framework for personalized cancer immunotherapy. J Clin Invest 125(9):3413–3421
41. Arriola E, Ottensmeier C (2016) TG4010: a vaccine with a therapeutic role in cancer. Immunotherapy 8(5):511–519
42. Alfonso S, Valdes-Zayas A, Santiesteban ER, Flores YI, Areces F, Hernandez M et al (2014) A randomized, multicenter, placebo-controlled clinical trial of racotumomab-alum vaccine as switch maintenance therapy in advanced non-small cell lung cancer patients. Clin Cancer Res 20(14):3660–3671

43. Deloch L, Derer A, Hartmann J, Frey B, Fietkau R, Gaipl US (2016) Modern radiotherapy concepts and the impact of radiation on immune activation. Front Oncol 6:141
44. Giaccone G, Bazhenova LA, Nemunaitis J, Tan M, Juhasz E, Ramlau R et al (2015) A phase III study of belagenpumatucel-L, an allogeneic tumour cell vaccine, as maintenance therapy for non-small cell lung cancer. Eur J Cancer 51(16):2321–2329
45. Deng L, Liang H, Burnette B, Beckett M, Darga T, Weichselbaum RR et al (2014) Irradiation and anti-PD-L1 treatment synergistically promote antitumor immunity in mice. J Clin Invest 124(2):687–695
46. Sharabi AB, Nirschl CJ, Kochel CM, Nirschl TR, Francica BJ, Velarde E et al (2015) Stereotactic radiation therapy augments antigen-specific PD-1-mediated antitumor immune responses via cross-presentation of tumor antigen. Cancer Immunol Res 3(4):345–355
47. Hinrichs CS, Rosenberg SA (2014) Exploiting the curative potential of adoptive T-cell therapy for cancer. Immunol Rev 257(1):56–71
48. Yee C (2013) Adoptive T-cell therapy for cancer: boutique therapy or treatment modality? Clin Cancer Res 19(17):4550–4552
49. Morgan RA, Dudley ME, Rosenberg SA (2010) Adoptive cell therapy: genetic modification to redirect effector cell specificity. Cancer J 16(4):336–341
50. Gross G, Waks T, Eshhar Z (1989) Expression of immunoglobulin-T-cell receptor chimeric molecules as functional receptors with antibody-type specificity. Proc Natl Acad Sci U S A 86(24):10024–10028
51. Lee DW, Kochenderfer JN, Stetler-Stevenson M, Cui YK, Delbrook C, Feldman SA et al (2015) T cells expressing CD19 chimeric antigen receptors for acute lymphoblastic leukaemia in children and young adults: a phase 1 dose-escalation trial. Lancet 385(9967):517–528
52. Michot JM, Bigenwald C, Champiat S, Collins M, Carbonnel F, Postel-Vinay S et al (2016) Immune-related adverse events with immune checkpoint blockade: a comprehensive review. Eur J Cancer 54:139–148
53. Kantoff PW, Higano CS, Shore ND, Berger ER, Small EJ, Penson DF et al (2010) Sipuleucel-T immunotherapy for castration-resistant prostate cancer. N Engl J Med 363(5):411–422
54. Lesterhuis WJ, Haanen JB, Punt CJ (2011) Cancer immunotherapy–revisited. Nat Rev Drug Discov 10(8):591–600
55. El Ghazal R, Yin X, Johns SC, Swanson L, Macal M, Ghosh P et al (2016) Glycan sulfation modulates dendritic cell biology and tumor growth. Neoplasia 18(5):294–306
56. Tammela T, Alitalo K (2010) Lymphangiogenesis: molecular mechanisms and future promise. Cell 140(4):460–476
57. Felts RL, Narayan K, Estes JD, Shi D, Trubey CM, Fu J et al (2010) 3D visualization of HIV transfer at the virological synapse between dendritic cells and T cells. Proc Natl Acad Sci U S A 107(30):13336–13341